Sulfonyl Ynamides as Useful Tools for N-Heterocyclic Chemistry

Zur Erlangung des akademischen Grades eines

DOKTORS DER NATURWISSENSCHAFTEN

(Dr. rer. nat.)

von der KIT-Fakultät für Chemie und Biowissenschaften

des Karlsruher Instituts für Technologie (KIT)

genehmigte

DISSERTATION

von

Tim Wezeman MSc.

aus

Essen, Deutschland

KIT-Dekan: Prof. Dr. Willem Klopper

Referent: Prof. Dr. Stefan Bräse

Korreferent: Prof. Dr. Joachim Podlech

Tag der mündlichen Prüfung: 22. Juli 2016

Band 62
Beiträge zur organischen Synthese
Hrsg.: Stefan Bräse

Prof. Dr. Stefan Bräse
Institut für Organische Chemie
Karlsruher Institut für Technologie (KIT)
Fritz-Haber-Weg 6
D-76131 Karlsruhe

Bibliografische Information der Deutschen Bibliothek

Die Deutsche Nationalbibliothek verzeichnet diese Publikation in der
Deutschen Nationalbibliografie; detaillierte bibliografische Daten sind
im Internet über http://dnb.d-nb.de abrufbar.

ISBN 978-3-8325-4397-6
ISSN 1862-5681

Logos Verlag Berlin GmbH
Comeniushof, Gubener Str. 47,
10243 Berlin
Tel.: +49 030 42 85 10 90
Fax: +49 030 42 85 10 92
INTERNET: http://www.logos-verlag.de

We don't make mistakes, we just have happy accidents.
Robert Norman "Bob" Ross

voor de fratsers die het leven zo leuk maken

Die vorliegende Arbeit wurde in der Zeit von September 2013 bis August 2016 am Institut für Organische Chemie des Karlsruher Instituts für Technologie (KIT), im Nanotechnology and Molecular Science Synthesis Laboratory am Queensland University of Technology (QUT) in Brisbane, Australien und bei Bayer CropScience in Frankfurt-Höchst, unter der Anleitung von Prof. Dr. Stefan Bräse durchgeführt. Während des Zeitraums vom 1. Dezember 2014 bis 31. März 2015 lag die wissenschaftliche Betreuung bei Dr. Kye-Simeon Masters (QUT) und während des Zeitraums vom 15. Mai 2015 bis 18. September 2015 lag die wissenschaftliche Betreuung bei Dr. Stephen D. Lindell (Bayer CropScience). Die Arbeit wurde gefördert durch die Europäischen Union, im Rahmen des FP7 Marie Skłodowska-Curie ITN "ECHONET" (project number 316379).

Hiermit erkläre ich, die vorliegende Arbeit selbstständig verfasst und keine anderen als die angegebenen Quellen und Hilfsmittel benutzt zu haben. Die Dissertation wurde bisher an keiner anderen Hochschule oder Universität eingereicht.

Tim Wezeman   Karlsruhe, 6. Juni 2016

Table of Contents

Abstract

Sulfonyl ynamides are highly versatile and synthetically useful reagents. Therefore, exploring their synthetic limitations can offer a fertile hunting ground for new reactions. The synthesis of a wide array of sulfonyl ynamides can be realized via copper-catalyzed amidative cross-couplings of sulfonyl amides with bromo acetylenes or by elimination of dichloroenamide precursors. Additionally the terminal ynamides can be further diversified via Sonogashira couplings. Solid-supported ynamides can be prepared by coupling of carboxylic acid-bearing ynamides to a Rink resin or attachment of dichloroenamides to a Merrifield triazene resin using CuAAC reactions. The use of these ynamides as synthons in heterocyclic organic chemistry was investigated.

Ynamides can be reacted with electrophilically-activated amides in order to access highly functionalized 4-aminoquinolines. A straightforward amide activation procedure with triflic anhydride and 2-chloropyridine, in combination with the modular ynamide synthesis, allowed for the development of a library of 4-aminoquinolines with ease. The tosylated and benzylated 4-aminoquinolines can be deprotected using potassium diphenylphosphide with subsequent hydrogenation, so that the 4-amino group can be further derivatized for additional libraries.

Moreover, 4-aminopyrazoles are accessible by reacting the sulfonyl ynamides with sydnones under copper catalysis. Choice of the copper catalyst was of crucial importance as hydrolysis of the ynamide was a persistent side-reaction. Using copper sulfate and a reducing agent several 4-aminopyrazoles were prepared, but under these conditions only C-4 unsubstituted sydnones were compatible. *In situ* prepared 3-azacyclohexyne was found to tolerate a much wider array of C-4 substituted sydnones, producing a mixture of both the 3,4- and 4,3-regio-isomers in good yields. The synthesis of a 3-azacyclooctyne could be verified by direct trapping with a sydnone as isolation of the cyclic ynamide has proven a challenge so far.

Additional investigations into heterocyclic methodology led to the development of highly sophisticated, non-symmetrical and axially-chiral dibenzo-1,3-diazepines, -oxazepines and -thiazepines from simple, commercially available anilines. The anilines were coupled to their corresponding reaction partners via a chloromethyl intermediate and the 7-membered ring was subsequently formed using direct arylation.

Kurzzusammenfassung – Abstract in German

Sulfonyl-Inamide sind vielfältige und synthetisch nützliche Reagenzien. Die Erweiterung ihrer synthetischen Einsetzbarkeit kann deshalb eine fruchtbare Quelle für die Entwicklung neuer Reaktionen darstellen. Zugänglich sind Sulfonyl-Inamide durch Kupfer-katalysierte amidierende Kreuzkupplungen von Sulfonylamiden mit Bromacetylenen oder durch Eliminierung von Dichlorenamid-Vorläufern. Unsubstituierte Inamide könnten via Sonogashira-Kreuzkupplungen weiter diversifiziert werden. Harzgebundene Inamide konnten durch die Kupplung von Inamiden mit Carbonsäurefunktion an ein Rink-Amid-Harz oder durch CuAAC-Reaktionen von Dichlorenamiden an ein Merrifield-Triazen-Harz synthetisiert werden. Im Anschluss wurde ihr synthetischer Wert untersucht.

Die modular zugänglichen Inamide reagierten mit Trifluormethansulfonsäure-anhydrid und 2-Chlorpyridin-aktivierten Amiden unter Bildung von hochfunktionalisierten 4-Amino-chinolinen. Tosylierte und benzylierte 4-Aminochinoline konnten anschließend mit Kalium-diphenylphosphid und durch Hydrierung entschützt werden, wodurch die erhaltene 4-Aminofunktionalität zur synthetischen Diversifizierung zur Verfügung steht.

4-Aminopyrazole sind durch eine Kupfer-katalysierte Cycloaddition der Sulfonyl-Inamide mit Sydnonen zugänglich. Die Wahl des Kupferkatalysators war dabei von entscheidender Bedeutung um Nebenreaktionen zu minimalisieren. Mehrere 4-Aminopyrazole konnten durch Verwendung einer Kombination aus Kupfersulfat und Reduktionsmittel synthetisiert werden, wobei jedoch nur C-4-unsubstituierte Sydnone kompatibel waren. Ein in situ hergestelltes 3-Azacyclohexin tolerierte hingegen ein viel breiteres Spektrum an substituierten Sydnonen, wobei Mischungen aus den Regio-isomeren in guten Ausbeuten erhalten wurden. Die Synthese eines 3-Azacyclooctins konnte nur indirekt durch die Reaktion mit einem Sydnon bewiesen werden, da die Isolierung des zyklischen Inamid bisher nicht erreicht werden konnte.

Zusätzliche Untersuchungen zur heterozyklischen Methodik führte zur Entwicklung der asymmetrischen und axial-chiralen Dibenzo-1,3-diazepine, -oxazepine und -thiazepine ausgehend von einfachen, handelsüblichen Anilinen. Die Aniline wurden dabei mit den entsprechenden Reaktionspartnern über eine Chlormethyl-Zwischenstufe gekuppelt. Eine anschließende direkte Arylierung unter Palladiumkatalyse mit einem CPhos-Katalysator bildete den 7-Ring.

Chapter 1. Introduction

Cyclic compounds that contain atoms of more than one different element as member of the ring are considered heterocyclic.[1] Although some heterocycles are inorganic compounds, often at least one carbon is present in the ring. In heterocyclic organic chemistry compounds containing oxygen, nitrogen or sulfur atoms are most commonly encountered. Saturated cyclic heterocycles often behave electronically similar to their acyclic counterparts, e.g. piperidine acts as an amine and 1,4-dioxane as an ether. However, unstrained and unsaturated compounds like pyridines, pyrazoles, thiophenes as well as benzene-fused heterocycles such as quinolines, benzofurans and indoles have gotten a lot more attention, since their aromaticity and heteroatomic π system participation are of great interest.

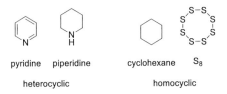

pyridine   piperidine   cyclohexane S_8

heterocyclic   homocyclic

Figure 1.1 Examples of heterocyclic and homocyclic compounds.

Since the beginning of the development of heterocyclic chemistry in the 1800s the field has made some great advances. Some notable[2] discoveries are the synthesis of alloxan (**1.1**) from uric acid in 1818,[3] the discovery of urea (**1.2**) by WÖHLER in 1828, the synthesis of furfural (**1.3**) by DÖBEREINER in 1832, the discovery of pyrrole (**1.4**) in 1834, the postulation of systematic chemical nomenclature for heterocyclic compounds by HANTZSCH[4] and WIDMAN,[5] the synthesis of synthetic dyes such as indigo (**1.5**) in the late 1800s and early 1900s, the isolation of porphyrins **1.6** and chlorophyll derivatives **1.7** from crude oil in 1936,[6] and the formulation of CHARGAFF's rules stating the ratios of pyrimidine and purine bases in double-stranded DNA.[7]

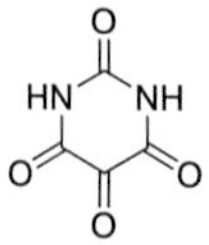

1.1
Alloxan
Brugnatelli, 1818

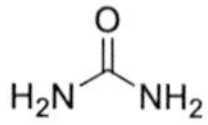

1.2
Urea
Wöhler, 1828

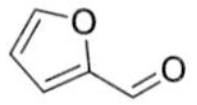

1.3
Furfural
Döbereiner, 1832

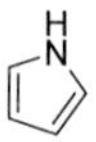

1.4
Pyrrole
Runge, 1834

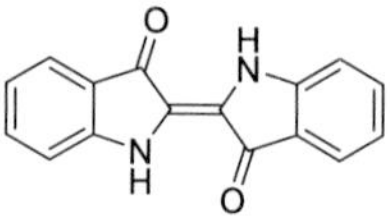

1.5
Indigo
von Baeyer, 1878

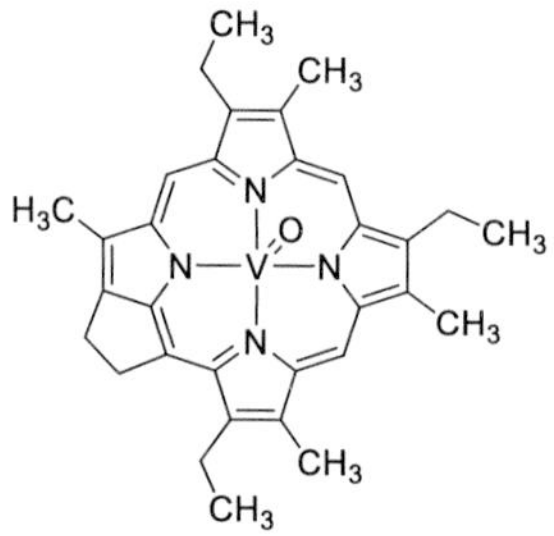

1.6
Porphyrin
Treibs, 1936

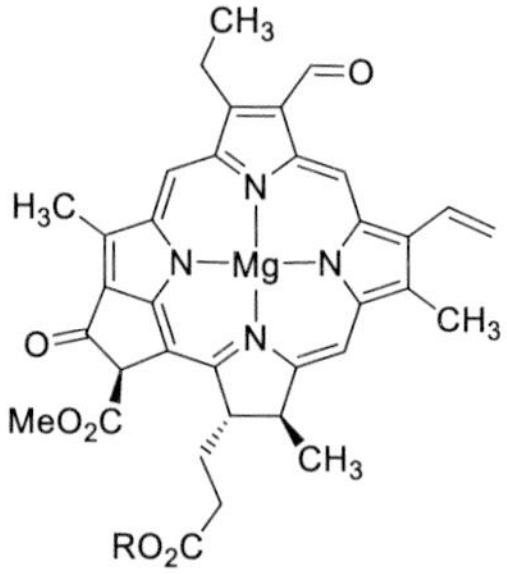

1.7
Chlorophyll a

Figure 1.2 Historically important heterocyclic discoveries.

The overall impact that the development of heterocyclic organic chemistry has had over the last centuries on the modern world is beyond measure. Many small molecules, whether discovered in nature or entirely man-made, have a heterocyclic component.[8] The majority of the most commonly prescribed drugs contain pyrazoles, pyridines, thiophenes or other heterocyclic moieties.[9,10]

Developing new reactions and new synthetic methods, as well as exploring the boundaries and limitations of existing methods, is the pinnacle of chemical innovation. An upcoming field in organic synthesis employs the use of aminoalkynes. These compounds, named ynamides (see Figure **1.3**), consist of a compound with a nitrogen atom directly adjacent to an alkyne. Since the nitrogen's lone pair activates the triple bond considerably, electron-withdrawing groups are often placed on the nitrogen atom in order to temper its reactivity and obtain more stable compounds. Although ynamides have gotten some considerable attention over the last years, a lot of their chemistry is still undiscovered.[11-13] Due to the nitrogen's tendency to form an N-allene type group (see Figure **1.3**), it displays a dual reactivity: C-1 is nucleophilic and C-2 electrophilic. The goal of this dissertation was to explore this dual reactivity of the ynamides and use it as a tool to prepare libraries of heterocyclic compounds that could be screened for their biological activity.

Figure 1.3 Ynamide isomerization to the corresponding *N*-allene.

First, Chapter 2 discusses some of the more practical and modular ynamide syntheses and related ongoing efforts to develop solid-supported ynamides that may be used for combinatorial chemistry. These modular approaches to sulfonyl ynamides are extremely valuable since they allow a direct access to a wide range of C-1 substituted ynamides, which can be used in the facile synthesis of libraries of heterocycles that can be submitted to biological screenings.

One of such syntheses of a heterocyclic library is discussed in Chapter 3. Here 4-amino-quinolines, which are considered to be privileged structures in medicinal chemistry due to their prominent role in malaria treatment, are prepared in a highly modular manner. The ynamides can react with electrophilically-activated amides to produce a wide range of highly functional-ized 4-aminoquinolines and, in contrast to previous synthetic routes, using this procedure allows for easy access to substitutions at all positions on the quinoline core structure. Since the developed tosylated and benzylated 4-aminoquinolines could be deprotected to yield the free amino analogs, these compounds can easily be further derivatized to yield libraries for biological testing.

Another common *N*-heterocyclic motif found both in nature and in modern drugs is the 5-mem-bered pyrazole. These are synthetically accessible via copper-catalyzed cycloadditions between sydnones – 5-membered meso-ionic heteroaromatic compounds – and electron-deficient alkynes. In order to expand the synthetic usability of this reaction to more reactive and electron-rich substrates, Chapter 4 details the reactions between sulfonyl ynamides and sydnones. Using copper catalysis, 4-aminopyrazoles were easily and regioselectively accessible, but the use of certain copper additives led to fast ynamide degradation.

In order to circumvent the issue of the copper-accelerated hydrolysis of terminal sulfonyl ynamides, Chapter 4 discusses the synthesis and use of strained cyclic ynamides. Both a six-membered 3-azacyclohexyne and an eight-membered version were prepared and yielded regio-isomers of the desired pyrazoles.

Lastly, Chapter 5 discusses recent investigations into heterocyclic methodology that led to the development of highly sophisticated, non-symmetrical and axially-chiral dibenzo-1,3-diaze-pines, -oxazepines and -thiazepines from simple, commercially available anilines. Beside their potential biological applications, these compounds could be used to make *N*-heterocyclic carbenes and serve as a novel type of ligands in catalysis.

To summarize, this thesis gives an overview of recent developments in the field of heterocyclic organic synthesis, discussing the synthesis and use of ynamides to form amino-substituted heterocycles with ease, and the synthesis of novel dibenzo-1,3-diazepines.

References

[1]   IUPAC, *Compendium of Chemical Terminology, 2nd ed. (the "Gold Book")*; Blackwell Scientific Publications: Oxford, 1997.

[2]   E. Campaigne, *J. Chem. Educ.*, **1986**, *63* (10), 860, *Adrien Albert and the rationalization of heterocyclic chemistry.*

[3]   G. Brugnatelli, *Giornale di Fisica, Chimica, Storia Naturale, Medicina, ed Arti*, **1818**, (1), 117-129, *Sopra i cangiamenti che avvengono nell' ossiurico (ac. urico) trattato coll' ossisettonoso (ac. nitroso).*

[4]   A. Hantzsch, J. H. Weber, *Ber. Dtsch. Chem. Ges.*, **1887**, *20* (2), 3118-3132, *Ueber Verbindungen des Thiazols (Pyridins der Thiophenreihe).*

[5]   O. Widman, *J. Prakt. Chem.*, **1888**, *38* (1), 185-201, *Zur Nomenclatur der Verbindungen, welche Stickstoffkerne enthalten.*

[6]   A. Treibs, *Angew. Chem.*, **1936**, *49* (38), 682-686, *Chlorophyll- und Häminderivate in organischen Mineralstoffen.*

[7]   E. Chargaff, R. Lipshitz, C. Green, *J. Biol. Chem.*, **1952**, *195* (1), 155-160, *Composition of the desoxypentose nucleic acids of four genera of sea-urchin.*

[8]   E. Vitaku, D. T. Smith, J. T. Njardarson, *J. Med. Chem.*, **2014**, *57* (24), 10257-10274, *Analysis of the Structural Diversity, Substitution Patterns, and Frequency of Nitrogen Heterocycles among U.S. FDA Approved Pharmaceuticals.*

[9]   N. A. McGrath, M. Brichacek, J. T. Njardarson, *J. Chem. Educ.*, **2010**, *87* (12), 1348-1349, *A Graphical Journey of Innovative Organic Architectures That Have Improved Our Lives.*

[10]   M. Baumann, I. R. Baxendale, *Beilstein J. Org. Chem.*, **2013**, *9*, 2265-2319, *An overview of the synthetic routes to the best selling drugs containing 6-membered heterocycles.*

[11]   K. A. DeKorver, H. Y. Li, A. G. Lohse, R. Hayashi, Z. J. Lu, Y. Zhang, R. P. Hsung, *Chem. Rev.*, **2010**, *110* (9), 5064-5106, *Ynamides: A Modern Functional Group for the New Millennium.*

[12]   G. Evano, A. Coste, K. Jouvin, *Angew. Chem. Int. Ed.*, **2010**, *49* (16), 2840-2859, *Ynamides: Versatile Tools in Organic Synthesis.*

[13]   X.-N. Wang, H.-S. Yeom, L.-C. Fang, S. He, Z.-X. Ma, B. L. Kedrowski, R. P. Hsung, *Acc. Chem. Res.*, **2014**, *47* (2), 560-578, *Ynamides in Ring Forming Transformations.*

Chapter 2. Sulfonyl ynamides

The research presented in this chapter has been conducted at the Institute of Organic Chemistry at the Karlsruhe Institute of Technology in Karlsruhe, Germany (with Prof. Dr. Stefan BRÄSE) and at Bayer CropScience GmbH in Frankfurt-Höchst, Germany (with Dr. Stephen D. LINDELL). Dr. Stephen D. LINDELL and Dr. Sabilla ZHONG are thanked for their help during my work on this chapter. Parts of this work have been published as T. Wezeman, S. Zhong, M. Nieger and S. Bräse, *Angew. Chem. Int. Ed.* **2016**, *55*, 3823 – 3827.

2.0 Abstract & Graphical abstract

A robust and modular synthesis of a diverse set of sulfonyl ynamides has been investigated. Using copper-catalyzed amidative cross-couplings of bromo-acetylenes to sulfonyl amides several sulfonyl ynamides are accessible. The terminal ynamides can then be diversified by subjecting them to Sonogashira couplings. Alternatively, a dichloroenamide precursor can be synthesized by reacting sulfonyl amides with trichloroethane. In the presence of phenyllithium these precursors undergo elimination and lithium-halogen exchange. The resulting highly reactive lithiated species can be quenched with different electrophiles to obtain sulfonyl ynamides in a modular way. Finally, the synthesis of a solid-supported ynamide system has been investigated. Since copper-catalyzed amidative cross-couplings on solid-supported sulfonamides failed, sulfonyl ynamides were coupled to the Rink resin via an amide bond. However, as the loading of the ynamides proved insufficient for subsequent library syntheses, an alkyne-bearing T1 triazene linker was constructed on commercially available Merrifield resin. This linker was then coupled to the azido dichloroenamide using a copper-catalyzed azide alkyne cycloaddition.

2.1 From ynamines to ynamides

An accidental synthesis

It could be argued that the most interesting discoveries are often unexpected. When ZAUGG *et al.* in 1958 tried to alkylate phenothiazine **2.1** with propargyl bromide and sodium hydride in dimethylformamide they accidentally prepared the first known ynamine **2.2**.[1] Only a few years later VIEHE developed a more practical route to ynamines (Scheme **2.1**).[2] Due to the electron-donating ability of the nitrogen the triple bond is polarized, leading to quite reactive species. Although the community of synthetic organic chemists did see their usefulness,[3-5] the ynamines remained laboratory curiosities for several decades, due to their instability and difficult handling.

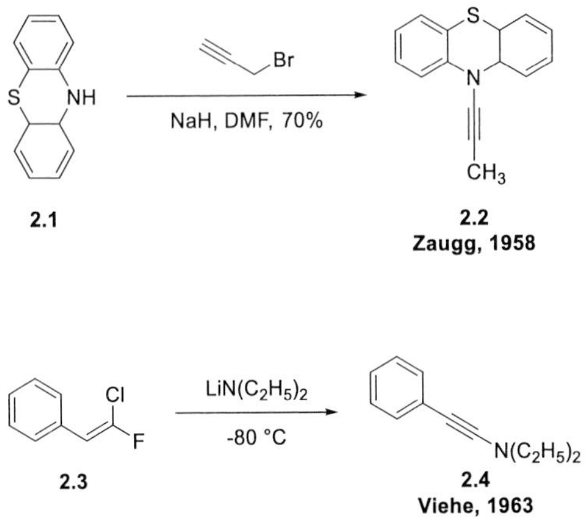

2.1   **2.2**
                                                                   Zaugg, 1958

2.3   **2.4**
                                                                   Viehe, 1963

Scheme 2.1 The synthesis of the first ynamines.

Thankfully the reactivity of the ynamines can be toned down by the introduction of electron-withdrawing groups on the nitrogen. These more thermally stable ynamides are often convenient to handle and offer a wide scope of possibilities for synthetic efforts. The first such ynamide was reported in 1972 and prepared by elimination of chloroenamide **2.6** (Scheme **2.2**).[6]

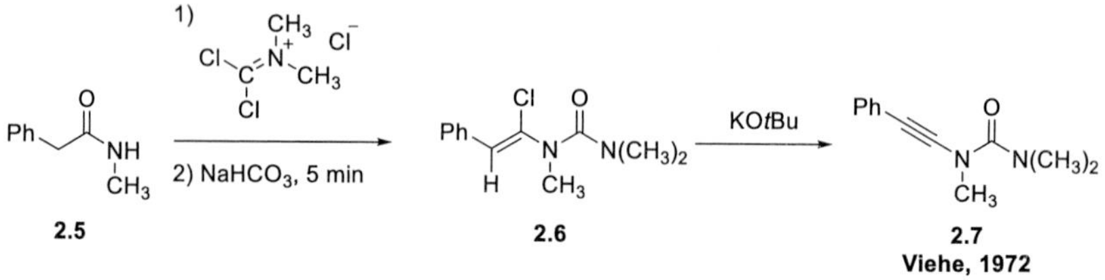

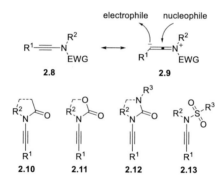

Scheme 2.2 The first synthesis of an ynamide in 1972.

Since the 1970s the field of ynamides has grown tremendously. Several commonly used electron-withdrawing groups include lactams **2.10**, carbamates **2.11**, ureas **2.12** and sulfonamides **2.13** (Figure **2.1**). For simplification all electron-deficient ynamines are typically called ynamides.

Figure 2.1 Ynamide reactivity and four common ynamide motifs.

Versatility is key

Ynamides can truly be called versatile reagents. Scheme **2.3** features a small excerpt out of the wide range of reactions possible with ynamides. For clarity Scheme **2.3** only shows intermolecular reactions, while there are also large amounts of intramolecular reactions published, which are less relevant for this chapter. By subjecting ynamides to copper catalysis isoxazoles[7] **2.14d** and triazoles[8] **2.14e** can be prepared in a straightforward manner. Furthermore Pauson-Khand reactions,[9] rhodium-catalyzed Diels-Alder reactions to anilines[10] **2.14a** and BF_3·Et_2O promoted hetero [2 + 2] cycloadditions to amides[11] **2.14b** have been investigated. To name some more exotic examples: nickel-catalyzed [3 + 2 + 2] cycloadditions to 7-membered rings **2.14h**,[12] multicomponent reactions to pyridines[13] **2.14g** (where the ynamide nitrogen is removed as a titanium complex before forming the final pyridine) and thermal isomerizations to nitriles[14] **2.14f** are also possible.

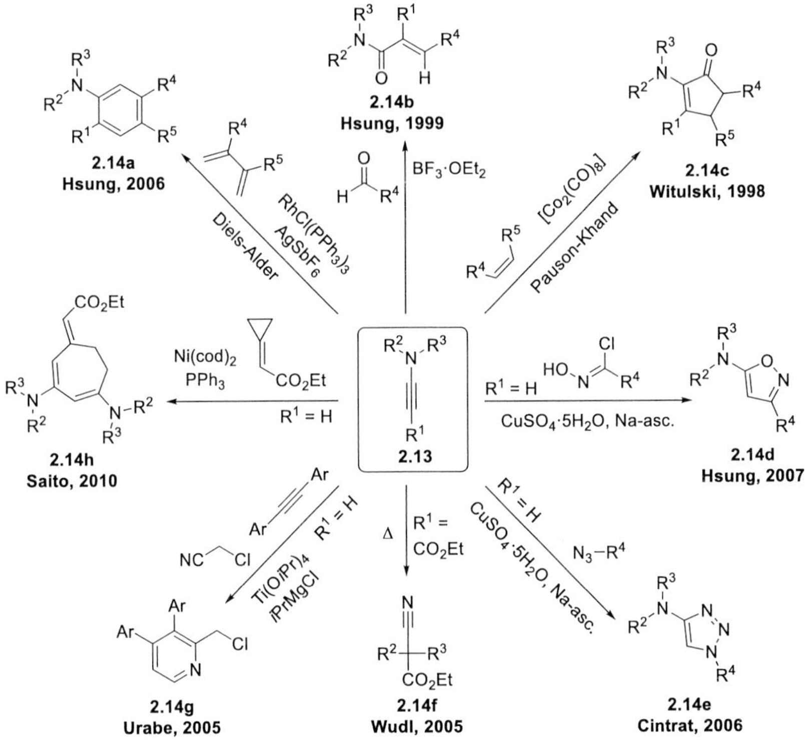

Scheme 2.3 Small excerpt of the versatility that ynamides encompass.[7-14]

Since the field of ynamide chemistry has grown so extensively, this chapter is hardly the place to comprehensively summarize all findings. Several excellent reviews on the chemistry of ynamines and ynamides have been published, discussing both ynamide synthesis and reactions with ynamides[3,5,15-20] including a special issue of Tetrahedron in 2006.[21]

2.2 Synthetic approaches to ynamides

There is more than one way that leads to Rome. The same can be said about most chemical reactions; often a certain compound can be synthesized by different routes. In order to explore what exciting chemistry can be done with sulfonyl ynamides – the subject of Chapters 3 and 4 – one first needs to address the synthesis of the ynamides.

A wide range of possibilities

Ynamides have been made by isomerization from propargyl species, as with ynamine **2.2**, shown in Scheme **2.1**, by coupling alkynyliodonium salts to amines, amidative cross-couplings or elimination of halo-enamides.

Perhaps the synthetically least useful of these methods is the isomerization of propargyl species. Although in principle ynamides are accessible via this route, HSUNG et al. showed that depending on the nature of the electron-withdrawing groups one could end up with allenamides **2.15** or the desired ynamides **2.17** under identical conditions.[22,23]

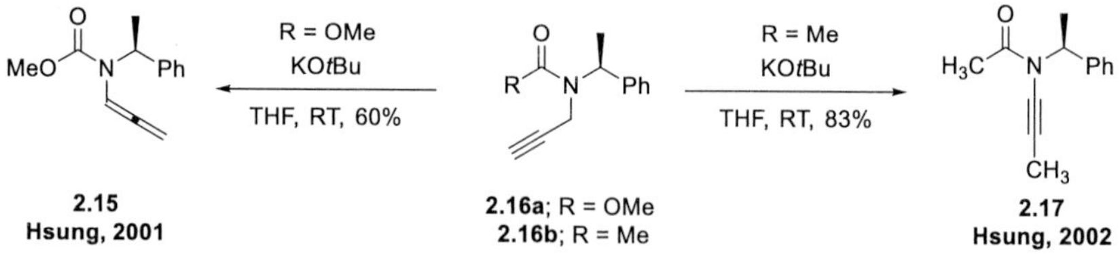

| **2.15** | **2.16a**; R = OMe | **2.17** |
| Hsung, 2001 | **2.16b**; R = Me | Hsung, 2002 |

Scheme **2.4** Ynamides by isomerization.

Alternatively ynamides can be prepared by coupling of alkynyliodonium salts to lithiated amides, parallel to what STANG pioneered in 1994 with his synthesis of ynamines **2.19**.[24]

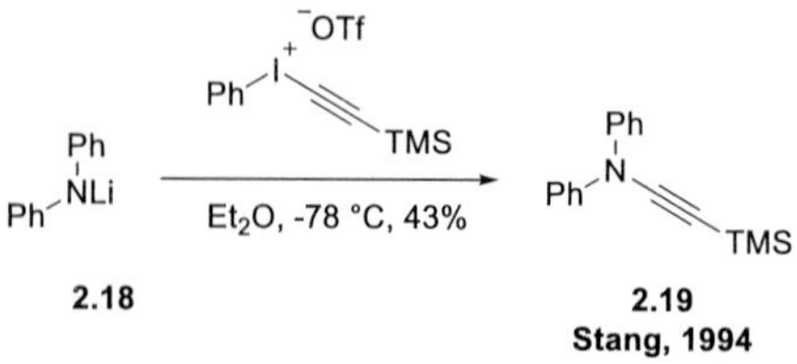

| **2.18** | **2.19** |
| | Stang, 1994 |

Scheme **2.5** Ynamines synthesis from alkynyliodonium salts.

The mechanism is believed to proceed by a nucleophilic attack in β-position to the iodine and subsequent 1,2-migration.[9] Although this method generally allows for the construction of bis-ynamides, this does not occur in every case.[25]

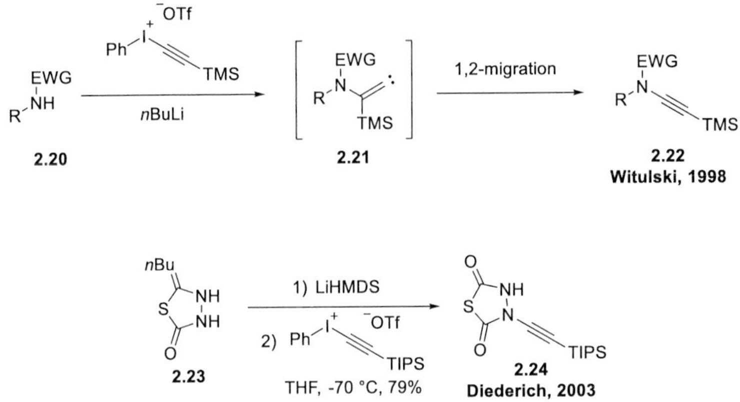

Scheme 2.6 Ynamides via alkynyliodonium salts.

Although both the isomerization and alkynyliodonium approach have proven to deliver ynamides in synthetically useful yields, both methods are unsuitable for the preparation of a library of ynamides, due to their lack of synthetic flexibility. However, the elimination of halo-enamides or amidative cross-coupling show great promise in this regard.

Amidative cross-coupling

The first example of a synthesis of an ynamide **2.27** by amidative cross-coupling dates back to 1985, when DOMIANO and colleagues accidentally found that their copper(I) acetylide reacted with the amide instead of displacing the iodine.[26]

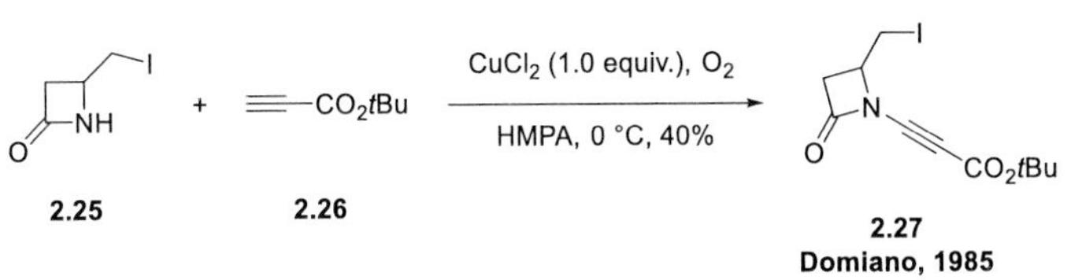

Scheme 2.7 First ynamide synthesis via amidative coupling, using stoichiometric amounts of copper chloride.

Almost twenty years later HSUNG *et al.* reported the first catalytic version, using 5 mol% CuCN or CuI and a simple ligand, that delivered ynamides **2.30** in satisfactory yields.[27]

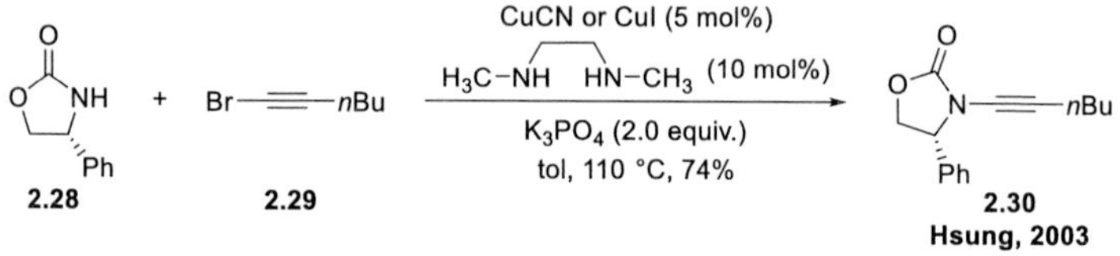

Scheme 2.8 Ynamide synthesis via catalytic amidative coupling.

Over the years several alterations to this original protocol have been reported, such as the use of stoichiometric copper iodide to make the reaction proceed at room temperature[28] and studies about the water content of the base used.[29] Most notable are the changes of the catalytic system to more environmentally friendly $CuSO_4 \cdot 5H_2O$ with 1,10-phenanthroline as ligand[10,30-32] or to inexpensive $FeCl_3 \cdot 6H_2O$.[33]

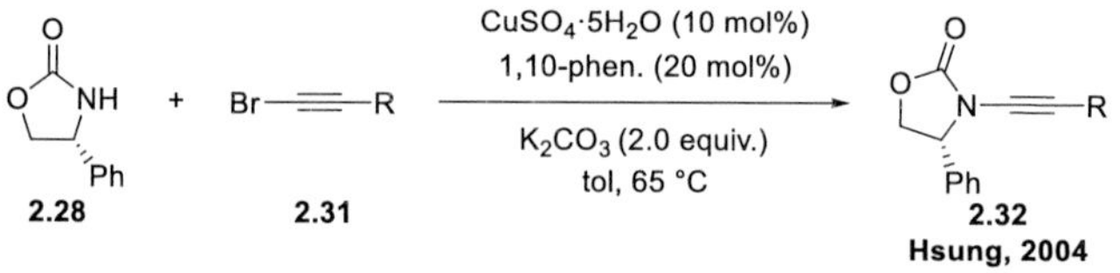

Scheme 2.9 Amidative coupling of bromo-acetylene **2.31** using catalytic copper sulfate.

Due to the relative ease and inexpensive nature of the copper sulfate-catalyzed reaction (Scheme **2.9**) it was decided to use the same method to prepare a small collection of sulfonyl ynamides. First sulfonyl amides **2.35** and **2.37** were prepared via either reductive amination or by reacting benzylamines **2.36** with tosyl chloride in a nucleophilic manner (Scheme **2.10**). The advantage of the reductive amination approach[34] is the increased flexibility in terms of the substitution on the benzyl group and the availability of methyl 4-sulfamoylbenzoate **2.33** as reaction partner. At a later stage the methyl ester can be deprotected with lithium hydroxide, thus opening up the carboxylic acid for subsequent couplings (see Chapter 2.4).

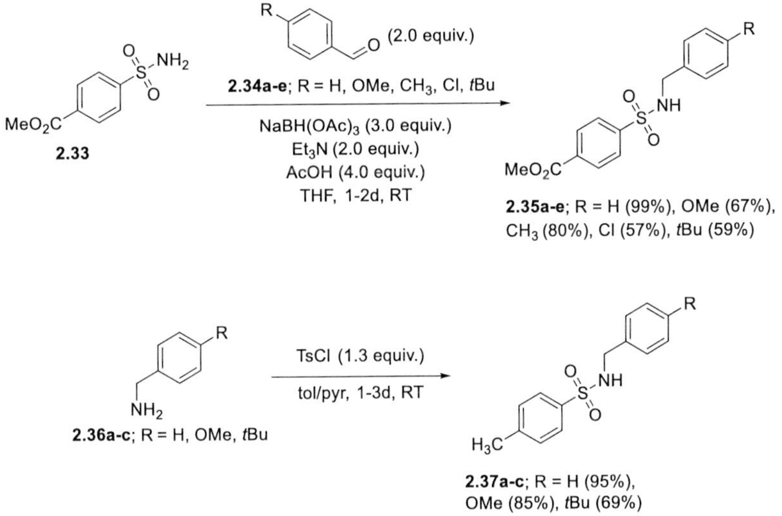

Scheme 2.10 Synthesis of sulfonyl amides via reductive amination or nucleophilic attack on tosyl chloride.

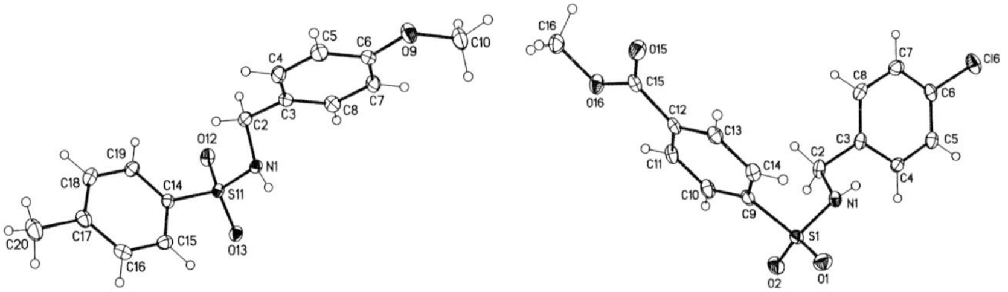

Figure 2.2 Molecular structures of **2.37b** (left) and **2.35b** (right).

Several bromo-acetylenes were prepared by bromination of terminal acetylenes under silver nitrate catalysis[35,36] allowing access to the coupling partners on large scale (> 5 g) in excellent yields. Care must be taken during workup since the acetylenes **2.39** were found to be volatile.

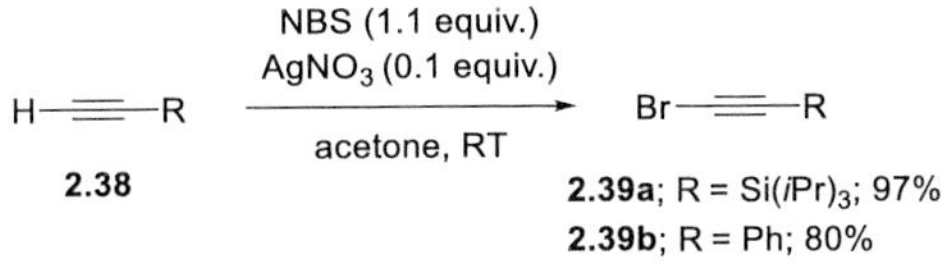

Scheme 2.11 Preparation of the bromo-acetylenes for the amidative coupling.

Now, with all the precursors at hand a small set of sulfonyl ynamides could be prepared (Scheme **2.12**) in acceptable to good yields. It must be noted that in order for the couplings to work as desired it was found helpful to grind the copper sulfate and 1,10-phenanthroline thoroughly before use.

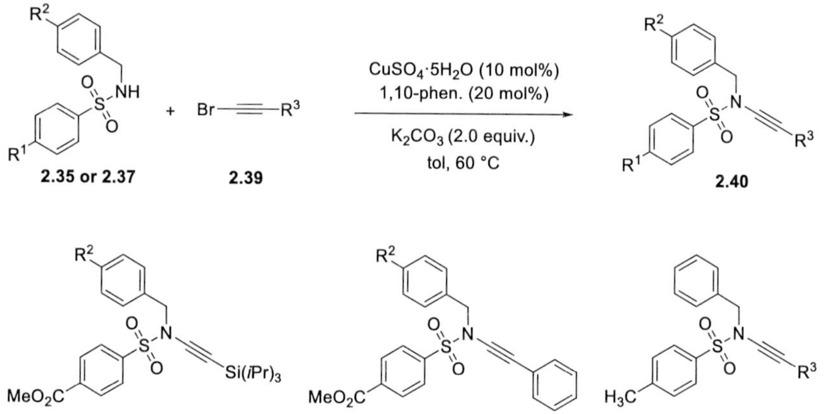

2.35 or 2.37   **2.39**

CuSO$_4$·5H$_2$O (10 mol%)
1,10-phen. (20 mol%)

K$_2$CO$_3$ (2.0 equiv.)
tol, 60 °C

2.40

2.40a-e; R^2 = H (quant.), OMe (29%),
CH$_3$ (78%), Cl (52%), *t*Bu (48%)

2.41a-c; R^2 = H (54%),
OMe (55%), CH$_3$ (74%)

2.42a; R^3 = Ph; 34%
2.42b; R^3 = Si(*i*Pr)$_3$; 86%

Scheme 2.12 Amidative coupling of bromo-acetylene using catalytic amounts of copper sulfate.

Terminal sulfonyl ynamides are easily accessible from the triisopropylsilane-protected precursors using tetra-*n*-butylammonium fluoride in THF. A simple and direct access to these more water-sensitive compounds is necessary since not all chemistry is compatible with internal ynamides. The synthesis of isoxazoles using nitrile oxides for example, requires the formation of a copper(I) acetylide intermediate, which does not happen with bis-substituted alkynes.[7]

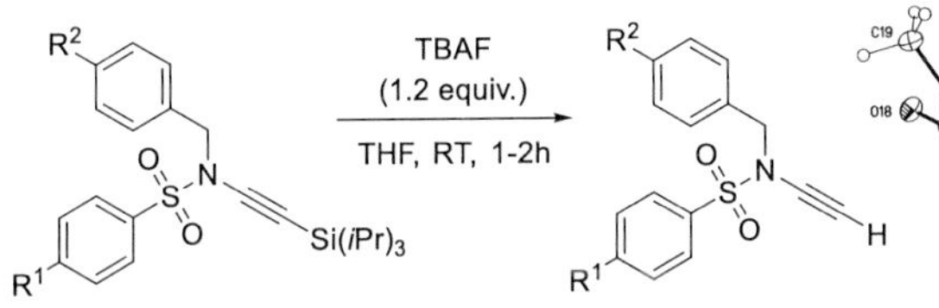

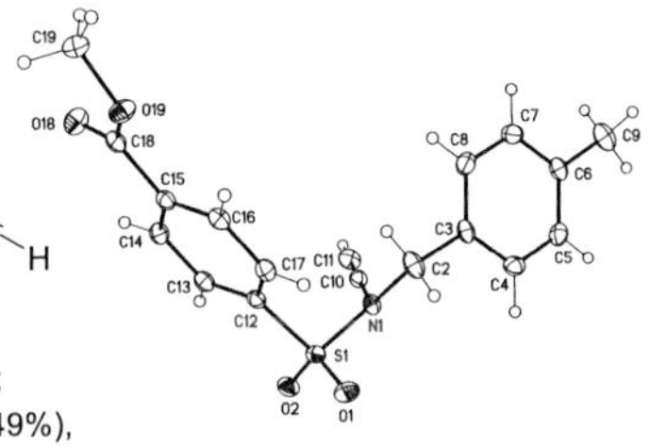

2.40a-e; R^1 = CO$_2$Me;
R^2 = H, OMe, CH$_3$, Cl, *t*Bu
2.42b; R^1 = CH$_3$; R^2 = H

2.43a-e; R^1 = CO$_2$Me;
R^2 = H (58%), OMe (49%),
CH$_3$ (76%), Cl (56%), *t*Bu (61%)
2.44; R^1 = CH$_3$; R^2 = H; 88%

2.43c

Scheme 2.13 Removal of the triisoproylsilane to yield terminal sulfonyl ynamides. Molecular structure of **2.43c**.

Although the amidative copper-catalyzed coupling of acetyl bromides to amides yields the desired ynamides without any issues, this synthetic route was found less convenient for larger scale reactions. Besides the formation of larger quantities of black residues that make purification complicated, larger amounts of acetylene bromides need to be synthesized. More inconveniently, for every different substitution on the ynamide a different bromo-acetylene has to be prepared. This slows down the process and does not allow for a very modular synthesis.

In order to solve this issue two different solutions were investigated. The first one involves changing from the copper-catalyzed amidative cross-coupling to the elimination of halo-enamides. The second uses Sonogashira cross-coupling reactions to modify terminal ynamides and prepare internal ynamides.

Elimination of haloenamides

Ynamides are also accessible via elimination of halogenated enamides. These reactions typically employ strong bases like *n*-butyllithium to abstract a proton and eliminate the halogen under triple bond formation. In this manner ynamide derivatives of the trichloroenamide analogs of purines and pyrimidines were prepared.[37] In 2000, BRÜCKNER developed a route to *β,β*-dichloroenamide **2.48** from formamides, which could be eliminated to the terminal ynamides,[38] quenched with zinc bromide[39] – to be then directly used in subsequent one-pot Negishi reactions – or even directly quenched with a suitable electrophile.[40]

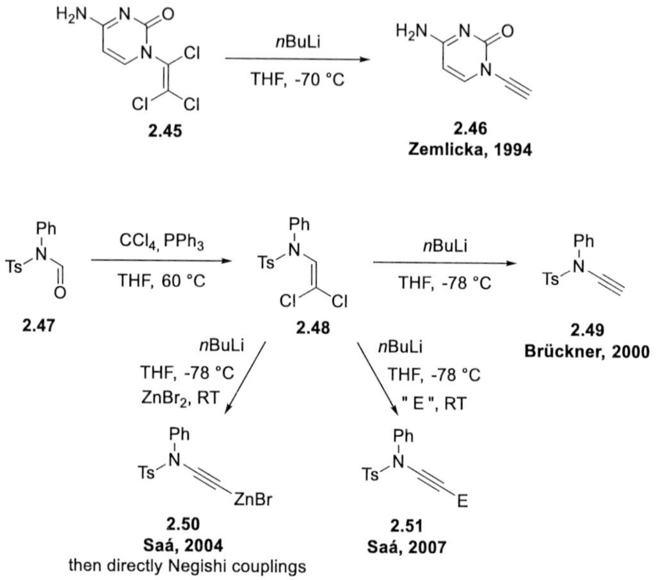

Scheme 2.14 Ynamides via elimination procedures.

Recognizing that the quenching with a suitable electrophile would allow for the most modular and direct route to sulfonyl ynamide analogs, it was decided to synthesize β,β-dichloroenamide **2.48**. However, around that time ANDERSON et al. published a similar route to E-α,β-dichloroenamides **2.52** in just one step.[41] Intrigued by this new method, that employed readily available trichloroethylene as synthon for the alkyne, it was decided to try these conditions.

The sulfonyl amides **2.37a-c** were coupled to 1,2-dichloroethyne, formed *in situ* from the trichloroethylene to result in the E-α,β-dichloroenamides **2.52** in excellent yields. The E-conformation of the chloro-substituents was confirmed using X-ray crystallography by both the ANDERSON group in their initial report and by the X-ray images of dichloroenamides **2.52**. Most notably, the dichloroenamides were easily accessible in multi-gram scale via this method, as purification could often be done by simple recrystallizations.

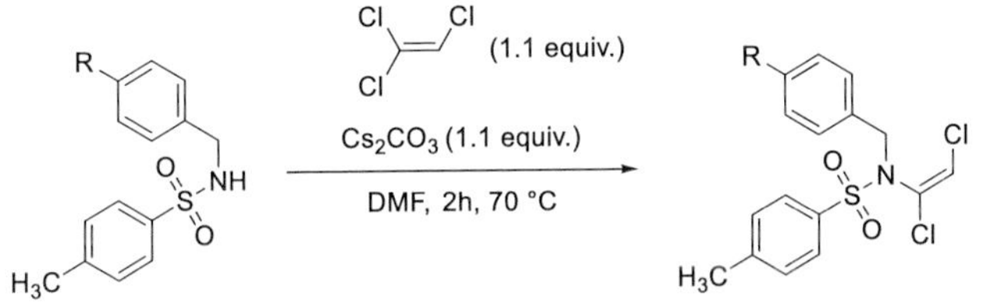

2.37a-c

2.52a; R = H; 35 g; 93%
2.52b; R = OMe; 2 g; 81%
2.52c; R = *t*Bu; 10 g; 67%

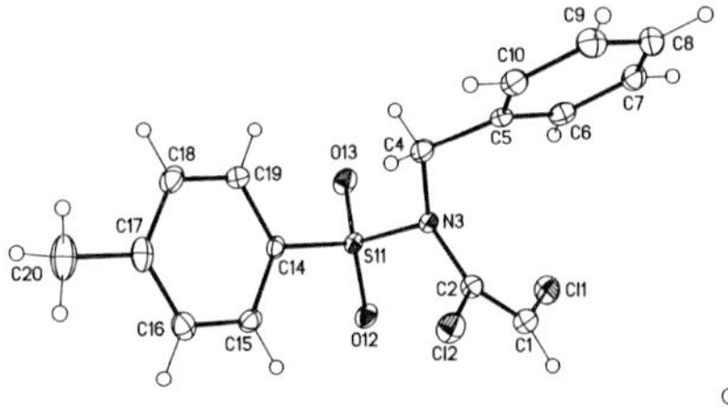

2.52a

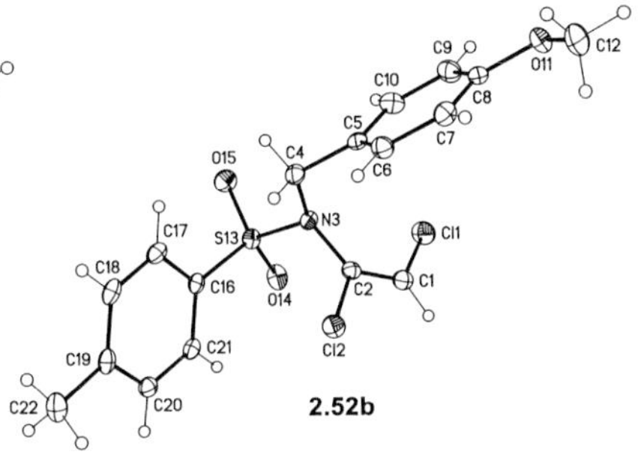

2.52b

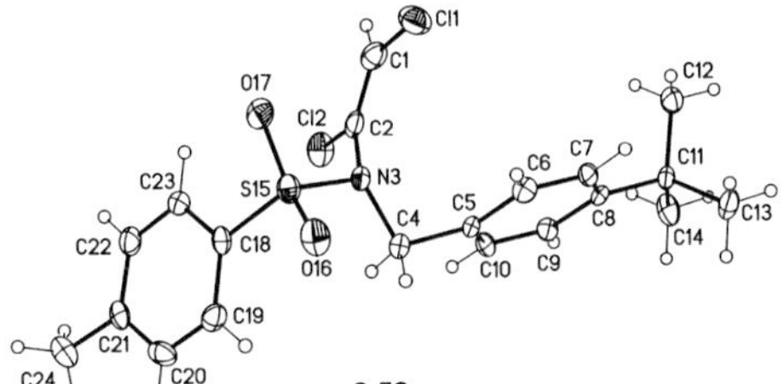

2.52c

Scheme 2.15 Multi-gram scale synthesis of *E*-*α,β*-dichloroenamides and their molecular structures.

With the dichloroenamides at hand, the synthesis of sulfonyl ynamides shown in Scheme **2.16** commenced. The first equivalent of phenyllithium is required for the elimination to the chloroethyne species **2.53**, which then undergoes lithium-halogen exchange with the second equivalent of phenyllithium. This lithiated ynamide **2.54** is reactive enough to react with electrophiles and produce the desired terminal **2.56 - 2.57** or internal ynamides **2.55 - 2.58**. Since the terminal ynamides arise from quenching with water, care must be taken to work under fully water-free conditions when internal ynamides are desired. Especially the electrophiles have to be dry, since they are added at the most crucial moment; otherwise mixtures of internal and terminal ynamides are made.

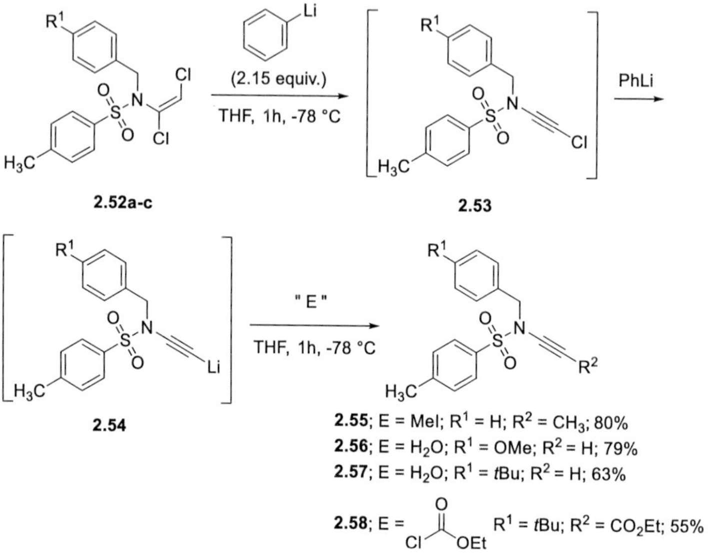

Scheme 2.16 Multi-gram scale synthesis of *E*-α,β-dichloroenamides.

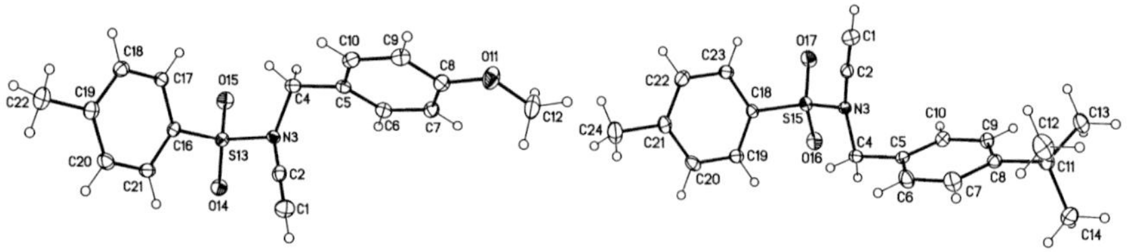

Figure 2.3 Molecular structures of ynamides **2.56** and **2.57**.

Diversification via Sonogashira chemistry

Although the two ynamide syntheses investigated yield terminal and internal ynamides in a convenient manner, an additional possibility to diversify the ynamides was examined. As discussed above, several different ynamides are accessible via the copper-catalyzed amidative cross-coupling, such as cyclopropane derived ynamide **2.59**.

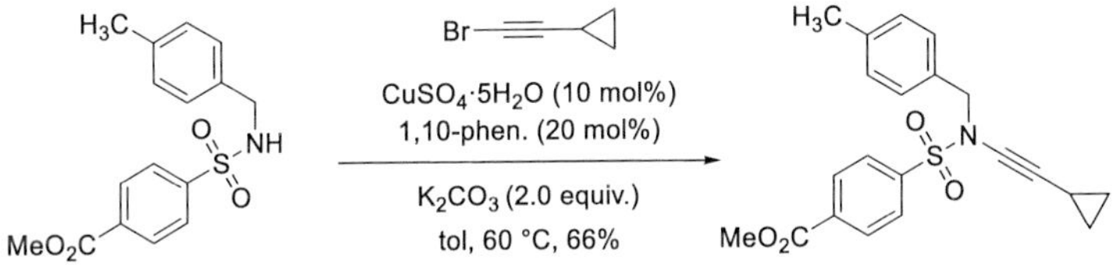

2.35c **2.59**

Scheme 2.17 Amidative cross-coupling requires different bromo-acetylenes, making this route less ideal for diversification.

However, this route is not the most convenient route to create analogs because it requires preparation of different bromo-acetylenes. Instead, with the terminal ynamides at hand, Sonogashira reactions were attempted, as it was known that ynamides should undergo these reactions quite well.[42] Although initial test reactions were performed by myself, the three Sonogashira products **2.60 - 2.62** and cyclopropyl ynamide **2.59** described in this chapter were synthesized and characterized by Dr. Sabilla ZHONG. Their synthetic and analytical data can be found online,[43] but for convenience the Sonogashira procedure is also mentioned as general procedure **GP2.6** in the experimental section.

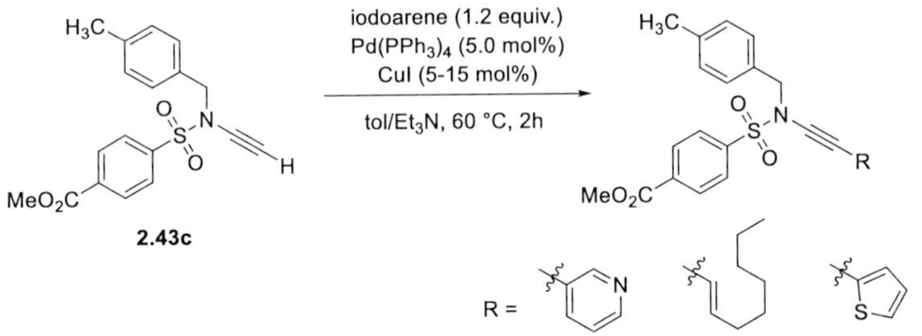

2.43c

2.60; 52% **2.61**; 59% **2.62**; 43%

Scheme 2.18 Diversification via Sonogashira reactions.

2.3 Solid-phase organic synthesis

Solid-supported chemistry has been initially developed by MERRIFIELD to serve as platform for peptide synthesis.[44,45] However, after four decades the field has grown tremendously and encompasses combinatorial small molecule library syntheses,[46] total syntheses of natural products,[47] organometallic reactions[48,49] and novel ways to cope with otherwise inconvenient, unpleasant or unstable reagents.[50] The best studied and most frequently used polystyrene resins include the Rink, Wang, Merrifield and tritylchloride resins (Figure **2.4**), but many more are available.

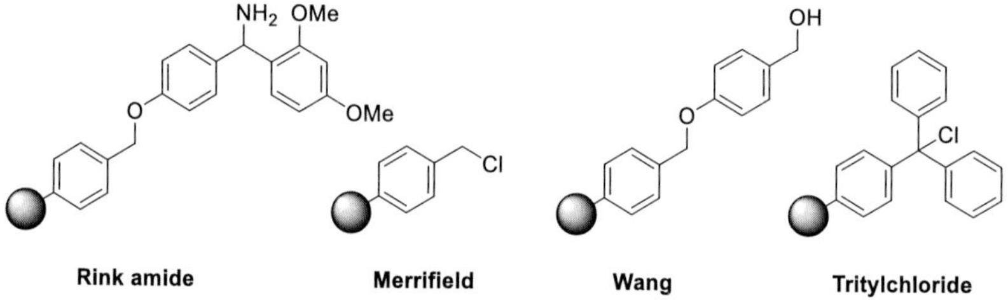

Figure 2.4 Common resins used for reactions on solid support.

Key advantage of solid-phase organic synthesis (SPOS) is that, since the product is bound to the polymeric support, the purification is simplified enormously. After the reaction has been driven to completion, typically by using large excess of reagents, concentrated mixtures and elevated temperatures − the resins can typically withstand up to 120 °C − the beads are transferred to a filter and all side-products and excess reagents are washed away. After this reaction sequence has been repeated and the final product is bound to the resin, the desired product can be cleaved off the beads. Cleaving conditions often involve strong acids but the exact conditions depend highly on the resin and linker used. Since over the years many different linkers and resins have been developed that are compatible with different chemistry and cleave under different circumstances, the choice of which resin to use can be complex. Choices include safety-catch linkers,[51] traceless linkers[52] and linkers that cleave while installing a new functional group, such as the triazene linkers.[53] To shed some light on the plethora of options, several books[54-57] and reviews[58,59] have been published discussing linker strategies.

A preparative limitation to SPOS is that reaction control can be a challenge. Since conventional methods such as thin layer chromatography, GC and LCMS are not suitable, alternatives have been developed. These include certain coloring tests, such as the Kaiser test (see Figure **2.5**), that can be used to detect amine conversion[60] and *in situ* spectroscopic methods such as IR and Raman can be used to track the presence of certain functional groups.[61]

Figure 2.5 Example of the Kaiser ninhydrin coloring test. From left to right: Blank – no resin, coloring reagents only; Resin with NH_2 – dark blue resin and solution; Resin after successful amide coupling – beads are colorless and the solution is reddish; Resin after incomplete amide coupling – beads are still blue and the solution is purple.

Alternatively, if the loading is high enough, a few beads can be taken, the product cleaved off and subsequently analyzed with mass spectrometry. Additional issues in SPOS reactions can arise when insoluble reagents are used or insoluble side-products formed.

Development of novel aluminium heating block for SPOS

Most small-scale reaction on solid support that require heating are performed in a sealable vial that is heated in a microwave or with a sand-bath. Advantage of the microwave-assisted heating is the precise temperature control, downside is that only one reaction can be done at a time. Since the resins are friction-sensitive the use of stirring bars is ill-advised: the beads can easily be ground into a powder. Typically a heated sand-bath is placed on a shaking plate and gently shaken. Since the resin-bound reactions are diffusion limited, the rate of shaking is irrelevant as long as everything mixes nicely. Although the use of the sand-bath allows us to run multiple reactions at the same time, the temperature control difficult. The top of the sand-bath often is much cooler than the lower half, leading to inconsistent reaction protocols. The use of aluminium heating blocks circumvents this issue, since here the heating is a lot more homogenous. However, conventional aluminium heating blocks used in organic synthesis labs feature only vertical pockets that are accessible from the top. This results in the sealed vials standing upright, which does not allow for good mixing when shaking the heating block. Therefore a heating block with the pockets on the sides, so that the reaction vessels could be resting horizontally in the block, was designed. By using a circular block it is guaranteed that all pockets have exactly the same temperature, since they are equally distanced from the thermometer in the center of the block. After discussions with Patrick HODAPP the new SPOS-suitable heating block was designed (Figure **2.6**) and constructed with aid of Dr. Patrick WEIS.

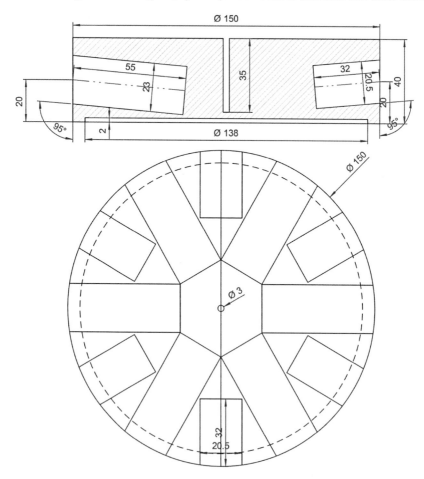

Figure 2.6 Detailed scheme of the aluminium SPOS heating block.

The SPOS block (see Figure **2.7**) features six pockets for large (20 mL) reaction vessels and six for smaller (10 mL) reaction vessels. The pockets are drilled at a slight angle to the horizontal plane, so the vessel is prevented from sliding out during shaking. A thermometer can be inserted through the small hole on the top of the block. Additionally a thin cut-out on the bottom of the block prevents the entire block from sliding of the heater during the shaking.

Figure 2.7 SPOS-suitable aluminium heating block.

By simply rotating the circular SPOS block all pockets are easily accessible. Figure **2.8** shows a direct comparison between the conventional aluminium blocks and the SPOS block. The use of the SPOS block drastically improved the mixing and heating of solid-supported reactions.

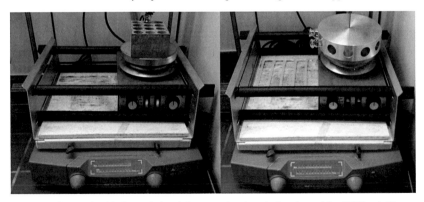

Figure 2.8 Heating blocks on shaker: left: conventional vertical pockets; right: SPOS-suitable horizontal pockets

2.4 Exploring solid-supported ynamide syntheses

Combining a solid-phase platform with a highly multi-functional group, such as the sulfonyl ynamides, offers great opportunities to synthesize large libraries with ease. Surprisingly, to date no reports of solid-supported ynamides have been published (based on detailed SciFinder search in May 2016).

Copper catalysis using the Rink amide Resin

Initial attempts to prepare solid-supported ynamides were designed in analogy to the copper-catalyzed ynamide synthesis discussed in Chapter 2.2. The route was based on the Rink amide resin, which is stable under a wide range of temperatures and conditions and cleaves only under strongly acidic conditions. Since this resin is widely used in solid-supported peptide synthesis it was Fmoc-protected. Facile deprotection with 20% piperidine in DMF yielded the free amine that could directly be coupled to 4-sulfamoylbenzoic acid under typical peptide coupling conditions to yield the sulfamoyl linker **2.63**. This linker was subjected to reductive amination with benzaldehyde to yield a solid-supported benzylated tosylamine **2.64**. Subsequently the resin was subjected to 1,10-phenanthroline and copper sulfate catalysis, used in solution phase, to prepare the silane-protected ynamides **2.64** → **2.65**. However, after many attempts to characterize the resin after the copper-catalyzed installment of the alkyne it became clear that this route was unpractical. Due to residual copper and black side-products – that can easily be filtered off over Celite® in the solution phase synthesis – the beads were too dark for coloring tests or on-bead spectroscopic Raman analysis and too inhomogeneous for ^{13}C-gel-NMR. Even after excessive washing with copper complexing agents such as Cupral solution the beads remained unanalyzable – even elemental analysis gave only inconclusive results. In an attempt to circumvent this issue it was decided to prepare a carboxylic acid containing ynamide in solution phase and subsequently couple the acid using conventional peptide chemistry to the Rink amide resin. Deprotection of the silane group with TBAF yielded the desired solid-supported ynamide **2.66**. Small-scale cleavage of **2.66** showed that the ynamide was bound to the beads, but also revealed that the ynamide loading of the beads was rather low. At this stage several attempts to react oxime chlorides and azides to the solid-supported ynamide were performed. Although the success of several of these reactions could be confirmed by mass spectrometric methods, the products were only obtained in very small quantities and were impure. Since it was unclear why the resin was yielding these rather poor results, it was assumed that there must be issues with either the coupling or deprotection step. In hindsight one could probably conclude that the use of a terminal ynamide on the solid support might not be the best idea, due to its sensitivity to water. Since the beads can be difficult to dry, the terminal sulfonyl ynamide has ample opportunity to react with water to the corresponding sulfonyl amide.

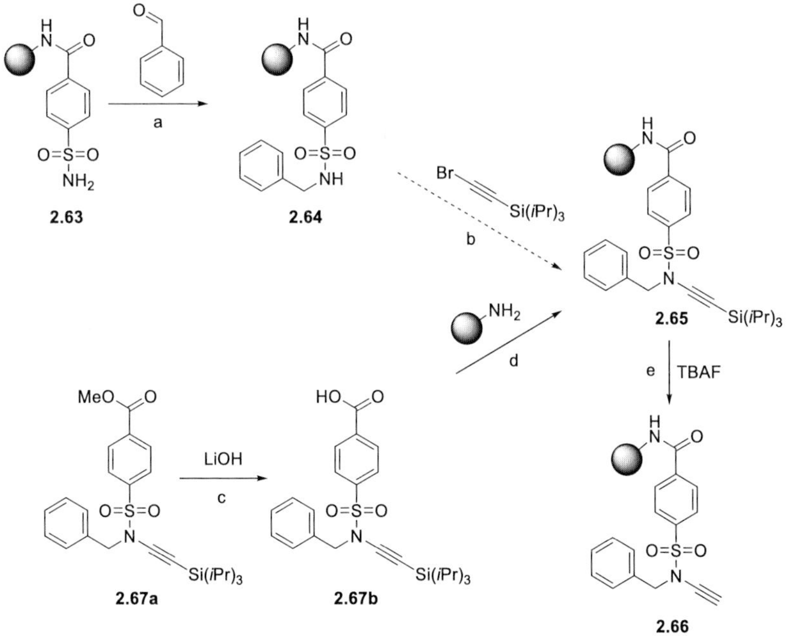

Scheme 2.19 Rink amide-based sulfonyl ynamide synthesis. Conditions: a) benzaldehyde (3.0 equiv.), Et₃N (3.0 equiv.), NaBH(OAc)₃ (2.0 equiv.), acetic acid (4.0 equiv.), THF, 80 °C, 16h; b) (bromoethynyl)triisopropylsilane (2.0 equiv.), CuSO₄·5H₂O (0.20 equiv.), 1,10-phenanthroline (0.40 equiv.), K₂CO₃ (2.0 equiv.), toluene, 85 °C, 16h; c) lithium hydroxide (2.0 equiv.), water/THF, RT, 2h, 81%; d) hydroxybenzotriazole (3.0 equiv.), N,N'-diisopropylcarbodiimide (3.0 equiv.), 4-(dimethylamino)pyridine (3.0 equiv.), 4-(N-benzyl-N-(triisopropylsilyl)sulfamoyl)benzoic acid (1.0 equiv.), Rink amide resin (1.0 equiv.), DMF, 100 °C, 3h; e) TBAF (1M in THF, 1.0 equiv.), THF, RT, 1h.

Ynamide construction on T1 linker – Issues with ditosylation

Due to the issues encountered with the copper-catalyzed ynamide installment a different approach was explored. Since the BRÄSE group has extensive expertise with the T1 triazene linker, which has the added advantage of allowing for a higher loading and offers a wide range of cleavage possibilities that allow for extra diversification,[62] it was decided to switch resins. The piperazine linker **2.68** could be easily made from the Merrifield resin. Next a nitrile was installed under the typical diazotization conditions that lead to the T1 linker. Reduction of nitrile **2.69** to the corresponding amine was achieved in a very facile manner with lithium aluminium hydride and could be confirmed by the loss of the nitrile bands in the IR spectra. In order to finish the synthesis, the plan was to first tosylate and then follow the dichloroenamide route to finally install non-terminal ynamides. However, since SPOS reactions usually rely on excess of reagents and elevated temperatures compared to their solution phase alternatives, only the ditosylated product **2.72** was obtained, as confirmed by elemental analysis.

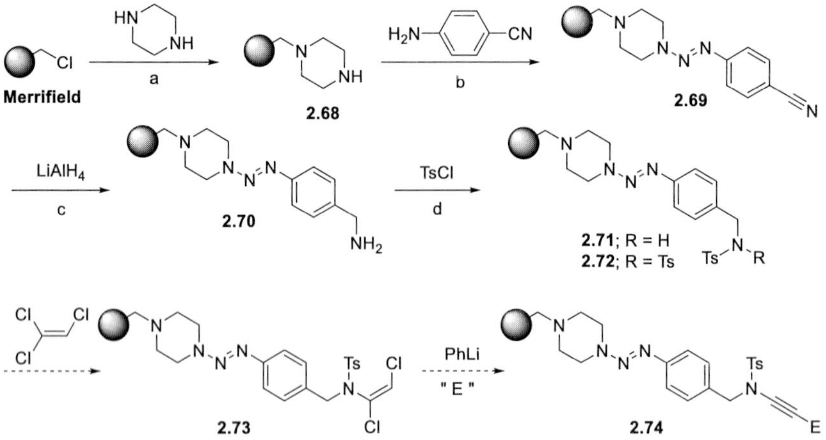

Scheme 2.20 Merrifield-based T1 linker synthesis and subsequent issues with ditosylation. Conditions: a) piperazine (10 equiv.) Et₃N (1.0 equiv.), DMF, 75 °C, 4d. 95%; b) 4-aminobenzonitrile (3.0 equiv.), BF₃·OEt₂ (5.0 equiv.), isoamylnitrite (5.0 equiv.) THF, −20 °C, then the resin THF/pyridine, 12h, 69%; c) LiAlH₄ (2.0 equiv.), THF, 16h, 100%; d) tosylchloride (5.0 equiv.), toluene/pyridine, RT → 50 °C, 2d, 87%.

Using diazotation to link the dichloroenamide to the resin

In order to circumvent the ditosylation issue it was decided to synthesize the dichloroenamide in solution phase and then diazotize this directly onto the piperazine linker. For this purpose, 4-bromobenzylamine was mono-tosylated and subsequently transformed into azide **2.77** by nucleophilic displacement with sodium azide. Next the dichloroenamide **2.78** was prepared in excellent yield under the typical conditions. To reduce the azide **2.78** to an amine **2.80** first a straightforward Staudinger reduction was attempted. However, even vigorous heating in acidic water for prolonged times did not yield the desired amine. Luckily an alternative route employing dichloroindium hydride[63] – *in situ* formed from indium chloride and triethylsilane – could be found, that was able to deliver the amine in reasonable yield.

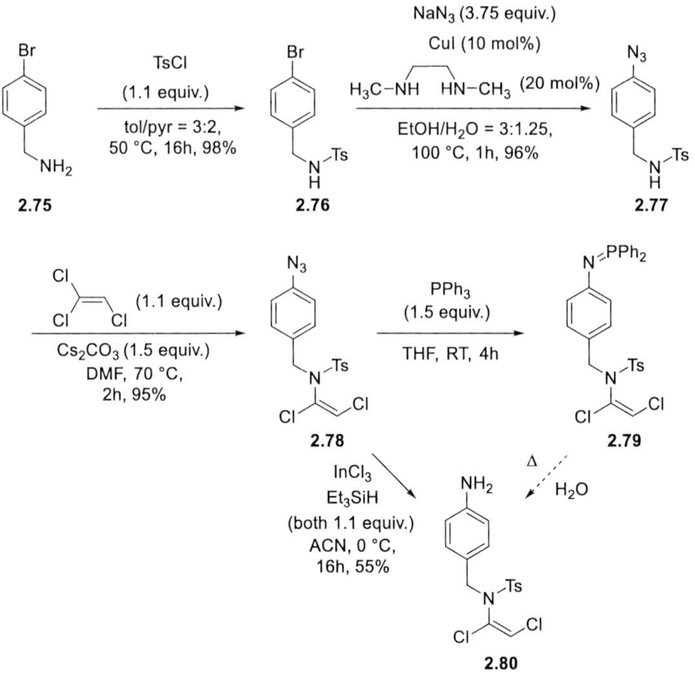

Scheme 2.21 Synthesis of dicholorenamide precursor suitable to be attached to the piperazine linker via diazotation of the amine.

With the amine **2.80** at hand, the coupling to the solid support was attempted. However, first the diazo reagent had to be prepared from the amine. Typically this salt crystallizes out during its preparation, but in this case it did not. This meant that the diazo salt could not be washed and purified before being used and so the crude mixture was used. While cleaving from the beads did yield some product as analyzed by LCMS, it was only in unpractically small amounts, rendering this route ineffective as well.

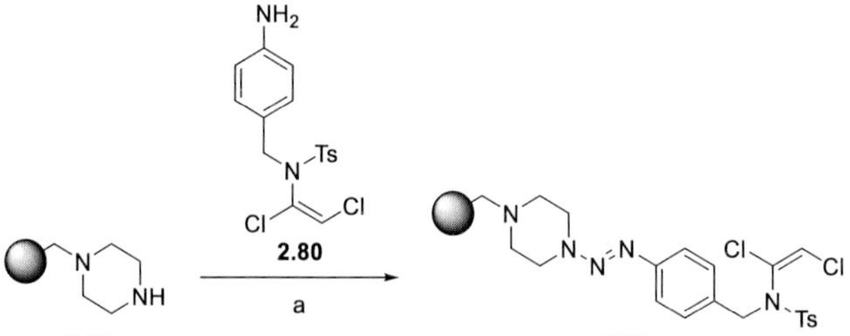

Scheme 2.22 Attempt to directly couple the dichloroenamide to the piperazine linker by diazotation of the primary amine. Conditions: a) dichloroenamide **2.80** (4.0 equiv.), BF$_3$·OEt$_2$ (6.0 equiv.), isoamylnitrite (6.0 equiv.), THF, -20 °C, then the resin in THF/pyridine, 12h, only traces of product observed after cleaving with TFA (10 equiv.).

Using CuAAC to link the dichloroenamide to the resin

Once it was realized that linking the dichloroenamide-amine via diazotation was unpractical, it was decided to take a drastically different approach. Instead of reducing the azide to yield the amine, it could be linked to a suitable alkyne resin using a conventional copper-catalyzed alkyne azide cycloaddition (CuAAC). The alkyne linker **2.83** was accessible in a fairly straightforward manner by coupling of the piperazine linker to the diazonium salt of iodoaniline, Sonogashira coupling with TMS-acetylene and subsequent deprotection of the TMS group with TBAF. This route worked wonderfully, as the installment of the acetylene could be monitored by IR spectroscopy and the presence of the TMS group could be confirmed by ^{13}C-gel-NMR. Unfortunately the residual palladium from the Sonogashira reaction did leave the beads too darkly tinted for Raman measurements. Cleavage of the final dichloroenamide after the CuAAC showed that the overall yield was 80%. Exploration of the correct conditions that can eliminate the dichloroenamide to the sulfonyl ynamides is currently ongoing.

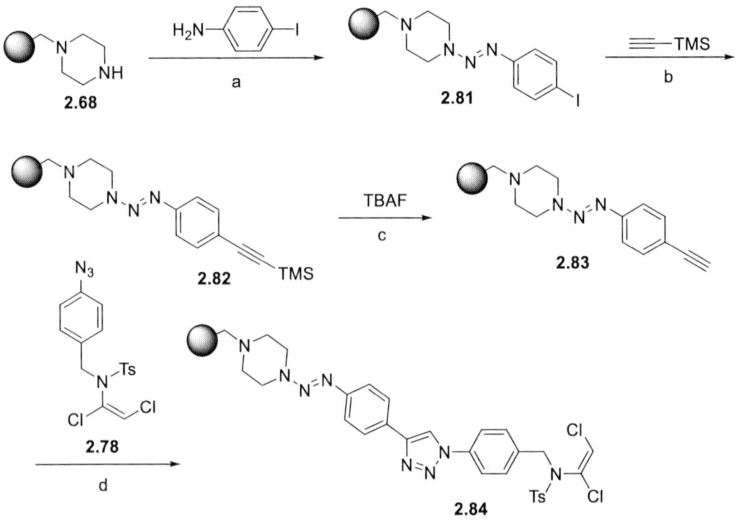

Scheme 2.23 Merrifield-based T1 linker with CuAAC access to solid-supported dichloroenamides. Conditions: a) 4-iodoaniline (4.0 equiv.), BF$_3$·OEt$_2$ (6.0 equiv.), isoamylnitrite (6.0 equiv.) THF, −20 °C, then the resin THF/pyridine, 12h, 99%; b) TMS-acetylene (3.0 equiv.), Et$_3$N (1.6 equiv.), CuI (0.70 equiv.), Pd(PPh$_3$)$_4$ (0.10 equiv.) DMF, RT, 2h, 96%; c) TBAF (1.75 equiv.), THF, RT, 2 min, 99%; d) dichloroenamide **2.78** (1.2 equiv.), CuI (0.40 equiv.), N-ethyl-N-isopropylpropan-2-amine (1.2 equiv.), RT, 80% (overall yield, as determined by cleaving of the resin with 10 equiv. of TFA in CH$_2$Cl$_2$).

2.5 Conclusion and outlook

Two different methods to prepare sulfonyl ynamides have been investigated. Using the copper-catalyzed amidative cross-couplings of bromo-acetylenes with sulfonyl amides several sulfonyl ynamides are accessible in solution phase without major issues. Main drawback is the requirement of different bromo-acetylenes when different internal ynamides are desired. In order to circumvent this nuisance, it was shown that terminal ynamides can be prepared from TIPS-protected ynamides and subsequently subjecting to Sonogashira reactions.

Attempts to use the copper-catalyzed amidative coupling on a Rink amide resin failed, due to the formation of large amounts of black side-products. Therefore ynamides carrying a carboxylic acid moiety were prepared and coupled to the solid support via traditional amide coupling. Although this worked, the amount of terminal ynamide coupled to the solid support was rather disappointing as it did not allow for the isolation of useful quantities of product after CuAAC reactions with azides.

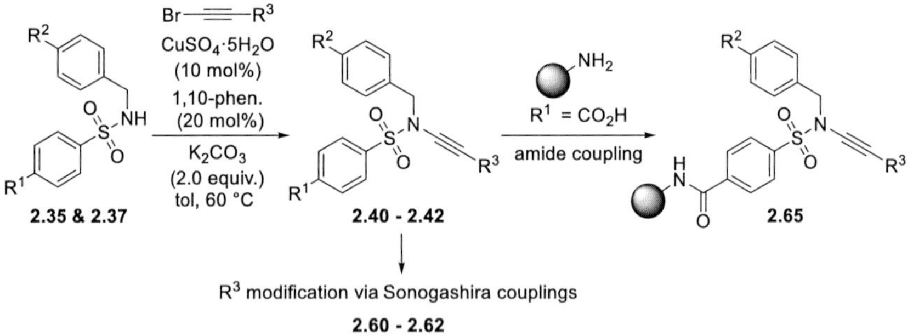

Scheme 2.24 Copper-catalyzed amidative cross-coupling of bromo-acetylenes to sulfonamides and subsequent immobilization on the Rink amide resin.

Alternatively, a dichloroenamide precursor can be made in solution phase by reacting sulfonyl amides with trichloroethane. In the presence of phenyllithium these precursors undergo elimination and lithium-halogen exchange. The resulting highly reactive lithiated species can be quenched with different electrophiles to obtain sulfonyl ynamides in a modular way.

In order to apply the dichloroenamide route to a solid-phase platform a piperazine-based T1 triazene linker was prepared by coupling of the diazo salt of 4-aminobenzonitrile to the piperazine resin. Subsequent reduction of the nitrile with lithium aluminium hydride afforded the free amine. However, since the subsequent tosylation only afforded the undesired ditosylated product, an attempt was made to couple a preformed dichloroenamide to the piperazine linker. For this purpose, a dichloroenamide bearing a free amino moiety on the benzyl group was prepared in five steps. Unfortunately this diazo coupling failed due to issues with the diazotation of the more complex amine. Therefore a T1 triazene phenylacetylene linker was synthesized via Sonogashira coupling of an acetylene to T1 triazene iodobenzene linker. Finally an azido-bearing dichloroenamide could be attached to the solid support via CuAAC.

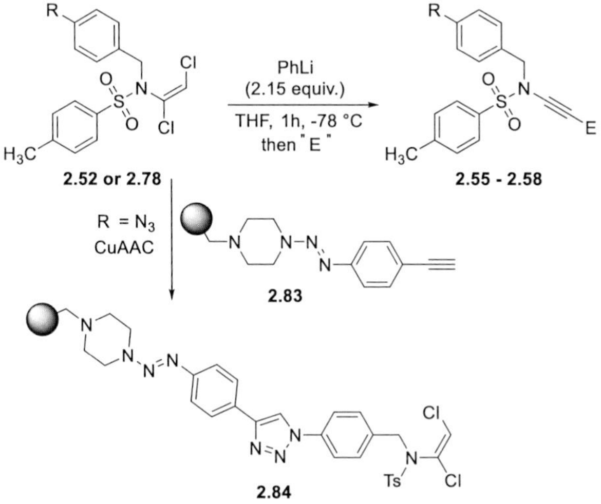

Scheme 2.25 Dichloroenamides are convenient precursors for sulfonyl ynamides. CuAAC of azide substituted dichloroenamide to a T1 triazene phenylacetylene Merrifield linker.

Since the CuAAC based installment of dichloroenamide **2.78** to the solid support (Scheme **2.23**) proved quite efficient the next step is to perform this reaction on larger scale and test suitable conditions for the elimination of the dichloroenamides to their respective ynamides. Typically this is done using slightly more than two equivalents of phenyllithium at −78 °C, but better conditions may be found. The choice for the piperazine linker instead of the common benzylamine linker was made with these conditions in mind: The piperazine linker has shown to be able to withstand *n*-butyllithium better than its benzylamine analog.[64]

A suggestion to make on-bead, *in situ* spectroscopic analysis possible would be to couple an aniline bearing a protected alkyne **2.85** to the solid support instead of doing the Sonogashira reaction, that is known to cause dark residues. With the final solid-supported sulfonyl ynamides at hand the combinatorial construction of compounds can commence and the reaction progress monitored by *in situ* spectroscopic methods.

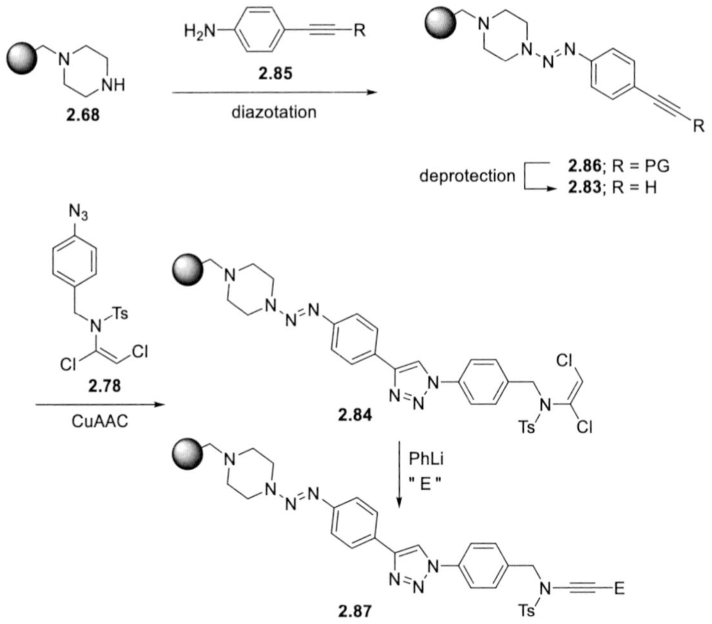

Scheme 2.26 Suggested synthesis of protected alkyne linker **2.86** via diazotation of aniline **2.85** to the Merrifield piperazine linker and subsequent elimination of dichloroenamide to obtain the substituted sulfonyl ynamide.

2.6 Experimental

General remarks

Nuclear magnetic resonance spectroscopy (NMR): ^{1}H-NMR spectra were recorded using the following devices: *Bruker* Avance 300 (300 MHz), *Bruker* Avance 400 (400 MHz) or a *Bruker* Avance DRX 500 (500 MHz); ^{13}C-NMR spectra were recorded using the following devices: *Bruker* Avance 300 (75 MHz), *Bruker* Avance 400 (100 MHz) or a *Bruker* Avance DRX 500 (125 MHz). All measurements were carried out at room temperature. The following solvents from *Eurisotop* were used: chloroform-d_1, methanol-d_4, acetone-d_6, DMSO-d_6. Chemical shifts δ were expressed in parts per million (ppm) and referenced to chloroform (^{1}H: δ = 7.26 ppm, ^{13}C: δ = 77.00 ppm), methanol (^{1}H: δ = 3.31 ppm, ^{13}C: δ = 49.00 ppm), acetone (^{1}H: δ = 2.05 ppm, ^{13}C: δ = 30.83 ppm) or DMSO (^{1}H: δ = 2.50 ppm, ^{13}C: δ = 39.43 ppm).[65] The signal structure is described as follows: s = singlet, d = doublet, t = triplet, q = quartet, quin = quintet, bs = broad singlet, m = multiplet, dt = doublet of triplets, dd = doublet of doublets, td = triplet of doublets, ddd = doublet of doublet of doublets. The spectra were analyzed according to first order and all coupling constants are absolute values and expressed in Hertz (Hz). The multiplicities of the signals of ^{13}C-NMR spectra were determined using characteristic chemical shifts and DEPT (Distortionless Enhancement by Polarization Transfer) and are described as follows: "+" = primary or tertiary (positive DEPT-135 signal), "–" = secondary (negative DEPT-135 signal), "C$_q$" = quarternary carbon atoms (no DEPT signal).

Infrared spectroscopy (IR): IR spectra were recorded on a *Bruker* IFS 88 using ATR (Attenuated Total Reflection). The intensities of the peaks are given as follows: vs = very strong (0 - 10% transmission), s = strong (11 - 40% transmission), m = medium (41 - 70% transmission), w = weak (71 - 90% transmission), vw = very weak (91 - 100% transmission).

Mass spectrometry (EI-MS, ESI-MS, FAB-MS): Mass spectra were recorded on a *Finnigan* MAT 95 using EI-MS (Electron ionization mass spectrometry) at 70 eV or FAB-MS (Fast Atom Bombardment Mass Spectroscopy) with 3-NBA as matrix. In special cases an *Agilent Technologies* 6230 TOF-LC/MS was used to record ESI-TOF-MS spectra in the positive mode. The molecular fragments are stated as the ratio of mass over charge *m/z*. The intensities of the

signals are given in percent relative to the intensity of the base signal (100%). The molecular ion is abbreviated $[M]^+$ for EI-MS and ESI-MS, the protonated molecular ion is abbreviated $[M+H]^+$ for FAB-MS.

Thin layer chromatography (TLC): Analytical thin layer chromatography was carried out using silica coated aluminium plates (silica 60, F_{254}, layer thickness: 0.25 mm) with fluorescence indicator by *Merck*. Detection proceeded under UV light at $\lambda = 254$ nm. For development phosphomolybdic acid solution (5% phosphomolybdic acid in ethanol, dip solution); potassium permanganate solution (1.00 g potassium permanganate, 2.00 g acetic acid, 5.00 g sodium bicarbonate in 100 mL water, dip solution) was used followed by heating in a hot air stream. Preparative thin layer chromatography was carried out using either the silica coated aluminium plates (silica 60, F_{254}, layer thickness: 0.25 mm) or the silica coated PSC glass plates (silica 60, F_{254}, layer thickness: 2.0 mm) by *Merck*.

Analytical balance: Used device: *Sartorius* Basic, model LA310S, range from 0.1 mg to 310.0 g.

Microwave: Reactions heated using microwave irradiation were carried out in a single mode *CEM* Discover LabMate microwave operated with *CEM*'s Synergy software. This instrument works with a constantly focused power source (0 - 300W). Irradiation can be adjusted via power- or temperature control. The temperature was monitored with an infrared sensor.

Solvents and reagents: Solvents of technical quality were distilled prior to use. Solvents of p.a. (*per analysi*) quality were commercially purchased (*Acros, Fisher Scientific, Sigma Aldrich*) and used without further purification. Absolute solvents were either commercially purchased as absolute solvents stored over molecular sieves under argon atmosphere or freshly prepared by distillation of p.a. quality over a drying agent (dichloromethane from calcium hydride, THF from sodium using benzophenone as indicator, and diethyl ether from sodium) and stored under argon atmosphere. Reagents were commercially purchased (*ABCR, Acros, Alfa Aesar, Fluka, J&K, Sigma Aldrich, TCI, VWR*) and used without further purification if not stated otherwise.

Reactions: For reactions with air- or moisture-sensitive reagents, the glassware with a PTFE coated magnetic stir bar was heated under high vacuum with a heat gun, filled with argon and closed with a rubber septum. All solution phase reactions were performed using the typical

Schlenk procedures with argon as inert gas. All reactions on solid support were done in closed vials, round bottom flasks or capped filter syringes. Large-scale reactions on solid support, e.g. > 2 g, were generally heated in an oil bath and stirred with a mechanical stirrer, all other reactions on solid support were shaken, not stirred. For the small-scale solid-supported reactions that required heating a special heating block was designed and used, as described in Chapter 2.3. Liquids were transferred with V2A-steel cannulas. Solids were used as powders if not indicated otherwise. Reactions at low temperatures were cooled in flat Dewar flasks from *Isotherm*. The following cooling mixtures were used: 0 °C (Ice/Water), 0 to −10 °C (Ice/Water/NaCl), −10 to −78 °C (acetone/dry ice *or* isopropanol/dry ice). In general, solvents were removed at preferably low temperatures (40 °C) under reduced pressure. If not stated otherwise, solutions of ammonium chloride, sodium chloride and sodium hydrogen carbonate are aqueous, saturated solutions. Reaction progress of liquid phase reaction was checked with thin-layer chromatography (TLC). Crude products were purified according to literature procedures by preparative TLC or flash column chromatography using Silica gel 60 (0.063×0.200 mm, 70-230 mesh ASTM) (*Merck*), Geduran® Silica gel 60 (0.040×0.063 mm, 230-400 mesh ASTM) (*Merck*) or Celite® (*Fluka*) and sea sand (calcined, purified with hydrochloric acid, *Riedel-de Haën*) as stationary phase. Eluents (mobile phase) were p.a. and volumetrically measured.

General remark: Procedures and spectroscopic data on compounds mentioned in this Chapter that were prepared and analyzed by Dr. Sabilla ZHONG can be found in the electronic supplementary information of "T. Wezeman, S. Zhong, M. Nieger and S. Bräse, *Angew. Chem. Int. Ed.* **2016**, *55*, 3823 – 3827".

General procedure GP2.1: Reductive amination of sulfam ides

A solution of the sulfamide (1.00 equiv.), aryl aldehyde (2.00 equiv.), Et$_3$N (3.00 equiv.), and NaBH(OAc)$_3$ (3.00 equiv.) in THF was treated dropwise with acetic acid (4.00 equiv.). After consumption of the starting material, the mixture was quenched with saturated aqueous NaHCO$_3$ solution and extracted with CH$_2$Cl$_2$. The combined organic layers were dried over Na$_2$SO$_4$ and the solvent was removed under reduced pressure. The crude product was purified by column chromatography.

General procedure GP2.2: Synthesis of ynamides by amidative coupling

A solution of the bromoalkyne (1.20 - 1.30 equiv.) in toluene was added to a mixture of the amide (1.00 equiv.), CuSO$_4$·5H$_2$O (0.20 equiv.), 1,10-phenanthroline (0.40 equiv.), and K$_2$CO$_3$ (2.00 equiv.) in toluene and heated in a sealed vial to 80 °C. After consumption of the starting material, the mixture was diluted with CH$_2$Cl$_2$ and washed with H$_2$O. The organic layer was dried over Na$_2$SO$_4$ and the solvent was removed under reduced pressure. The crude product was purified by column chromatography.

General procedure GP2.3: Deprotection of TIPS-alkynes to terminal alkynes

A TBAF solution (1M in THF, 1.20 equiv.) was added dropwise to a solution of the TIPS-protected alkyne (1.00 equiv.) in THF. The mixture was stirred at RT until complete consumption of the starting material. The solvent was removed under reduced pressure and the crude product was purified by column chromatography.

General procedure GP2.4: Synthesis of 1,2-dichloroenamides from sulfamides

A solution of the sulfamide (1.00 equiv.), caesium carbonate (1.10 equiv.) and trichloroethylene (1.10 equiv.) in DMF was stirred at 70 °C for 2h. After completion was observed by TLC, the reaction was cooled to RT and quenched with water. The mixture was extracted with EtOAc four times, washed with brine and the combined organic layers were dried over Na_2SO_4. The solvent was removed under reduced pressure and the crude product was purified by column chromatography.

General procedure GP2.5: Synthesis of ynamides from 1,2-dichloroenamides

An oven-dried and argon-flushed flask was charged with the 1,2-dichloroenamide (1.00 equiv.) in anhydrous THF (4.0 mL per mmol substrate) and cooled to at –78 °C whilst stirring. A solution of phenyllithium (1.8M in *t*butyl-ether, 2.15 equiv.) was added dropwise and the reaction was left to stir for 1h. After complete conversion to the lithiated ynamide, as confirmed by TLC, the reaction was quenched by dropwise addition of an electrophile. The reaction was then stirred for 1h at –78 °C and warmed to RT before the addition of water. The mixture was extracted with EtOAc, washed with brine and the combined organic layers were dried over Na_2SO_4. The solvent was removed under reduced pressure and the crude product was purified by column chromatography.

General procedure GP2.6: Diversification with Sonogashira couplings

A vial containing alkyne (1.00 equiv.), iodoarene (1.20 equiv.), and Pd(PPh$_3$)$_4$ (5.0 mol%), and toluene (3 mL / mmol) was degassed and Et$_3$N (6 mL / mmol) added. After stirring for 10 min at RT, CuI (5 - 15 mol%) was added and the mixture was stirred at 60 °C for 2h. Afterwards, the suspension was diluted with EtOAc and filtered through Celite®. The solvent was removed under reduced pressure and the crude product was purified by column chromatography.

Methyl 4-sulfamoylbenzoate (2.33)

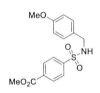

Thionyl chloride (3.25 mL, 44.7 mmol, 1.50 equiv.) was added dropwise to a solution of 4-sulfamoylbenzoic acid (6.00 g, 29.8 mmol) in dry methanol (50 mL). The reaction mixture was stirred for 2h at RT under argon atmosphere before the solvent was evaporated. The resulting solid was then triturated three times with EtOAc to isolate the pure white product (5.78 g, 26.9 mmol, 90%). **¹H-NMR** (300 MHz, CDCl₃): δ = 8.12 (d, J = 8.6 Hz, 2H), 7.93 (d, J = 8.6 Hz, 2H), 4.79 (s, 2H), 3.90 (s, 3H) ppm. – **¹³C-NMR** (101 MHz, Acetone-d_6): δ = 166.2, 148.9, 134.0, 130.7, 127.1, 52.8 ppm. – **IR** (ATR): ṽ = 3356 (w), 3266 (w), 1724 (w), 1535 (w), 1433 (w), 1399 (w), 1338 (w), 1320 (w), 1280 (w), 1162 (m), 1106 (w), 1090 (w) cm⁻¹. – **MS** (EI), m/z (%): 134.8 (20), 184.0 (100), 215.0 (45) [M]⁺. – **HRMS** (EI, C₈H₉NO₄³²S): calc. 215.0247; found 215.0245.

Methyl 4-(N-(4-methoxybenzyl)sulfamoyl)benzoate (2.35b)

This compound was synthesized according to **GP2.1** using methyl 4-sulfamoylbenzoate (1.00 g, 4.65 mmol, 1.00 equiv.) and 4-methoxybenzaldehyde (1.13 mL, 1.27 g, 9.29 mmol, 2.00 equiv.) in 10 mL THF for 1d. The crude product was purified by column chromatography (cHex/EtOAc = 3:1) and the sulfamide was obtained as a colorless solid (1.04 g, 3.12 mmol, 67%). R_f (cHex/EtOAc = 3:1) = 0.19. – **m.p.** = 88 - 92 °C. – **¹H-NMR** (300 MHz, CDCl₃): δ = 8.09 (d, ^3J = 8.6 Hz, 2H), 7.85 (d, ^3J = 8.6 Hz, 2H), 7.02 (d, ^3J = 8.7 Hz, 2H), 6.72 (d, ^3J = 8.7 Hz, 2H), 4.68 (t, ^3J = 5.8 Hz, 1H, NH), 4.11 (d, ^3J = 5.9 Hz, 2H, CH_2), 3.97 (s, 3H, CO₂CH_3), 3.77 (s, 3H, OCH_3) ppm. – **¹³C-NMR** (126 MHz, CDCl₃): δ = 165.6 (C_q, CO_2Me), 159.4 (C_q), 144.0 (C_q), 133.8 (C_q), 130.3 (+, 2 × CH_{Ar}), 129.2 (+, CH_{Ar}), 127.8 (+, CH_{Ar}), 127.1 (+, 2 × CH_{Ar}), 114.1 (+, 2 × CH_{Ar}), 55.3 (+, CH_3), 52.7 (+, CH_3), 46.9 (−, CH_2) ppm. – **IR** (ATR): ṽ = 3265 (m), 2953 (w), 1762 (m), 1611 (w), 1511 (m), 1433 (m), 1400 (m), 1319 (m), 1276 (s), 1246 (m), 1157 (s), 1105 (m), 1089 (m), 1064 (m), 1028 (m) cm⁻¹. – **MS** (EI), m/z (%): 335.1 (4) [M]⁺, 215.2 (44) [C₈H₉NO₄S]⁺, 184.2 (100). – **HRMS** (EI, C₁₆H₁₇NO₅³²S): calc. 335.0822; found 335.0821.

Methyl 4-(N-(4-chlorobenzyl)sulfamoyl)benzoate (2.35d)

 This compound was synthesized according to **GP2.1** using methyl 4-sulfamoylbenzoate (1.00 g, 4.65 mmol, 1.00 equiv.) and 4-chlorobenzaldehyde (1.31 g, 9.29 mmol, 2.00 equiv.) in 10 mL THF for 1d. The crude product was purified by column chromatography (cHex/EtOAc = 3:1) and the sulfamide was obtained as colorless solid (895 mg, 2.63 mmol, 57%). R_f (cHex/EtOAc = 3:1) = 0.19. – **m.p.** = 131 - 133 °C. – **^{1}H-NMR** (400 MHz, CDCl$_3$): δ = 8.13 (d, ^3J = 8.6 Hz, 2H, CH$_{Ar}$), 7.98 (d, ^3J = 8.6 Hz, 2H, CH$_{Ar}$), 7.22 (d, ^3J = 8.6 Hz, 2H, CH$_{Ar}$), 7.12 (d, ^3J = 8.6 Hz, 2H, CH$_{Ar}$), 5.14 (t, ^3J = 6.2 Hz, 1H, NH), 4.13 (d, ^3J = 6.2 Hz, 2H, CH$_2$), 3.96 (s, 3H, CO$_2$CH$_3$) ppm. – **^{13}C-NMR** (100 MHz, CDCl$_3$): δ = 165.6 (C$_q$, CO$_2$Me), 143.8 (C$_q$, CS), 134.4 (C$_q$), 133.91 (C$_q$), 133.8 (C$_q$), 130.3 (+, 2 × CH$_{Ar}$), 129.2 (+, 2 × CH$_{Ar}$), 128.9 (+, 2 × CH$_{Ar}$), 127.0 (+, 2 × CH$_{Ar}$), 52.7 (+, CO$_2$CH$_3$), 46.6 (–, CH$_2$) ppm. – **IR** (ATR): ṽ = 3228 (w), 2958 (vw), 1704 (m), 1598 (w), 1573 (vw), 1490 (w), 1471 (vw), 1433 (w), 1397 (w), 1332 (w), 1282 (m), 1160 (m), 1107 (m), 1090 (m), 1055 (m), 1012 (w) cm^{-1}. – **MS** (EI), m/z (%): 341/339 (0.5/1) [M]$^+$, 142/140 (29/100) [C$_7$H$_7$ClN]$^+$. – **HRMS** (EI, C$_{15}$H$_{14}$NO$_4$^{35}Cl^{32}S): calc. 339.0327; found 339.0325.

N-benzyl-4-methylbenzenesulfonamide (2.37a)

Benzylamine (100 mL, 915 mmol, 1.00 equiv.) was dissolved in pyridine (250 mL) and toluene (150 mL) in a 2L three-neck round bottom flask equipped with a mechanical stirrer. A suspension of para-toluenesulfonyl chloride (183 g, 961 mmol, 1.05 equiv.) in toluene (100 mL) is added dropwise over 1h via a dropping funnel. The reaction was stirred at RT for 3d before it was quenched with addition of a 4M HCl solution. The formed crystals were filtered, triturated (Hept/EtOAc = 4:1) and dried overnight. The remaining liquid was then extracted with Et$_2$O (3 × 250 mL) and the combined organic layers were washed with 1M HCl solution and water before being concentrated under vacuum. The crude product was triturated (Hept/EtOAc = 4:1) and the fractions combined to yield slightly yellowish crystals (227 g, 870 mmol, 95%). R_f (cHex/EtOAc = 2:1) = 0.71. – **^{1}H-NMR** (400 MHz, CDCl$_3$): δ = 7.78 (d, J = 7.9 Hz, 2H), 7.32 (d, J = 7.9 Hz, 2H), 7.26 - 7.29 (m, 3H), 7.16 - 7.23 (m, 2H),

4.62 (br. s., 1H), 4.14 (d, J = 6.0 Hz, 2H), 2.45 (s, 3H) ppm. – ^{13}C-NMR (100 MHz, CDCl$_3$): δ = 132.4, 129.8, 128.7, 127.9, 127.8, 127.2, 47.3, 21.5 ppm. – **IR** (ATR): ṽ = 3266 (w), 1597 (w), 1494 (w), 1452 (w), 1420 (vw), 1380 (w), 1157 (m) cm^{-1}. – **MS** (EI), m/z (%): 261.2 (1) [M]$^+$, 106.0 (100). – **HRMS** (EI, C$_{14}$H$_{15}$NO$_2$^{32}S): calc. 261.0818; found 261.0817.

N-(4-methoxybenzyl)-4-methylbenzenesulfonamide (2.37b)

4-(Methoxy)benzylamine (8.51 g, 8.10 mL, 62.0 mmol, 1.00 equiv.) was stirred with 35 mL pyridine and 25 mL toluene in a 500 mL round bottom flask. A solution of *para*-toluenesulphonyl chloride (15.7 g, 82.2 mmol, 1.33 equiv.) in toluene (40 mL) was then added dropwise over 20 min via a dropping funnel. The reaction was stirred at RT for 3d before it was quenched with addition of water. The reaction mixture was then extracted with EtOAc and the combined organic layers were washed with 1M HCl solution, sat. aq. NaHCO$_3$ and subsequently dried over Mg$_2$SO$_4$ and concentrated under vacuum. The crude product is triturated (cHex:Et$_2$O = 4:1) to yield a slightly tan solid (15.40 g, 52.85 mmol, 85%). ^{1}H-NMR (400 MHz, CDCl$_3$): δ = 7.74 (d, ^3J = 8.1 Hz, 2H, CH_{Ar}), 7.30 (d, ^3J = 8.3 Hz, 2H, CH_{Ar}), 7.10 (d, ^3J = 8.6 Hz, 2H, CH_{Ar}), 6.79 (d, ^3J = 8.3 Hz, 2H, CH_{Ar}), 4.75 (t, ^3J = 5.8 Hz, 1H, CH_2), 4.04 (d, ^3J = 6.0 Hz, 1H, NH), 3.77 (s, 3H, CH_3), 2.43 (s, 3H, CH_3) ppm. – ^{13}C-NMR (101 MHz, CDCl$_3$): δ = 159.3 (C_q), 143.4 (C_q), 137.0 (C_q), 129.7 (C_q), 129.2 (+, 2 × CH_{Ar}), 128.3 (+, 2 × CH_{Ar}), 127.2 (+, 2 × CH_{Ar}), 114.1 (+, 2 × CH_{Ar}), 55.3 (+, CH_3), 46.8 (−, CH_2), 21.5 (+, CH_3) ppm. – **IR** (ATR): ṽ = 3247 (w), 1612 (vw), 1513 (m), 1460 (w), 1427 (m), 1319 (m), 1288 (m), 1250 (w), 1178 (s), 1155 (s), 1058 (m), 1028 (m) cm^{-1}. – **MS** (EI): m/z (%): 291.2 (14) [M]$^+$, 135.1 (100). – **HRMS** (EI, C$_{15}$H$_{17}$NO$_3$^{32}S): calc. 291.0924; found 291.0926.

N-(4-(*tert*-butyl)benzyl)-4-methylbenzenesulfonamide 2.37c]

4-(Tert-butyl)benzylamine (9.27 g, 10.0 mL, 55.1 mmol, 1.00 equiv.) was stirred with pyridine (30 mL) and toluene (20 mL) in a 500 mL round bottom flask. A solution of *para*-toluenesulfonyl chloride (12.6 g, 66.1 mmol, 1.20 equiv.) in toluene (30 mL) was then added dropwise over

20 min via a dropping funnel. The reaction was stirred at RT for 2h before it was quenched by addition of water. The reaction mixture was then extracted with EtOAc and the combined organic layers were washed with 1M HCl solution, sat. aq. NaHCO$_3$ and subsequently dried over Mg$_2$SO$_4$ and concentrated under vacuum. The crude product is filtered through a short plug of silica (3 cm, CH$_2$Cl$_2$/EtOAc = 1:1) and concentrated under reduced pressure to yield an off-white solid (12.0 g, 37.8 mmol, 69%). ^{1}H-NMR (400 MHz, CDCl$_3$): δ = 7.71 (d, ^3J = 8.1 Hz, 2H, 2 × CH_{Ar}), 7.25 (d, ^3J = 4.8 Hz, 4H, 4 × CH_{Ar}), 7.10 (d, ^3J = 8.3 Hz, 2H, 2 × CH_{Ar}), 5.05 (br. s., 1H, NH), 4.06 (d, ^3J = 6.0 Hz, 2H, CH_2), 2.39 (s, 3H, CH_3), 1.26 (s, 9H, 3 × CH_3) ppm. – ^{13}C-NMR (101 MHz, CDCl$_3$): δ = 150.8 (C_q), 143.2 (C_q), 137.0 (C_q), 133.3 (C_q), 129.6 (+, 2 × CH$_{Ar}$), 127.6 (+, 2 × CH$_{Ar}$), 127.1 (+, 2 × CH$_{Ar}$), 125.4 (+, 2 × CH$_{Ar}$), 46.8 (−, CH$_2$), 34.4 (C_q, C(CH$_3$)$_3$), 31.2 (+, CH$_3$), 21.4 (+, CH$_3$) ppm. – IR (ATR): ṽ = 3247 (m), 2957 (w), 1595 (w), 1514 (w), 1465 (m), 1418 (m), 1324 (m), 1268 (w), 1159 (s), 1090 (m), 1056 (m), 1017 (m) cm^{-1}. – MS (EI): m/z (%): 317.1 (2) [M]$^+$, 307.3 (56), 250.2 (17), 147.2 (100). – HRMS (EI, C$_{18}$H$_{23}$NO$_2$^{32}S): calc. 317.1444; found 317.1446.

(Bromoethynyl)triisopropylsilane (2.39a)

Br──═──Si(iPr)$_3$ N-Bromosuccinimide (5.37 g, 30.2 mmol, 1.10 equiv.) and silver nitrate (466 mg, 2.74 mmol, 0.10 equiv.) were added to a solution of ethynyltriisopropylsilane (5.00 g, 27.4 mmol) in acetone (83 mL). The reaction mixture was stirred for 3h at RT under argon atmosphere before the solvent was evaporated. The reaction was then quenched with water, extracted with pentane, washed with brine, and dried with anhydrous Na$_2$SO$_4$. Evaporation under reduced pressure (250 mbar) yielded the product as clear oil (6.91 g, 26.5 mmol, 97%). – ^{1}H-NMR (300 MHz, CDCl$_3$): δ = 1.07 (s, 18H) ppm. – ^{13}C-NMR (75 MHz, CDCl$_3$): δ = 83.4, 61.7, 18.5, 11.3 ppm. – IR (ATR): ṽ = 2942 (w), 2864 (w), 2119 (m), 1461 (w), 1383 (w), 1070 (w) cm^{-1}. – MS (EI): m/z (%): 137.1 (25), 149.0 (74), 161.0 (59), 189.1 (50) 191.1 (51), 219.1 (100), 260.2 (25) [M]$^+$, 262.2 (26). – HRMS: (EI, C$_{11}$H$_{21}$^{79}Br^{28}Si): calc. 260.0590; found 260.0588.

Methyl 4-(_N_-benzyl-_N_-((triisopropylsilyl)ethynyl)sulfamoyl) benzoate (2.40a)

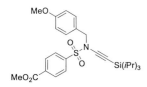

This compound was synthesized following **GP2.2** using methyl 4-(_N_-benzylsulfamoyl)benzoate (2.00 g, 6.55 mmol, 1.00 equiv.) and (bromoethynyl)triisopropylsilane (1.65 mL, 7.86 mmol, 1.20 equiv.) in 12 mL toluene at 85 °C for 3d. The crude product (3.10 g) was used in subsequent steps without further purification. ^{1}**H-NMR** (300 MHz, CDCl$_3$): δ = 8.12 (d, ^3J = 8.6 Hz, 2H, 2 × CH_{Ar}), 7.89 (d, ^3J = 8.6 Hz, 2H, 2 × CH_{Ar}), 7.31 - 7.24 (m, 5H, 5 × CH_{Ar}), 4.55 (s, 2H, CH_2), 3.97 (s, 3H, CO$_2$CH_3), 0.96 (s, 21H, Si(CH(CH_3)$_2$)$_3$) ppm. – ^{13}**C-NMR** (100 MHz, CDCl$_3$): δ = 165.5 (CO$_2$Me), 159.1 (C_{Ar}), 141.4 (C_{Ar}), 134.5 (C_{Ar}), 134.0 (C_{Ar}), 130.1 (2 × C_{Ar}), 128.9 (2 × C_{Ar}), 128.5 (2 × C_{Ar}), 127.7 (2 × C_{Ar}), 95.8 (C≡C), 70.8 (C≡C), 55.9 (CH$_2$), 52.7 (CO$_2$CH$_3$), 18.5 (3 × CH(CH$_3$)$_2$), 11.3 (CH(CH$_3$)$_2$) ppm. – **IR** (ATR): ṽ = 2939 (m) 2862 (m), 2161 (s), 1728 (vs), 1457 (w), 1372 (s), 1276 (s), 1170 (s), 1103 (s), 1086 (s), 1012 (m) cm^{-1}. – **MS** (EI): _m/z_ (%): 485.2 (25) [M–C$_3$H$_6$]$^+$, 442.1 (30) [M–C$_6$H$_{12}$]$^+$.

Methyl 4-(_N_-(4-methoxybenzyl)-_N_-((triisopropylsilyl)ethynyl)sulfamoyl)benzoate (2.40b)

This compound was synthesized following **GP2.2** using methyl 4-(_N_-(4-methoxybenzyl)sulfamoyl)benzoate (551 mg, 1.64 mmol, 1.00 equiv.) and (bromoethynyl)triisopropylsilane (515 mg, 1.97 mmol, 1.20 equiv.) in 3.0 mL toluene for 3d. The crude product was purified by column chromatography (_c_Hex/EtOAc = 10:1) and the ynamide was obtained as a yellow oil (248 mg, 0.480 mmol, 29%). R_f (_c_Hex/EtOAc = 5:1) = 0.40. – ^{1}**H-NMR** (400 MHz, CDCl$_3$): δ = 8.12 (d, ^3J = 8.6 Hz, 2H, CH_{Ar}), 7.89 (d, ^3J = 8.6 Hz, 2H, CH_{Ar}), 7.18 (d, ^3J = 8.6 Hz, 2H, CH_{Ar}), 6.77 (d, ^3J = 8.6 Hz, 2H, CH_{Ar}), 4.48 (s, 2H, CH_2), 3.97 (s, 3H, CO$_2$CH_3), 3.77 (s, 3H, OCH_3), 0.94 - 0.97 (m, 21H, Si(CH(CH_3)$_2$)$_3$) ppm. – ^{13}**C-NMR** (100 MHz, CDCl$_3$): δ = 165.5 (C_q, CO$_2$Me), 159.8 (C_q, COMe), 141.4 (C_q, CS), 134.4 (C_q), 130.4 (+, 2 × CH_{Ar}), 130.0 (+, 2 × CH_{Ar}), 127.7 (+, 2 × CH_{Ar}), 126.0 (C_q), 113.8 (+, 2 × CH_{Ar}), 95.7 (C_q, C≡C), 70.8 (C_q, C≡C), 55.4 (–, CH$_2$), 55.3 (+, OCH$_3$), 52.7 (+, CO$_2$CH$_3$), 18.5 (+, 3 × CH(CH$_3$)$_2$), 11.2 (+, 3 × CH(CH$_3$)$_2$) ppm. – **IR** (ATR): ṽ = 2941 (w), 2863 (w), 2163 (w), 1789 (m), 1684 (w),

1611 (w), 1513 (m), 1461 (w), 1436 (w), 1399 (w), 1356 (m), 1276 (m), 1247 (m), 1171 (m), 1106 (m), 1085 (m), 1032 (m) cm^{-1}. – **MS** (EI), m/z (%): 515.0 (1) [M]$^+$, 472 (5) [M–iPr]$^+$, 121 (100) [C$_8$H$_9$O]$^+$. – **HRMS** (EI, C$_{27}$H$_{37}$NO$_5$^{32}S^{28}Si): calc. 515.2156; found 515.2157.

Methyl 4-(N-(4-chlorobenzyl)-N-((triisopropylsilyl)ethynyl)sulfamoyl)benzoate (2.40d)

This compound was synthesized following **GP2.2** using methyl 4-(N-(4-chlorobenzyl)sulfamoyl)benzoate (115 mg, 0.340 mmol, 1.00 equiv.) and (bromoethynyl)triisopropylsilane (107 mg, 0.408 mmol, 1.20 equiv.) in 1.0 mL toluene for 3d. The crude product was purified by column chromatography (cHex/EtOAc = 5:1) and the ynamide was obtained as a slightly orange solid (92.8 mg, 0.178 mmol, 52%). R_f (cHex/EtOAc = 10:1) = 0.20. – **m.p.** = 43 - 45 °C. – **^{1}H-NMR** (400 MHz, CDCl$_3$): δ = 8.15 (d, ^3J = 8.6 Hz, 2H, 2 × CH_{Ar}), 7.91 (d, ^3J = 8.6 Hz, 2H, 2 × CH_{Ar}), 7.25 (d, ^3J = 8.3 Hz, 2H, 2 × CH_{Ar}), 7.21 (d, ^3J = 8.3 Hz, 2H, 2 × CH_{Ar}), 4.50 (s, 2H, CH_2), 3.98 (s, 3H, CO$_2$CH_3), 0.95 (s, 21H, Si(CH(CH$_3$)$_2$)$_3$), ppm. – **^{13}C-NMR** (100 MHz, CDCl$_3$): δ = 165.4 (C_q, CO$_2$Me), 141.1 (C_q), 134.6 (C_q), 134.5 (C_q), 132.5 (C_q), 130.22 (+, 2 × CH_{Ar}), 130.15 (+, 2 × CH_{Ar}), 128.7 (+, 2 × CH_{Ar}), 127.7 (+, 2 × CH_{Ar}), 95.3 (C_q, C≡C), 71.1 (C_q, C≡C), 55.0 (–, CH$_2$), 52.8 (+, CO$_2$$CH_3$), 18.4 (+, 3 × CH(CH$_3$)$_2$), 11.2 (+, 3 × CH(CH$_3$)$_2$) ppm. – **IR** (ATR): ṽ = 2944 (m), 2864 (m), 2157 (m), 1720 (m), 1601 (vw), 1493 (w), 1461 (w), 1437 (w), 1400 (w), 1375 (m), 1271 (s), 1170 (s), 1103 (m), 1082 (m), 1008 (m) cm^{-1}. – **MS** (EI), m/z (%): 521/519 (3/6) [M]$^+$, 478/476 (71/99) [M–iPr]$^+$, 207 (24), 179 (21), 127/125 (41/100) [C$_7$H$_6$Cl]$^+$.– **HRMS** (EI, C$_{26}$H$_{34}$NO$_4$^{35}Cl^{32}S^{28}Si): calc. 519.1661; found 519.1663.

Methyl 4-(N-(4-(t-butyl)benzyl)-N-((triisopropylsilyl)ethynyl)sulfamoyl)benzoate (2.40e)

This compound was synthesized following **GP2.2** using methyl 4-(N-(4-($tert$-butyl)benzyl)sulfamoyl)benzoate (1.56 g, 4.32 mmol, 1.00 equiv.) and (bromoethynyl)triisopropylsilane (0.95 mL, 4.75 mmol, 1.10 equiv.) in 10.0 mL toluene for 2d. The crude product was purified by column chromatography (cHex/EtOAc = 9:1) and the ynamide was obtained as a solid (1.21 g,

2.07 mmol, 48%). **^{1}H-NMR** (400 MHz, CDCl$_3$): δ = 8.09 (d, ^3J = 8.7 Hz, 2H, 2 × CH_{Ar}), 7.87 (d, ^3J = 8.7 Hz, 2H, 2 × CH_{Ar}), 7.24 (d, ^3J = 6.8 Hz, 2H, 2 × CH_{Ar}), 7.16 (d, ^3J = 8.4 Hz, 2H, 2 × CH_{Ar}), 4.51 (s, 2H, CH_2), 3.96 (s, 3H, CO$_2$CH_3), 1.26 (s, 9H, 3 × CH_3), 0.94 (s, 21H, Si(CH(CH_3)$_2$)$_3$) ppm. – **^{13}C-NMR** (100 MHz, CDCl$_3$): δ = 165.5 (CO$_2$Me), 151.6 (C_{Ar}), 141.5 (C_{Ar}), 134.4 (C_{Ar}), 130.81 (C_{Ar}), 129.9 (2 × C_{Ar}), 128.8 (2 × C_{Ar}), 127.7 (2 × C_{Ar}), 125.4 (2 × C_{Ar}), 95.9 (C≡C), 70.7 (C≡C), 55.6 (CH$_2$), 52.6 (CH$_2$), 34.5 (CCH$_3$), 31.2 (3 ×CCH$_3$), 18.5 (3 × CH(CH$_3$)$_2$), 11.3 (CH(CH$_3$)$_2$) ppm. – **IR** (ATR): ṽ = 2940 (m), 2862 (m), 2160 (m), 1732 (s), 1461 (w), 1373 (s), 1169 (s), 1104 (s) cm^{-1}. – **MS** (EI), m/z (%): 542.4 (85) [M–C$_3$H$_5$]$^+$, 516.3 (100) [M–C$_5$H$_7$]$^+$, 498.3 (90), 442.2 (85). – **HRMS** (EI, C$_{30}$H$_{44}$NO$_4$^{32}S^{28}Si, [M–C$_3$H$_5$]$^+$): calc. 542.2755; found 542.2753.

Methyl 4-(*N*-benzyl-*N*-(phenylethynyl)sulfamoyl)benzoate (2.41a)

This compound was synthesized following **GP2.2** using methyl 4-(*N*-benzylsulfamoyl)benzoate (1.99 g, 6.52 mmol, 1.00 equiv.) and (bromoethynyl)benzene (1.17 mL, 9.78 mmol, 1.5 equiv.) in 15 mL toluene at 85 °C for 1d. The crude product was purified by column chromatography (*c*Hex/EtOAc = 9:1) and the ynamide was obtained as a white solid (1.43 g, 3.53 mmol, 54%). **^{1}H-NMR** (300 MHz, CDCl$_3$): δ = 8.15 (d, ^3J = 8.1 Hz, 2H, 2 × CH_{Ar}), 7.93 (d, ^3J = 8.1 Hz, 2H, 2 × CH_{Ar}), 7.29 (s, 10H, 10 × CH_{Ar}), 4.63 (s, 2H, CH_2), 3.97 (s, 3H, CO$_2$CH_3) ppm. – **^{13}C-NMR** (75 MHz, CDCl$_3$): δ = 171.2 (C_q, CO$_2$Me), 143.1 (C_q), 136.1 (C_q), 134.7 (C_q), 132.7 (+, CH_{Ar}), 130.1 (+, 2 × CH_{Ar}), 129.1 (+, 2 × CH_{Ar}), 128.9 (+, 2 × CH_{Ar}), 128.7 (+, 2 × CH_{Ar}), 128.2 (+, 2 × CH_{Ar}), 128.1 (+, 2 × CH_{Ar}), 127.5 (+, CH_{Ar}), 127.4 (+, CH_{Ar}), 99.9 (C_q, C≡C), 52.8 (–, CH_2), 49.8 (+, CO$_2$$CH_3$) ppm. – **IR** (ATR): ṽ = 3060 (w), 1954 (w), 1730 (s), 1693 (s), 1601 (m), 1495 (m), 1453 (m), 1342 (s), 1275 (s), 1167 (s), 1106 (s), 1015 (s) cm^{-1}. – **MS** (EI): m/z (%): 405.1 (47) [M]$^+$, 206.1 (40). – **HRMS** (EI, C$_{22}$H$_{17}$NO$_4$^{32}S): calc. 405.1029; found 405.1031.

Methyl 4-(*N*-(4-methoxybenzyl)-*N*-(phenylethynyl) sulfamoyl)benzoate (2.41b)

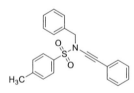

This compound was synthesized following **GP2.2** using methyl 4-(*N*-(4-methoxybenzyl)sulfamoyl)benzoate (1.50 g, 4.48 mmol, 1.00 equiv.) and (bromoethynyl)benzene (1.17 mL, 9.78 mmol, 1.50 equiv.) in 15 mL toluene at 85 °C for 1d. The crude product was purified by column chromatography (*c*Hex/EtOAc = 9:1) finally recrystallized (*c*Hex/Et$_2$O = 2:1) several times and the ynamide was obtained as a slightly yellow solid (1.07 g, 2.65 mmol, 55%). ^{1}**H-NMR** (300 MHz, CDCl$_3$): δ = 8.14 (d, ^3J = 8.7 Hz, 2H, 2 × CH_{Ar}), 7.95 (d, ^3J = 8.6 Hz, 2H, 2 × CH_{Ar}), 7.22 - 7.25 (m, 7H, 7 × CH_{Ar}), 6.80 (d, ^3J = 8.7 Hz, 2H, 2 × CH_{Ar}), 4.56 (s, 2H, CH_2), 3.92 (s, 3H, CO$_2$CH_3), 3.73 (s, 3H, COCH_3) ppm. – ^{13}**C-NMR** (75 MHz, CDCl$_3$): δ = 165.4 (*C*O$_2$H), 159.9 (*C*$_{Ar}$), 141.4 (*C*$_{Ar}$), 134.3 (*C*$_{Ar}$), 130.9 (3 × *C*$_{Ar}$), 129.1 (3 × *C*$_{Ar}$), 128.1 (3 × *C*$_{Ar}$), 127.4 (2 × *C*$_{Ar}$), 122.2 (*C*$_{Ar}$), 113.7 (3 × *C*$_{Ar}$), 82.2 (*C*≡C), 71.7 (C≡*C*), 55.4 (*C*H$_2$), 55.2 (CO$_2$*C*H$_3$), 52.6 (O*C*H$_3$) ppm. – **IR** (ATR): ṽ = 2952 (w), 2836 (w), 1723 (s), 1611 (m), 1512 (s), 1434 (w), 1399 (w), 1346 (w), 1275 (s), 1246 (s), 1166 (s), 1104 (s), 1084 (s) cm^{-1}. – **MS** (EI): *m/z* (%): 454.1 (10) [M+H$_2$O]$^+$, 435.1 (4) [M]$^+$, 254.1 (100). – **HRMS** (EI, C$_{24}$H$_{21}$NO$_5$^{32}S): calc. 435.1135; found 435.1135.

_N_-Benzyl-4-methyl-*N*-(phenylethynyl)benzenesulfonamide (2.42a)

This compound was synthesized following **GP2.2** using *N*-benzyl-4-methylbenzenesulfonamide (1.51 g, 5.75 mmol, 1.00 equiv.) and (bromoethynyl)benzene (1.25 g, 6.89 mmol, 1.20 equiv.) in 28 mL toluene for 3d. The crude product was purified by column chromatography (*c*Hex/EtOAc = 10:1) and the ynamide was obtained as an orange solid (2.05 g, 5.72 mmol, 99%). *R*$_f$ (*c*Hex/EtOAc = 10:1) = 0.30. – ^{1}**H-NMR** (400 MHz, CDCl$_3$): δ = 7.82 (d, *J* = 8.3 Hz, 2H, CH_{Ar}), 7.29 - 7.39 (m, 7H, CH_{Ar}), 7.27 (br. s., 5H, CH_{Ar}), 4.95 (s, 2H, CH_2), 2.46 (s, 3H, CH_3) ppm. – ^{13}**C-NMR** (100 MHz, CDCl$_3$): δ = 144.6 (*C*$_q$), 134.6 (*C*$_q$), 134.4 (*C*$_q$), 131.1 (+, 2 × CH_{Ar}), 129.7 (+,2 × CH_{Ar}), 128.9 (+, 2 × CH_{Ar}), 128.5 (+, 2 × CH_{Ar}), 128.3 (+, CH_{Ar}), 128.2 (+, 2 × CH_{Ar}), 127.7 (+, 2 × CH_{Ar}), 127.6 (+, CH_{Ar}), 122.8 (*C*$_q$), 82.6 (*C*$_q$, C≡*C*), 71.3

(C_q, $C\equiv C$), 55.7 (–, CH_2), 21.6 (+, CH_3) ppm. – **IR** (ATR): $\tilde{\nu}$ = 3031 (w), 1694 (s), 1594 (m), 1493 (m), 1453 (m), 1344 (s), 1168 (s), 1116 (s) cm^{-1}. – **MS** (EI), m/z (%): 361 (81) [M]$^+$, 224.1 (100), 206.4 (95) [M–$C_7H_7O_2S$]$^+$, 179.7 (37), 155.3 (16) [$C_7H_7O_2S$]$^+$. – **HRMS** (EI, $C_{22}H_{19}NO_2{}^{32}S$): calc. 361.1131; found 361.1131.

N-Benzyl-4-methyl-*N*-((triisopropylsilyl)ethynyl)benzene sulfonamide (2.42b)

This compound was synthesized following **GP2.2** using *N*-benzyl-4-methylbenzenesulfonamide (500 mg, 1.92 mmol, 1.00 equiv.) and (bromoethynyl)triisopropylsilane (0.60 mL, 2.87 mmol, 1.50 equiv.) in 3.0 mL toluene for 1d. The crude product was purified by column chromatography (*c*Hex/EtOAc = 9:1) and the ynamide was obtained as a white solid (728 mg, 1.65 mmol, 86%). **¹H-NMR** (300 MHz, CDCl₃): δ = 7.72 (d, ^3J = 8.3 Hz, 2H, 2 × CH_{Ar}), 7.29 - 7.20 (m, 7H, 7 × CH_{Ar}), 4.46 (s, 2H, CH_2), 2.40 (s, 3H, CH_3), 0.98 - 0.85 (m, 21H, Si($CH(CH_3)_2)_3$) ppm. – **¹³C-NMR** (75 MHz, CDCl₃): δ = 144.5 (C_{Ar}), 134.6 (C_{Ar}), 134.3 (C_{Ar}), 129.5 (2 × C_{Ar}), 128.9 (2 × C_{Ar}), 128.4 (2 × C_{Ar}), 128.2 (2 × C_{Ar}), 127.7(C_{Ar}), 96.3 ($C\equiv C$), 70.1 ($C\equiv C$), 55.3 (CH_2), 21.5 (CH_3), 18.4 (3 × $CH(CH_3)_2$), 11.2 ($CH(CH_3)_2$) ppm. – **IR** (ATR) $\tilde{\nu}$ = 2940 (s), 2862 (s), 2161 (vs), 1595 (w), 1495 (s), 1485 (m), 1363 (s), 1168 (vs), 1088 (m), 1014 (m) cm^{-1}. – **MS** (EI): m/z (%): 441.5 (8) [M]$^+$, 398.4 (100), 91.1 (47). – **HRMS**: (EI, $C_{25}H_{35}NO_2{}^{32}S^{28}Si$): calc. 441.2152; found 441.2150.

Methyl 4-(*N*-benzyl-*N*-ethynylsulfamoyl)benzoate (2.43a)

This compound was synthesized following **GP2.3** from methyl 4-(*N*-benzyl-*N*-((triisopropylsilyl)ethynyl)sulfamoyl) benzoate (223 mg, 0.459 mmol, 1.00 equiv.) in 4.0 mL THF for 30 min. The crude product was purified by column chromatography (*c*Hex/EtOAc = 10:1) and the ynamide was obtained as a colorless solid (88.1 mg, 0.267 mmol, 58%). R_f (*c*Hex/EtOAc = 5:1) = 0.25. – **m.p.** = 99 - 102 °C. – **¹H-NMR** (400 MHz, CDCl₃): δ = 8.14 (d, ^3J = 8.6 Hz, 2H, CH_{Ar}), 7.89 (d, ^3J = 8.6 Hz, 2H, CH_{Ar}),

7.27 - 7.31 (m, 5H, CH$_{Ar}$), 4.55 (s, CH$_2$), 3.97 (s, 3H, CO$_2$CH$_3$), 2.72 (s, 1H, C≡CH) ppm. – ^{13}C-NMR (100 MHz, CDCl$_3$): δ = 165.4 (C$_q$, CO$_2$Me), 141.3 (C$_q$, CS), 134.6 (C$_q$), 133.8 (C$_q$), 130.2 (+, 2 × CH$_{Ar}$), 128.7 (+, 2 × CH$_{Ar}$), 128.59 (+, 2 × CH$_{Ar}$), 128.56 (+, CH$_{Ar}$), 127.6 (+, 2 × CH$_{Ar}$), 75.7 (C$_q$, C≡CH), 60.0 (+, C≡CH), 55.6 (–, CH$_2$), 52.7 (+, CO$_2$CH$_3$) ppm. – IR (ATR): ṽ = 3275 (w), 2921 (w), 2136 (w), 1725 (s), 1598 (w), 1496 (w), 1434 (m), 1400 (m), 1359 (s), 1277 (s), 1170 (s), 1106 (s), 1086 (s), 1042 (m), 1028 (m), 1013 (m) cm^{-1}. – MS (EI), m/z (%): 329 (6) [M]$^+$, 298.2 (2). – HRMS (EI, C$_{17}$H$_{15}$NO$_4$^{32}S): calc. 329.0716; found 329.0718.

Methyl 4-(N-ethynyl-N-(4-methoxybenzyl)sulfamoyl)benzoate (2.43b)

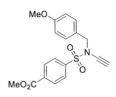

This compound was synthesized following **GP2.3** from methyl 4-(N-(4-methoxybenzyl)-N-((triisopropylsilyl)ethynyl)sulfamoyl)benzoate (45.3 mg, 87.8 µmol, 1.00 equiv.) in 1.0 mL THF for 1.5h. The crude product was purified by column chromatography (cHex/EtOAc = 5:1) and the ynamide was obtained as a colorless solid (15.6 mg, 43.4 µmol, 49%). R$_f$ (cHex/EtOAc = 5:1) = 0.19. – m.p. = 93 - 97 °C. – ^{1}H-NMR (400 MHz, CDCl$_3$): δ = 8.15 (d, ^3J = 8.7 Hz, 2H, CH$_{Ar}$), 7.90 (d, ^3J = 8.7 Hz, 2H, CH$_{Ar}$), 7.20 (d, ^3J = 8.8 Hz, 2H, CH$_{Ar}$), 6.80 (d, ^3J = 8.8 Hz, 2H, CH$_{Ar}$), 4.48 (s, CH$_2$), 3.97 (s, 3H, CO$_2$CH$_3$), 3.78 (s, 3H, OCH$_3$), 2.71 (s, 1H, C≡CH) ppm. – ^{13}C-NMR (100 MHz, CDCl$_3$): δ = 165.4 (C$_q$, CO$_2$Me), 159.8 (C$_q$, COMe), 141.4 (C$_q$, CS), 134.5 (C$_q$), 130.24 (+, 2 × CH$_{Ar}$), 130.18 (+, 2 × CH$_{Ar}$), 127.6 (+, 2 × CH$_{Ar}$), 125.8 (C$_q$), 113.9 (+, 2 × CH$_{Ar}$), 75.7 (C$_q$, C≡CH), 60.1 (+, C≡CH), 55.2 (+, OCH$_3$), 55.1 (–, CH$_2$), 52.7 (+, CO$_2$CH$_3$) ppm. – IR (ATR): ṽ = 3273 (w), 2932 (vw), 2135 (vw), 1715 (w), 1613 (w), 1517 (w), 1461 (vw), 1432 (w), 1399 (w), 1356 (w), 1280 (w), 1257 (w), 1169 (w), 1117 (w), 1085 (w), 1023 (w), 1010 (w) cm^{-1}. – MS (EI), m/z (%): 358 (8) [M]$^+$, 121 (100) [C$_8$H$_9$O]$^+$. – HRMS (EI, C$_{18}$H$_{17}$NO$_5$^{32}S): calc. 359.0822; found 359.0821.

N-Benzyl-N-ethynyl-4-methylbenzenesulfonamide (2.44)

This compound was synthesized following **GP2.3** from N-benzyl-4-methyl-N-((triisopropylsilyl)ethynyl)benzene sulfonamide (150 mg, 0.34 mmol, 1.00 equiv.) in 5.0 mL THF for 60 min. The crude product was purified by column chromatography (cHex/EtOAc = 5:1) and the terminal ynamide was obtained as a white crystals (85.1 mg, 0.299 mmol, 88%). **^{1}H-NMR** (300 MHz, CDCl$_3$): δ = 7.76 (d, ^3J = 8.3 Hz, 2H, 2 × CH$_{Ar}$), 7.38 - 7.31 (m, 7H, 7 × CH$_{Ar}$), 4.50 (s, 2H, CH$_2$), 2.68 (s, 1H, C≡CH), 2.45 (s, 3H, CH$_3$) ppm. – **^{13}C-NMR** (75 MHz, CDCl$_3$): δ = 144.7 (C$_{Ar}$), 134.6 (C$_{Ar}$), 134.2 (C$_{Ar}$), 129.7 (2 × C$_{Ar}$), 128.6 (2 × C$_{Ar}$), 128.5 (C$_{Ar}$), 128.3 (C$_{Ar}$), 127.9 (C$_{Ar}$), 127.7 (C$_{Ar}$), 127.2 (C$_{Ar}$), 76.2 (C≡C), 59.7 (C$_q$, C≡C), 55.2 (CH$_2$), 21.6 (CH$_3$). – **IR** (ATR): $\tilde{v}$ = 3267 (vs), 2922 (w), 2136 (s), 1595 (m), 1494 (w), 1454 (w), 1354 (s), 1161 (vs), 1086 (s), 1026 (s) cm^{-1}. – **MS** (EI): m/z (%): 366.5 (27), 285.2 (32) [M+H]$^+$, 214.2 (45), 155.1 (100). – **HRMS** (EI, C$_{16}$H$_{15}$NO$_2$^{32}S): calc. 285.0818; found: 285.0820.

(E)-N-benzyl-N-(1,2-dichlorovinyl)-4-methylbenzenesulfonamide (2.52a)

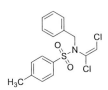

This compound was synthesized following **GP2.4** using N-benzyl-4-methylbenzenesulfonamide (27.70 g, 106.0 mmol, 1.00 equiv.), trichloroethylene (10.5 mL, 116.6 mmol, 1.10 equiv.) and Cs$_2$CO$_3$ (38.0 g, 116.6 mmol, 1.10 equiv.) in DMF (250 mL) at 70 °C for 2h. The crude product was purified by recrystallization (Et$_2$O) and the dichloroenamide was obtained as an off-white solid (35.0 g, 98.3 mmol, 93%). **^{1}H-NMR** (400 MHz, CDCl$_3$): δ = 7.86 (d, ^3J = 8.1 Hz, 2H, CH$_{Ar}$), 7.36 (d, ^3J = 8.3 Hz, 2H, CH$_{Ar}$), 7.28 - 7.34 (m, 4H, CH$_{Ar}$), 6.27 (s, 1H, CClH), 4.45 (br. s, 2H, CH$_2$) 2.47 (s, 3H, CH$_3$) ppm. – **^{13}C-NMR** (101 MHz, CDCl$_3$): δ = 144.7 (CCl), 135.1 (C$_q$), 133.3 (C$_q$), 129.8 (C$_q$), 129.7 (+, CH$_{Ar}$), 129.3 (+, CH$_{Ar}$), 128.7 (+, 2 × CH$_{Ar}$), 128.4 (+, CH$_{Ar}$), 128.4 (+, CH$_{Ar}$), 128.3 (+, CH$_{Ar}$), 127.8 (+, CH$_{Ar}$), 127.2 (+, CH$_{Ar}$), 121.7 (+, CClH), 51.8 (−, CH$_2$), 21.7 (+, CH$_3$) ppm. – **IR** (ATR): $\tilde{v}$ = 3077 (w), 2921 (w), 1594 (w), 1494 (w), 1455 (m), 1355 (m), 1167 (m), 1134 (w), 1084 (w), 1028 (m) cm^{-1}. – **MS** (EI): m/z (%): 355.1/357.1 (1/2) [M]$^+$, 256.1 (11),

155.0 (9), 106.1 (11), 91.1 (100). – **HRMS** (EI, $C_{16}H_{15}NO_2{}^{32}S^{35}Cl_2$): calc. 355.0195; found 355.0194.

((E)-N-(1,2-dichlorovinyl)-N-(4-methoxybenzyl)-4-methylbenzenesulfonamide (2.52b)

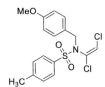

This compound was synthesized following **GP2.4** using N-(4-methoxybenzyl)-4-methylbenzenesulfonamide (2.00 g, 6.86 mmol, 1.00 equiv.), trichloroethylene (0.68 mL, 7.55 mmol, 1.10 equiv.) and Cs_2CO_3 (3.35 g, 10.3 mmol, 1.50 equiv.) in 15 mL DMF at 70 °C for 2h. The crude product was purified by recrystallization (Et$_2$O) and the dichloroenamide was obtained as colorless crystals (2.14 g, 5.53 mmol, 81%).
¹H-NMR (400 MHz, CDCl$_3$): δ = 7.85 (d, ^3J = 8.1 Hz, 2H, CH$_{Ar}$), 7.36 (d, ^3J = 8.6 Hz, 2H, CH$_{Ar}$), 7.23 (d, ^3J = 8.6 Hz, 2H, CH$_{Ar}$), 6.81 (d, ^3J = 8.6 Hz, 2H, CH$_{Ar}$), 6.26 (s, 1H, CClH), 4.65 (br. d, ^2J = 264.7 Hz, 2H, CH$_2$), 3.78 (s, 3H, OCH$_3$), 2.46 (s, 3H, CH$_3$) ppm. – **¹³C-NMR** (101 MHz, CDCl$_3$): δ = 159.6 (C_q, OCCH$_{Ar}$), 144.6 (C_q, CCl), 135.2 (C_q), 130.7 (+, 2 × CH$_{Ar}$), 129.7 (+, 2 × CH$_{Ar}$), 129.4 (C_q), 128.4 (+, 2 × CH$_{Ar}$), 125.2 (2 × C_q), 121.6 (+, CClH), 113.7 (+, 2 × CH$_{Ar}$), 55.1 (–, CH$_2$), 51.2 (+, OCH$_3$), 21.6 (+, CH$_3$) ppm. – **IR** (ATR): ṽ = 3077 (w), 2832 (w), 1611 (w), 1513 (m), 1454 (w), 1352 (m), 1305 (m), 1277 (w), 1249 (m), 1162 (m) cm^{-1}. – **MS** (EI): m/z (%): 385.1/387.1 (80/54) [M]$^+$, 350.2 (100). – **HRMS** (EI, $C_{17}H_{17}NO_3{}^{32}S^{35}Cl_2$): calc. 385.0301; found 385.0302.

(E)-N-(4-(tert-butyl)benzyl)-N-(1,2-dichlorovinyl)-4-methylbenzenesulfonamide (2.52c)

This compound was synthesized following **GP2.4** using N-(4-(tert-butyl)benzyl)-4-methylbenzenesulfonamide (11.3 g, 35.6 mmol, 1.00 equiv.), trichloroethylene (3.52 mL, 39.1 mmol, 1.10 equiv.) and Cs_2CO_3 (12.8 g, 39.1 mmol, 1.10 equiv.) in 50 mL DMF at 70 °C for 2h. The crude product was purified by recrystallization (Et$_2$O) and the dichloroenamide was obtained as off-white crystals (9.62 g, 23.5 mmol, 66%). **¹H-NMR** (400 MHz, CDCl$_3$): δ = 7.82 (d, J = 8.3 Hz, 2H, CH$_{Ar}$), 7.32 (d, J = 8.1 Hz, 2H, CH$_{Ar}$), 7.24 (dd, J = 24.9, 8.3 Hz, 4H, CH$_{Ar}$), 6.24 (s, 1H), 4.33 (bd, J = 207.6 Hz, 2H, CH$_2$), 2.43 (s, 3H),

1.26 (s, 9H). – ^{13}**C-NMR** (101 MHz, CDCl$_3$): δ = 151.4 (C_q), 144.6 (C_q, NCCl), 135.3 (C_q), 130.3 (C_q), 129.7 (+, 2 × CH_{Ar}), 129.6 (+, 2 × CH_{Ar}), 128.9 (C_q), 128.4 (+, 2 × CH_{Ar}), 125.2 (+, 2 × CH_{Ar}), 121.5 (+, HCCl), 51.6 (–, CH_2), 34.5 (C_q), 31.2 (+, 3 × CH_3), 21.6 (+, CH_3). – **IR** (ATR): ṽ = 3080 (w), 2963 (w), 1595 (w), 1451 (m), 1358 (m), 1166 (m), 1127 (m), 866 (m) cm^{-1}. – **MS** (EI): m/z (%): 411.1/413.1 (16/12) [M]$^+$, 376.3 (44), 312.3 (100). – **HRMS** (EI, $C_{20}H_{23}NO_2{}^{32}S^{35}Cl_2$): calc. 411.0821; found 411.0822.

N-benzyl-4-methyl-N-(prop-1-yn-1-yl)benzenesulfonamide (2.55)

This compound was synthesized following **GP2.5** using (E)-N-benzyl-N-(1,2-dichlorovinyl)-4-methylbenzenesulfonamide (10.0 g, 28.1 mmol, 1.00 equiv.), phenyllithium (23.0 mL, 41.4 mmol, 1.8M in $t$butyl-ether, 2.15 equiv.) and methyl iodide as electrophile (1.92 mL, 30.9 mmol, 1.10 equiv.) in 100 mL THF at –78 °C for 2.5h. The crude product was purified by column chromatography (cHex/EtOAc = 9:1) and the ynamide was obtained as off-white crystals (6.72 g, 22.5 mmol, 80%). ^{1}**H-NMR** (300 MHz, CDCl$_3$): δ = 7.69 (d, ^3J = 8.3 Hz, 2H), 7.23 (s, 6H), 4.39 (s, 2H), 2.38 (s, 3H), 1.75 (s, 3H) ppm. – ^{13}**C-NMR** (75 MHz, CDCl$_3$): δ = 144.4 (C_{Ar}), 134.9 (C_{Ar}), 134.8 (C_{Ar}), 129.9 (C_{Ar}), 129.5 (C_{Ar}), 128.8 (C_{Ar}), 128.7 (C_{Ar}), 128.3 (C_{Ar}), 128.2 (C_{Ar}), 127.9 (C_{Ar}), 127.4 (C_{Ar}), 126.6 (C_{Ar}), 72.3 (C≡C), 66.2 (C≡C), 55.5 (CH_2), 21.5 (CH_3), 3.3 (CH_3) ppm. – **IR** (ATR): ṽ = 2918 (v), 1703 (m), 1595 (w), 1495 (w), 1453 (w), 1349 (m), 1156 (m) cm^{-1}. – **MS:** (EI): m/z (%): 299.5 (100) [M]$^+$, 91.3 (79). – **HRMS** (EI, $C_{17}H_{17}NO_2{}^{32}S$): calc. 299.0975; found 299.0975.

N-ethynyl-N-(4-methoxybenzyl)-4-methylbenzenesulfonamide (2.56)

This compound was synthesized following **GP2.5** using ((E)-N-(1,2-dichlorovinyl)-N-(4-methoxybenzyl)-4-methylbenzenesulfonamide (1.75 g, 4.53 mmol, 1.00 equiv.), phenyllithium (5.54 mL, 9.97 mmol, 1.8M in $t$butyl-ether, 2.15 equiv.) and water as electrophile (10 mL) in 10 mL THF at –78 °C for 3h. The crude product was purified by recrystallization (Hept/EtOAc = 9:1) and the ynamide was obtained as slightly yellow crystals (1.12 g, 3.57 mmol, 79%). **^{1}-NMR** (400 MHz, CDCl$_3$): δ = 7.76 (d, ^3J = 8.3 Hz, 2H, CH_{Ar}), 7.27 - 7.42 (m, 2H, CH_{Ar}), 7.21 (d, ^3J = 8.5 Hz, 2H, CH_{Ar}), 6.82 (d, ^3J = 8.5 Hz, 2H, CH_{Ar}), 4.43 (s, 2H, CH_2), 3.79 (s, 3H, OCH_3), 2.67 - 2.70 (m, 1H, CH), 2.44 (s, 3H, CH_{Ar}) ppm. – **^{13}C-NMR** (101 MHz, CDCl$_3$): δ = 159.6 (C_q), 144.6 (C_q), 134.7 (C_q), 130.2 (+, 2 × CH_{Ar}), 129.7 (+, 2 × CH_{Ar}), 127.6 (+, 2 × CH_{Ar}), 126.2 (C_q), 113.8 (+, 2 × CH_{Ar}), 76.2 (C_q, C≡C), 59.7 (C_q, C≡C), 55.2 (–, CH_2), 54.8 (+, OCH$_3$), 21.6 (+, CH$_3$) ppm. – **IR** (ATR): ṽ = 3280 (w), 2131 (w), 1610 (w), 1512 (m), 1440 (w), 1353 (m), 1302 (w), 1187 (m), 1084 (m), 1027 (m) cm^{-1}. – **MS** (EI): *m/z* (%): 315.2 (100) [M]$^+$, 262.2 (77). – **HRMS** (EI, C$_{17}$H$_{17}$NO$_3$^{32}S): calc. 315.0924; found 315.0923.

N-(4-(*tert*-butyl)benzyl)-*N*-ethynyl-4-methylbenzenesulfonamide (2.57)

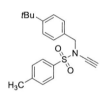

This compound was synthesized following **GP2.5** using (E)-*N*-(4-(t-butyl)benzyl)-*N*-(1,2-dichlorovinyl)-4-methylbenzene-sulfonamide (2.00 g, 4.85 mmol, 1.00 equiv.), phenyllithium (5.93 mL, 10.67 mmol, 1.8M in *t*butyl-ether, 2.15 equiv.) and water as electrophile (10 mL) in 10 mL THF at −78 °C for 2h. The crude product was purified by recrystallization (Et$_2$O) and the ynamide was obtained as white crystals (1.04 g, 3.06 mmol, 63%). **^{1}H-NMR** (400 MHz, CDCl$_3$): δ = 7.73 (d, ^3J = 8.3 Hz, 2H, CH_{Ar}), 7.27 - 7.37 (m, 4H, CH_{Ar}), 7.18 - 7.25 (m, 2H, CH_{Ar}), 4.48 (s, 2H, CH_2), 2.71 (s, 1H, C≡CH), 2.43 (s, 3H, CH_3), 1.30 (s, 9H, 3 × CH_3) ppm. – **^{13}C-NMR** (101 MHz, CDCl$_3$): δ = 151.3 (C_q), 144.5 (C_q), 134.8 (C_q), 131.2 (C_q), 129.6 (+, 2 × CH_{Ar}), 128.3 (+, 2 × CH_{Ar}), 127.7 (+, 2 × CH_{Ar}), 125.4 (+, 2 × CH_{Ar}), 76.5 (C_q, C≡C), 59.5 (C_q, C≡C), 55.1 (–, CH_2), 34.5 (C_q), 31.3 (+, 3 × CH_3), 21.6 (+, CH_3) ppm. – **IR** (ATR): ṽ = 3271 (w), 2950 (w), 2133 (w), 1595 (w), 1456 (w), 1355 (m), 1167 (m), 1089 (m), 1011 (m) cm^{-1}. – **MS** (EI): *m/z* (%): 341.3 (6) [M]$^+$, 147.1 (100). – **HRMS** (EI, C$_{20}$H$_{23}$NO$_2$^{32}S): calc. 341.1444; found 341.1445.

Ethyl 3-((N-(4-(tert-butyl)benzyl)-4-methylphenyl)sulfonamido)propiolate (2.58)

This compound was synthesized following **GP2.5** using (E)-N-(4-(tert-butyl)benzyl)-N-(1,2-dichlorovinyl)-4-methyl-benzenesulfonamide (2.00 g, 4.85 mmol, 1.00 equiv.), phenyllithium (5.93 mL, 10.7 mmol, 1.8M in tbutyl-ether, 2.15 equiv.) and ethyl chloroformate as electrophile (0.56 mL, 5.82 mmol, 1.20 equiv.) in 50 mL THF at −78 °C for 2h. The crude product was purified by column chromatography (Hept/EtOAc) and the ynamide was obtained as a thick yellow oil that slowly solidified (1.10 g, 2.67 mmol, 55%). ^{1}**H-NMR** (400 MHz, CDCl$_3$): δ = 7.63 (d, ^{3}J = 8.3 Hz, 2H, CH_{Ar}), 7.20 - 7.34 (m, 4H, CH_{Ar}), 7.10 - 7.20 (m, 2H, CH_{Ar}), 4.57 (s, 2H, CH_2), 4.16 (q, ^{3}J = 7.1 Hz, 2H, OCH_2CH$_3$), 2.38 (s, 3H, OCH$_2$CH_3), 1.26 (s, 9H, 3 × CH_3) ppm. – ^{13}**C-NMR** (101 MHz, CDCl$_3$): δ = 154.1 (C_q), 151.7 (C_q), 145.2 (C_q), 134.5 (C_q), 130.5 (C_q), 129.8 (+, 2 × CH_{Ar}), 128.4 (+, 2 × CH_{Ar}), 127.8 (+, 2 × CH_{Ar}), 125.5 (+, 2 × CH_{Ar}), 83.1 (C_q, C≡C), 68.0 (C_q, C≡C), 61.5 (−, CH_2), 55.4 (−, CH_2), 34.6 (C_q, C(CH$_3$)$_3$), 31.2 (+, CH_3), 21.6 (+, CH_3), 14.1 (+, CH$_2$CH_3) ppm. – **IR** (ATR): ṽ = 2961 (w), 2203 (m), 1705 (m), 1594 (w), 1372 (m), 1257 (m), 1186 (m), 1148 (m), 1109 (m), 1026 (m) cm^{-1}. – **MS** (EI): m/z (%): 413.3 (5) [M]$^{+}$, 368.3 (8), 258.2 (11), 202.1 (12), 155.0 (40), 147.1 (100). – **HRMS** (EI, C$_{23}$H$_{27}$NO$_4$^{32}S): calc. 413.1655; found 413.1657.

N-benzyl(sulfamoylbenzamide)-Rink Resin (2.64)

A suspension of sulfamide Rink Resin (900 mg, 0.549 mmol, 1.00 equiv.), benzaldehyde (0.17 mL, 1.65 mmol, 3.00 equiv.), Et$_3$N (0.23 mL, 1.65 mmol, 3.00 equiv.), and NaBH(OAc)$_3$ (234 mg, 1.10 mmol, 2.00 equiv.) in THF (4 mL) was treated dropwise with acetic acid (4.00 equiv.) and heated at 80 °C overnight. After cooling to RT, the beads were transferred to a filter and washed with 3 × DMF, 3 × water, 3 × acetone, 3 × CH$_2$Cl$_2$ and dried under high vacuum overnight. Full conversion was observed by Kaiser Test and ^{1}H-NMR after test-cleavage. – **IR** (ATR): ṽ = 3024 (w), 2921 (w), 1664 (w), 1605 (w), 1504 (w), 1450 (w), 1293 (w), 1207 (w) cm^{-1}. – **Elemental analysis** (C$_{122}$H$_{118}$N$_2$O$_3$S$_1$): calc. C86.59 H7.03 N1.66 S1.89; found C82.32 H7.34 N2.15 S1.01 and C82.07 H7.33 N2.16 S1.00.

4-(N-benzyl-N-((triisopropylsilyl)ethynyl)sulfamoyl)-benzamide-Rink Resin (2.65)

A suspension of Fmoc-protected Rink amide Resin (1580 mg, 0.915 mmol, 1.00 equiv.) was deprotected with piperidine (4 mL, 20% in in DMF). The resin was washed with $3 \times$ DMF and subsequently heated at 100 °C with hydroxybenzotriazole (420 mg, 2.74 mmol, 3.00 equiv.), N,N'-diisopropylcarbodiimide (0.43 mL, 2.74 mmol, 3.00 equiv.), and 4-(dimethylamino)pyridine (335 mg, 2.74 mmol, 3.00 equiv.) and 4-(N-benzyl-N-(triisopropylsilyl)- sulfamoyl)benzoic acid (431 mg, 0.915 mmol, 1.00 equiv.) in DMF (5 mL) for 3h with the aid of microwave irradiation. After cooling to RT, the beads were transferred to a filter and washed with $3 \times$ DMF, $3 \times$ water, $3 \times$ acetone, $3 \times$ CH$_2$Cl$_2$ and dried under high vacuum overnight. Full conversion was observed by Kaiser Test. – **IR** (ATR): ṽ = 3023 (w), 2920 (w), 2161 (w, C≡C stretch), 1672 (w), 1605 (w), 1505 (w), 1451 (w), 1369 (w), 1293 (w) cm^{-1}. – **Elemental analysis** (C$_{133}$H$_{138}$N$_2$O$_3$S$_1$Si$_1$): calc. C85.30 H7.43 N1.50 S1.71; found C81.13 H7.52 N2.57 S0.98 and C81.37 H7.61 N2.56 S0.84.

4-(N-benzyl-N-(ethynyl)sulfamoyl)-benzamide-Rink Resin (2.66)

A suspension of 4-(N-benzyl-N-((triisopropylsilyl)ethynyl)sulfamoyl)- benzamide Rink Resin (1.00 g, 0.610 mmol, 1.00 equiv.) was treated with TBAF (1.5 mL, 1M in THF, 1.00 equiv.) in THF (3 mL). After shaking for 1h the beads were transferred to a filter and washed with $3 \times$ DMF, $3 \times$ water, $3 \times$ acetone, $3 \times$ CH$_2$Cl$_2$ and dried under high vacuum overnight. – **IR** (ATR): ṽ = 3023 (w), 2920 (w), 2110 (vw, C≡C stretch), 1682 (w), 1603 (w), 1504 (w), 1450 (w), 1369 (w), 1290 (w) cm^{-1}.

4-(*N*-benzyl-*N*-((triisopropylsilyl)ethynyl)sulfamoyl)benzoic acid (2.67b)

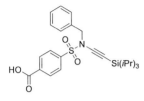

Lithium hydroxide (7.4 mg, 0.309 mmol, 2.00 equiv.) dissolved in water (1.0 mL) was added to a solution of methyl 4-(*N*-benzyl-*N*-((triisopropylsilyl)ethynyl)sulfamoyl)-benzoate (75.0 mg, 0.206 mmol, 1.00 equiv.) in THF (1.0 mL). The reaction was washed with a saturated aqueous solution of ammonium chloride after stirring for 2h at RT. The mixture was then extracted with CH_2Cl_2 and concentrated under reduced pressure. The crude product was purified by column chromatography (*c*Hex/EtOAc = 2:1) and the ynamide was obtained as a white solid (58.3 mg, 0.167 mmol, 81%). **^{1}H-NMR** (300 MHz, Acetone-d_6): δ = 8.26 (d, J = 8.7 Hz, 2H), 8.06 (d, J = 8.7 Hz, 2H), 7.43 - 7.13 (m, 5H), 4.65 (s, 2H), 0.97 (s, 21H) ppm. − **^{13}C-NMR** (101 MHz, Acetone-d_6): δ = 166.2 (CO_2H), 142.4 (C_{Ar}), 136.2 (C_{Ar}), 135.4 (C_{Ar}), 131.3 (C_{Ar}), 129.8 (C_{Ar}), 129.4 (C_{Ar}), 129.2 (C_{Ar}), 128.7 (C_{Ar}), 97.3 ($C{\equiv}C$), 70.8 ($C{\equiv}C$), 56.4 (CH_2), 18.8 (CH), 11.9 (CH_3). − **IR**: (ATR) v= 2939 (w), 2862 (w), 2164 (w), 1689 (w), 1376 (w), 1171 (w) cm^{-1}. − **MS**: (EI): *m/z* (%): 471.2 (5) [M]$^+$, 428.2 (100) [M−(CH(CH$_3$)$_2$)]$^+$. − **HRMS**: (EI, $C_{25}H_{25}NO_3{}^{32}S^{28}$Si) calc. 471.1137; found 471.1135.

Piperazine Merrifield Resin (2.68)

A 500 mL three-neck flask with condenser and mechanical stirrer was charged Merrifield resin (22.000 g, 21.72 mmol, loading = 0.99 mmol/g, 1.00 equiv.) and piperazine (18.71 g, 217 mmol, 10.0 equiv.) in DMF (200 mL) to give a suspension. Et$_3$N (3.03 mL, 21.72 mmol, 1.00 equiv.) was added and the reaction mixture was mechanically stirred at 75 °C for 4d. After cooling to RT, the beads were transferred to a filter and washed with 3 × DMF, 3 × water, 3 × acetone, 3 × CH_2Cl_2 and dried at 90 °C overnight. (23.028 g, loading = 0.8986 mmol/g, 95%) − **^{13}C-NMR** (100 MHz, gel, CDCl$_3$): δ = 145.1, 63.2, 54.1, 43.4 ppm. − **IR** (ATR): ṽ = 3057 (w), 3023 (w), 2918 (w), 2847 (w), 1712 (w), 1678 (w), 1600 (w), 1491 (w), 1450 (w), 1360 (w) cm^{-1}. − **Elemental analysis** ($C_{79}H_{85}N_2$): calc. C89.30 H8.06 N2.64; found C88.04 H7.89 N2.74 and C88.48 H7.92 N2.71.

T1-nitrile Merrifield Resin (2.69)

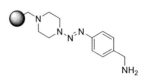 A 100 mL round-bottomed flask was charged with 4-aminobenzonitrile (3.236 g, 27.4 mmol, 3.00 equiv.) and boron trifluoride diethyl etherate (5.5 mL, 44.6 mmol, 4.88 equiv.) in THF (30 mL) ml to give a yellow solution. The reaction mixture was cooled to −40 °C and was stirred for 5 min before isoamylnitrite (6.01 mL, 44.7 mmol, 4.90 equiv.) was added. After stirring for 2h at −20 °C the precipitate was filtered and dissolved in a minimal amount of acetonitrile (40 mL). The piperazine Merrifield Resin (10.21 g, 9.13 mmol, 1.00 equiv.) was presoaked in a 250 mL round-bottomed flask with THF (90 mL) and Pyridine (10 mL) before ~1/3 of the dissolved diazonium salt is added. The reaction mixture was shaken for 30 min and the resin washed with clean toluene. This procedure is repeated 2 more times, but the third time the suspension is allowed to shake overnight. The beads were transferred to a filter and washed with 3 × DMF, 3 × water, 3 × acetone, 3 × CH₂Cl₂ and dried at 90 °C overnight. (11.0302 g, loading = 0.5758 mmol/g, 69%). – **IR** (ATR): ṽ = 3023 (w), 2918 (w), 2222 (w), 1676 (w), 1599 (w), 1491 (w), 1449 (w), 1342 (w) cm⁻¹. – **Elemental analysis** (C₈₆H₈₈N₅): calc. C86.68 H7.44 N5.88; found C85.61 H7.49 N3.69 and C85.51 H7.50 N3.71.

T1-amino Merrifield Resin (2.70)

A 250 mL round-bottomed flask with mechanical stirrer was charged with T1-nitrile Merrifield Resin (10.69 g, 8.97 mmol, 1.00 equiv.) in THF (50 mL) at 0 °C to give a colorless suspension. Lithium aluminium hydride (0.681 g, 17.94 mmol, 2.00 equiv.) was added slowly in small portions with strong stirring. After addition the reaction mixture was stirred at RT overnight. The beads were transferred to a filter and washed with 3 × sat. aq. NaK tartate, 3 × DMF, 3 × water, 3 × acetone, 3 × CH₂Cl₂ and dried at 90 °C overnight. (11.8542 g, loading = 0.5758 mmol/g, 100%). – **IR** (ATR): ṽ = 3413 (w), 3056 (w), 3023 (w), 2917 (w), 1599 (w), 1491 (w), 1450 (w), 1353 (w) cm⁻¹. – **Elemental analysis** (C₈₆H₈₈N₅): calc. C86.39 H 7.76 N 5.86; found C78.41 H7.40 N3.22 and C78.19 H7.33 N3.23.

N-(4-bromobenzyl)-4-methylbenzenesulfonamide (2.76)

To a solution of (4-bromophenyl)methanamine (5.0 mL, 38.4 mmol, 1.00 equiv.) in toluene (30 mL) and pyridine (20 mL) was slowly added a solution of 4-methylbenzenesulfonyl chloride (8.05 g, 42.2 mmol, 1.10 equiv.) in toluene (25 mL). The reaction mixture was stirred at RT for 3h and subsequently heated at 50 °C for 16h. The reaction mixture was quenched with water (25 mL) and stirred for 5 min at RT. The aqueous layer was extracted with EtOAc (2 × 200 mL) and the combined organic layers were washed with 1M HCl (1 × 50 mL) and sat. aq. NaCl (1 × 100 mL). The orange crude was dried over anhydrous magnesium sulfate, filtered through a SiO_2 filter column (2 cm, cHex/EtOAc = 1:1). The solvent was evaporated to yield an off-white solid that was used in the next step without further purification (12.8 g, 37.7 mmol, 98%). **^{1}H-NMR** (300 MHz, CDCl$_3$): δ = 7.73 (d, J = 8.1 Hz, 2H), 7.39 (d, J = 8.5 Hz, 2H), 7.30 (d, J = 7.9 Hz, 2H), 7.08 (d, J = 8.5 Hz, 2H), 4.74 (t, J = 5.8 Hz, 1H), 4.08 (d, J = 6.2 Hz, 2H), 2.44 (s, 3H) ppm. – **^{13}C-NMR** (75 MHz, CDCl$_3$): δ = 143.7, 136.7, 135.3, 132.1, 1131.2, 129.8, 129.5, 127.2, 121.9, 46.6, 26.9 ppm. – **IR** (ATR): $\tilde{v}$ = 3267 (m), 1596 (w), 1484 (w), 1447 (m), 1364 (m), 1316 (m), 1305 (m), 1154 (m) cm^{-1}. – **MS** (EI): m/z (%): 339.1/341.1 (2/2) [M]$^+$, 184/186 (100/92). – **HRMS** (EI, C$_{14}$H$_{14}$NO$_2$^{32}S^{79}Br): calc. 338.9923; found 338.9921.

N-(4-azidobenzyl)-4-methylbenzenesulfonamide (2.77)

A 20 mL closed vial was charged with N-(4-bromobenzyl)-4-methylbenzenesulfonamide (340 mg, 1.00 mmol, 1.0 equiv.), copper(I) iodide (25.0 mg, 0.131 mmol, 0.13 equiv.), sodium azide (245 mg, 3.77 mmol, 3.77 equiv.) and N1,N2-dimethylethane-1,2-diamine (0.022 mL, 0.200 mmol, 0.2 equiv.) and sodium ascorbate (9.90 mg, 0.050 mmol, 0.050 equiv.) in a mixture of EtOH (3 mL) and water (1.25 mL). After the reaction mixture was stirred at 100 °C for 1h it was cooled to RT and taken up in CH$_2$Cl$_2$. The aqueous layer was extracted with EtOAc (2 × 200 mL) and the combined organic layers were washed with sat. aq. NaCl (1 × 100 mL). The crude was dried over anhydrous sodium sulfate and filtered through a Celite$^{®}$ filter column (2 cm,

cHex/EtOAc = 1:1). The solvent was evaporated to yield a slightly yellow solid that was used in the next step without further purification (290 mg, 0.959 mmol, 96%). **^{1}H-NMR** (300 MHz, CDCl$_3$): δ = 7.64 (d, J = 8.1 Hz, 2H), 7.19 (d, J = 7.9 Hz, 2H), 7.08 (d, J = 8.3 Hz, 2H), 6.80 (d, J = 8.3 Hz, 2H), 5.21 (t, J = 5.3 Hz, 1H), 3.97 (d, J = 5.9 Hz, 2H), 2.34 (s, 3H) ppm. – **^{13}C-NMR** (75 MHz, CDCl$_3$): δ = 156.9, 143.6, 143.4, 139.5, 136.8, 132.6, 129.7, 119.1, 115.2, 58.4, 46.6, 21.5 ppm. – **IR** (ATR): ṽ = 3269 (m), 2921 (w), 2107 (m), 1597 (m), 1503 (m), 1457 (m), 1429 (m), 1280 (m) cm^{-1}. – **MS** (EI): m/z (%): 302.2 (13) [M]$^+$, 273.2 (42), 155.0 (37), 91.1 (100). – **HRMS** (EI, C$_{14}$H$_{14}$N$_4$O$_2$^{32}S): calc. 302.0832; found 302.0831.

(E)-N-(4-azidobenzyl)-N-(1,2-dichlorovinyl)-4-methylbenzenesulfonamide (2.78)

This compound was synthesized following **GP2.4** using N-(4-azidobenzyl)-4-methylbenzenesulfonamide (7.59 g, 25.1 mmol, 1.00 equiv.), trichloroethylene (2.48 mL, 27.6 mmol, 1.10 equiv.) and Cs$_2$CO$_3$ (12.3 g, 37.6 mmol, 1.50 equiv.) in 30 mL DMF at 70 °C for 2h. The crude product was filtered through a SiO$_2$ filter column (2 cm, cHex/EtOAc = 1:1). The solvent was evaporated to a yellow solid that was used in the next step without further purification (9.43 g, 23.7 mmol, 95%). **^{1}H-NMR** (300 MHz, CDCl$_3$): δ = 7.84 (d, J = 8.1 Hz, 2H), 7.27 - 7.40 (m, 4H), 6.95 (d, J = 8.5 Hz, 2H), 6.28 (s, 1H), 4.32 (d, J = 134.8 Hz, 2H), 2.46 (s, 3H) ppm. – **^{13}C-NMR** (75 MHz, CDCl$_3$): δ = 144.8, 140.2, 135.0, 130.8, 130.0, 129.8, 129.3, 128.4, 121.8, 118.9, 51.1, 21.6 ppm. – **IR** (ATR): ṽ = 3093 (w), 2106 (s), 1669 (w), 1595 (m), 1505 (m), 1351 (w), 1285 (w), 1162 (m) cm^{-1}. – **MS** (FAB): m/z (%): 397.1/399.0 (50/45) [M+H]$^+$, 278.0/280.0 (100/70). – **HRMS** (FAB, C$_{16}$H$_{15}$N$_4$O$_2$^{35}Cl$_2$^{32}S): calc. 397.0287; found 397.0285.

((E)-N-(4-aminobenzyl)-N-(1,2-dichlorovinyl)-4-methylbenzenesulfonamide (2.80)

A 100 mL Schlenk flask was charged with indium(III)chloride (0.612 g, 2.77 mmol, 1.10 equiv.) and dried for 1h at 130 °C under vacuum. After the flask is cooled to 0 °C, triethylsilane (0.44 mL, 2.77 mmol, 1.10 equiv.) and acetonitrile (5 mL) were added. The reaction mixture was stirred at 0 °C for 20 min before (E)-N-(4-azidobenzyl)-

N-(1,2-dichlorovinyl)-4-methylbenzenesulfonamide (1.00 g, 2.52 mmol, 1.00 equiv.) was added as solution in acetonitrile (5 mL). The reaction was stirred overnight and quenched with 1M HCl (2 × 10 mL). The aqueous layer was washed with EtOAc (1 × 20 mL) and the combined aqueous layers were neutralized with sat. aq. NaHCO$_3$ and extracted with EtOAc (3 × 20 mL). The crude was dried over anhydrous sodium sulfate and the solvent was evaporated to an off-white solid that was used in the next step without further purification (515 mg, 1.39 mmol, 55%). **^{1}H-NMR** (300 MHz, CDCl$_3$): δ = 7.87 (d, J = 8.1 Hz, 2H), 7.40 (d, J = 7.9 Hz, 2H), 7.16 (d, J = 8.3 Hz, 2H), 6.84 (d, J = 8.1 Hz, 2H), 6.28 (s, 3H), 2.48 (s, 3H) ppm. – **^{13}C-NMR** (75 MHz, CDCl$_3$): δ = 146.7, 143.9, 135.0, 132.1, 113.2, 129.8, 129.5, 127.2, 121.2, 116.9, 51.2, 21.4 ppm. – **MS** (ESI), m/z (%): 371.043/373.039 (100/70) [M]$^+$.

T1-iodo Merrifield Resin (2.81)

A 100 mL round-bottomed flask was charged with 4-iodoaniline (7.87 g, 35.9 mmol, 4.00 equiv.) and boron trifluoride diethyl etherate (6.65 mL, 53.9 mmol, 6.00 equiv.) in THF (70 mL) to give a purple solution. The reaction mixture was cooled to −20 °C and was stirred for 5 min before *iso*amylnitrite (7.24 mL, 53.9 mmol, 6.00 equiv.) was added. After stirring for 2h at −20 °C the precipitate was filtered and dissolved in a minimal amount of acetonitrile (40 mL). The piperazine Merrifield Resin (10.0468 g, 8.98 mmol, 1.00 equiv.) was presoaked in a 250 mL round-bottomed flask with THF (70 mL) and pyridine (10 mL) before ~1/2 of the dissolved diazonium salt is added. The reaction mixture was shaken for 2h and the resin washed with clean toluene. The remaining diazonium salt solution was added and the resin allowed to shake overnight. The beads were transferred to a filter and washed with 3 × DMF, 3 × water, 3 × acetone, 3 × CH$_2$Cl$_2$ and dried at 90 °C overnight. (12.101 g, loading = 0.738 mmol/g, 99%). – **IR** (ATR): ṽ = 3023 (w), 2917 (w), 1674 (w), 1600 (w), 1491 (w), 1450 (w), 1351 (w), 1305 (w) cm^{-1}. – **Elemental analysis** (C$_{85}$H$_{88}$I$_1$N$_4$): calc. C78.98 H6.86 N4.33; found C79.82 H7.00 N3.08 and C79.84 H6.99 N3.05.

T1-TMS-acetylene Merrifield Resin (2.82)

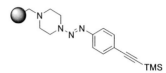

A 10 mL closed vial was charged with T1-iodo Merrifield Resin (11.36 g, 7.91 mmol, 1.00 equiv.), TMS-acetylene (3.35 mL, 23.73 mmol, 3.00 equiv.), Et₃N (1.76 mL, 12.66 mmol, 1.60 equiv.), copper(I) iodide (1.05 g, 5.54 mmol, 0.70 equiv.) and Pd(PPh₃)₄ (0.914 g, 0.791 mmol, 0.10 equiv.) in DMF (7 mL). The reaction mixture was stirred for 2h at RT and subsequently transferred to a filter and washed with Cupral in DMF followed by CH₂Cl₂ until the solvent was no longer colored. The resin is washed with 3 × DMF, 3 × water, 3 × acetone, 3 × CH₂Cl₂ and dried at 90 °C overnight. (11.120 g, loading = 0.7269 mmol/g, 96%) − **¹³C-NMR** (100 MHz, gel, CDCl₃): δ = 62.6, 52.1, 45.5, 0.10 ppm. Due to low sensitivity the alkyne carbons are not distinguishable from the background. − **IR** (ATR): ṽ = 3059 (w), 3023 (w), 2919 (w), 2848 (w), 2232 (vw, C≡C stretch), 1678 (w), 1600 (w), 1491 (w), 1450 (w), 1350 (w) cm⁻¹. − **Elemental analysis** (C₉₀H₉₇N₄Si): calc. C85.60 H7.74 N4.4; found C85.35 H7.58 N3.36 and C85.04 H7.54 N3.32.

T1-acetylene Merrifield Resin (2.83)

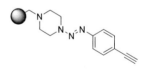

A suspension of T1-TMS-acetylene Merrifield Resin (3.027 g, 2.397 mmol, 1.00 equiv.) was treated with TBAF (4.19 mL, 1M in THF, 1.75 equiv.) in THF (3 mL). After shaking for 2 min the beads were transferred to a filter and washed with 3 × DMF, 3 × water, 3 × acetone, 3 × CH₂Cl₂ and dried at 90 °C vacuum overnight. (2.869 g, loading = 0.7630 mmol/g, 99%) − **¹³C-NMR** (100 MHz, gel, CDCl₃): δ = 62.6, 52.1, 45.5 ppm. Due to low sensitivity the alkyne carbons are not distinguishable from the background. − **IR** (ATR): ṽ = 3225 (vw, C≡C−H stretch), 3023 (w), 2919 (w), 1677 (w), 1599 (w), 1491 (w), 1449 (w), 1345 (w) cm⁻¹. The weak C≡C stretch vibration could not be identified. − **Elemental analysis** (C₈₇H₈₉N₄): calc. C87.76 H7.53 N4.71; found C86.71 H7.48 N3.39 and C86.66 H7.54 N3.40.

T1-dichloroenamide Merrifield Resin (2.84)

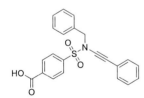

A suspension of T1-acetylene Merrifield Resin (1.00 g, 0.798 mmol, 1.00 equiv.) was treated with (E)-N-(4-azidobenzyl)-N-(1,2-dichlorovinyl)-4-methylbenzenesulfonamide (0.475 g, 1.197 mmol, 1.20 equiv.), copper(I) iodide (0.076 g, 0.399 mmol, 0.40 equiv.) and N-ethyl-N-isopropylpropan-2-amine (0.207 mL, 1.197 mmol, 1.20 equiv.) in dry THF (3 mL). The black suspension was shaken at RT overnight and transferred to a filter and washed with 3 × DMF, 3 × water, 3 × acetone, 3 × CH₂Cl₂ and dried at 90 °C vacuum overnight. Subsequently part of the resin (238 mg, 0.150 mmol) was cleaved with trifluoroacetic acid (0.12 mL, 1.50 mmol, 10.0 equiv.) in CH₂Cl₂ (4 mL). The crude product was purified by column chromatography (cHex/EtOAc = 4:1) resulting in a white powder (60.0 mg, 80%) – **IR** (ATR): ṽ = 3023 (w), 2914 (w), 1678 (w), 1600 (w), 1491 (w), 1450 (w), 1354 (w) cm⁻¹. – **Elemental analysis** (C₁₀₃H₁₀₃N₈Cl₂O₂S₁): calc. C77.91 H6.53 N7.06 S2.02; found C78.64 H6.87 N4.38 S0.93 and C78.31 H6.84 N4.37 S0.92.

4-(N-benzyl-N-(phenylethynyl)sulfamoyl)benzoic acid (2.85)

Lithium hydroxide (16.9 mg, 0.707 mmol, 2.00 equiv.) dissolved in water (6 mL) was added to a solution of methyl 4-(N-benzyl-N-(phenylethynyl)sulfamoyl)benzoate (143 mg, 0.354 mmol) in THF (6 mL). The reaction was washed with a saturated aqueous solution of ammonium chloride after stirring for two hours at RT. The mixture was then extracted with CH₂Cl₂ and concentrated under reduced pressure to yield a white solid in quantitative yield. **¹H-NMR** (300 MHz, Acetone-d_6): δ = 8.29 (d, ³J = 8.5 Hz, 2H, 2 × CH_{Ar}), 8.13 (d, ³J = 8.5 Hz, 2H, 2 × CH_{Ar}), 7.74 - 6.98 (m, 10H, 10 × CH_{Ar}), 4.72 (s, 2H, CH_2) ppm. – **¹³C-NMR** (101 MHz, Acetone-d_6): δ = 166.2 (C_q, CO_2H), 142.1 (C_q), 136.2 (C_q), 135.6 (C_q), 131.9 (+, 2 × CH_{Ar}), 131.4 (+, 2 × CH_{Ar}), 129.9 (+, 2 × CH_{Ar}), 129.5 (+, 2 ×CH_{Ar}), 129.3 (+, 2 × CH_{Ar}), 129.30 (+, CH_{Ar}), 128.9 (+, 2 × CH_{Ar}), 128.8 (C_q), 123.4 (+, CH_{Ar}), 83.2 (C_q, C≡C), 72.1 (C_q, C≡C), 56.7 (−, CH_2) ppm. – **IR** (ATR): ṽ = 3029 (w), 2262 (w), 1705 (s), 1599 (w), 1494 (w), 1398 (m), 1359 (m), 1290 (s), 1150 (s),

1021 (s), 1006 (s) cm^{-1}. – **MS** (EI): *m/z* (%): 391.1 (65) [M]$^+$, 206.1 (72). – **HRMS** (EI, C$_{22}$H$_{17}$NO$_4$^{32}S): calc. 391.0873; found 391.0871.

X-ray Crystallographic data

All X-ray crystallographic work in this chapter has been performed by Dr. Martin NIEGER, the Laboratory of Inorganic Chemistry, University of Helsinki (Finland). All X-ray images shown in this Chapter are depicted with displacement parameters drawn at 50 % probability level.

Methyl 4-(N-(4-chlorobenzyl)sulfamoyl)benzoate – SB800_HY – CCDC 1402826 – (2.35b)

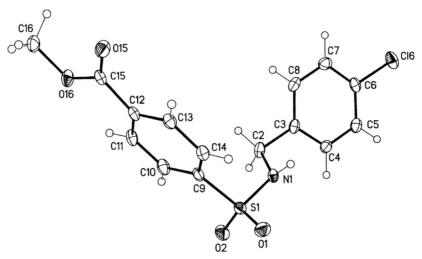

Crystal data

$C_{15}H_{14}ClNO_4S$	$Z = 2$
$M_r = 339.78$	$F(000) = 352$
Triclinic, $P\text{-}1$ (no.2)	$D_x = 1.502$ Mg m^{-3}
$a = 5.8872$ (3) Å	Cu $K\alpha$ radiation, $\lambda = 1.54178$ Å
$b = 10.5339$ (6) Å	Cell parameters from 9855 reflections
$c = 12.3822$ (7) Å	$\theta = 3.7–72.1°$
$\alpha = 78.417$ (1)°	$\mu = 3.72$ mm^{-1}
$\beta = 87.191$ (1)°	$T = 123$ K
$\gamma = 89.181$ (1)°	Needles, colourless
$V = 751.33$ (7) Å^3	$0.50 \times 0.30 \times 0.20$ mm

Data collection

Bruker D8 Venture diffractometer with Photon100 detector	2900 reflections with $I > 2\sigma(I)$
Radiation source: IµS microfocus	$R_{int} = 0.031$
rotation in ϕ and ω, n°, shutterless scans	$\theta_{max} = 72.3°, \theta_{min} = 3.7°$
Absorption correction: multi-scan SADABD (Sheldrick, 2015)	$h = -7\rightarrow7$
$T_{min} = 0.339, T_{max} = 0.523$	$k = -13\rightarrow13$
17280 measured reflections	$l = -15\rightarrow15$
2945 independent reflections	

Refinement

Refinement on F^2	Secondary atom site location: difference Fourier map
Least-squares matrix: full	Hydrogen site location: difference Fourier map
$R[F^2 > 2\sigma(F^2)] = 0.030$	H atoms treated by a mixture of independent and constrained refinement
$wR(F^2) = 0.080$	$w = 1/[\sigma^2(F_o^2) + (0.0394P)^2 + 0.4714P]$ where $P = (F_o^2 + 2F_c^2)/3$
$S = 1.05$	$(\Delta/\sigma)_{max} = 0.001$
2945 reflections	$\Delta\rangle_{max} = 0.43$ e Å^{-3}
204 parameters	$\Delta\rangle_{min} = -0.32$ e Å^{-3}
1 restraint	Extinction correction: SHELXL, Fc*=kFc[1+0.001xFc$^2\lambda^3$/sin(2θ)]$^{-1/4}$
Primary atom site location: structure-invariant direct methods	Extinction coefficient: 0.0159 (10)

N-(4-methoxybenzyl)-4-methylbenzene-sulfonamide − SB853_HY − (2.37b)

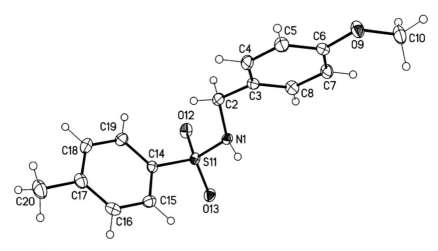

Crystal data

$C_{15}H_{17}NO_3S$	$F(000) = 616$
$M_r = 291.35$	$D_x = 1.363$ Mg m^{-3}
Monoclinic, $P2_1/c$ *(no.14)*	Mo $K\alpha$ radiation, $\lambda = 0.71073$ Å
$a = 13.0146$ (6) Å	Cell parameters from 9284 reflections
$b = 5.1785$ (3) Å	$\theta = 2.8–27.5°$
$c = 21.6705$ (10) Å	$\mu = 0.23$ mm^{-1}
$\beta = 103.505$ (2)°	$T = 123$ K
$V = 1420.12$ (12) Å^3	Blocks, colourless
$Z = 4$	$0.20 \times 0.10 \times 0.06$ mm

Data collection

Bruker D8 VENTURE diffractometer with Photon100 detector	3264 independent reflections
Radiation source: INCOATEC microfocus sealed tube	2896 reflections with $I > 2\sigma(I)$
Detector resolution: 10.4167 pixels mm^{-1}	$R_{int} = 0.030$
rotation in ϕ and ω, 0.5°, shutterless scans	$\theta_{max} = 27.5°$, $\theta_{min} = 2.2°$
Absorption correction: multi-scan *SADABS* (Sheldrick, 2014)	$h = -15 \rightarrow 16$
$T_{min} = 0.930$, $T_{max} = 0.987$	$k = -6 \rightarrow 6$
33653 measured reflections	$l = -28 \rightarrow 28$

Refinement

Refinement on F^2	Secondary atom site location: difference Fourier map
Least-squares matrix: full	Hydrogen site location: difference Fourier map
$R[F^2 > 2\sigma(F^2)] = 0.033$	H atoms treated by a mixture of independent and constrained refinement
$wR(F^2) = 0.085$	$w = 1/[\sigma^2(F_o^2) + (0.0403P)^2 + 0.9104P]$ where $P = (F_o^2 + 2F_c^2)/3$
$S = 1.03$	$(\Delta/\sigma)_{max} = 0.001$
3264 reflections	$\Delta\rangle_{max} = 0.40$ e Å^{-3}
187 parameters	$\Delta\rangle_{min} = -0.39$ e Å^{-3}
1 restraint	Extinction correction: *SHELXL2014*/7 (Sheldrick 2014, Fc*=kFc[1+0.001xFc$^2\lambda^3$/sin(2θ)]$^{-1/4}$
Primary atom site location: structure-invariant direct methods	Extinction coefficient: 0.0068 (9)

Methyl 4-(*N*-ethynyl-*N*-(4-methylbenzyl)sulfamoyl)benzoate – SB807_HY – CCDC 1402829 – (2.43c)

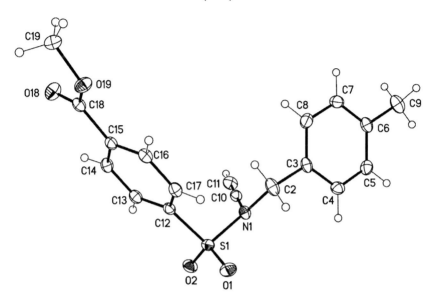

Crystal data

C₁₈H₁₇NO₄S	$Z = 2$
$M_r = 343.38$	$F(000) = 360$
Triclinic, $P\text{-}1$ (no.2)	$D_x = 1.397$ Mg m⁻³
$a = 7.6484$ (3) Å	Mo $K\alpha$ radiation, $\lambda = 0.71073$ Å
$b = 8.0925$ (3) Å	Cell parameters from 9983 reflections
$c = 14.0623$ (6) Å	$\theta = 2.6–27.5°$
$\alpha = 74.350$ (2)°	$\mu = 0.22$ mm⁻¹
$\beta = 77.105$ (2)°	$T = 123$ K
$\gamma = 84.161$ (2)°	Blocks, colourless
$V = 816.18$ (6) Å³	0.24 × 0.18 × 0.06 mm

Data collection

Bruker D8 Venture diffractometer with Photon100 detector	3247 reflections with $I > 2\sigma(I)$
Radiation source: IµS microfocus	$R_{int} = 0.036$
rotation in φ and ω, 1°, shutterless scans	$\theta_{max} = 27.6°$, $\theta_{min} = 2.6°$
Absorption correction: multi-scan *SADABS* (Sheldrick, 2015)	$h = -9 \to 9$
$T_{min} = 0.891$, $T_{max} = 0.996$	$k = -10 \to 10$
21522 measured reflections	$l = -18 \to 18$
3769 independent reflections	

Refinement

Refinement on F^2	Primary atom site location: structure-invariant direct methods
Least-squares matrix: full	Secondary atom site location: difference Fourier map
$R[F^2 > 2\sigma(F^2)] = 0.032$	Hydrogen site location: difference Fourier map
$wR(F^2) = 0.084$	H-atom parameters constrained
$S = 1.04$	$w = 1/[\sigma^2(F_o^2) + (0.0357P)^2 + 0.4287P]$ where $P = (F_o^2 + 2F_c^2)/3$
3769 reflections	$(\Delta/\sigma)_{max} = 0.001$
219 parameters	$\Delta\rangle_{max} = 0.30$ e Å⁻³
0 restraints	$\Delta\rangle_{min} = -0.45$ e Å⁻³

(E)−*N*-benzyl-*N*-(1,2-dichlorovinyl)-4-methybenzenesulfonamide − SB855_HY − (2.52a)

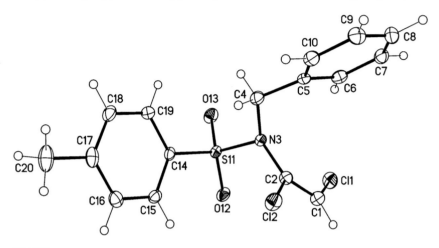

Crystal data

$C_{16}H_{15}Cl_2NO_2S$	$Z = 2$
$M_r = 356.25$	$F(000) = 368$
Triclinic, $P\text{-}1$ (no.2)	$D_x = 1.480$ Mg m^{-3}
$a = 8.2389$ (3) Å	Cu $K\alpha$ radiation, $\lambda = 1.54178$ Å
$b = 9.1762$ (4) Å	Cell parameters from 9848 reflections
$c = 10.7660$ (4) Å	$\theta = 4.4$–$72.1°$
$\alpha = 92.911$ (1)°	$\mu = 4.92$ mm^{-1}
$\beta = 100.098$ (1)°	$T = 123$ K
$\gamma = 91.758$ (1)°	Blocks, colourless
$V = 799.64$ (5) Å^3	$0.30 \times 0.20 \times 0.10$ mm

Data collection

Bruker D8 VENTURE diffractometer with Photon100 detector	3129 independent reflections
Radiation source: INCOATEC microfocus sealed tube	3052 reflections with $I > 2\sigma(I)$
Detector resolution: 10.4167 pixels mm^{-1}	$R_{int} = 0.025$
rotation in ϕ and ω, 1°, shutterless scans	$\theta_{max} = 72.1°$, $\theta_{min} = 4.2°$
Absorption correction: multi-scan *SADABS* (Sheldrick, 2014)	$h = -10 \rightarrow 10$
$T_{min} = 0.432$, $T_{max} = 0.608$	$k = -11 \rightarrow 11$
13173 measured reflections	$l = -13 \rightarrow 13$

Refinement

Refinement on F^2	Secondary atom site location: difference Fourier map
Least-squares matrix: full	Hydrogen site location: difference Fourier map
$R[F^2 > 2\sigma(F^2)] = 0.030$	H-atom parameters constrained
$wR(F^2) = 0.070$	$w = 1/[\sigma^2(F_o^2) + (0.0179P)^2 + 0.8151P]$ where $P = (F_o^2 + 2F_c^2)/3$
$S = 1.07$	$(\Delta/\sigma)_{max} = 0.001$
3129 reflections	$\Delta\rangle_{max} = 0.45$ e Å^{-3}
201 parameters	$\Delta\rangle_{min} = -0.33$ e Å^{-3}
0 restraints	Extinction correction: *SHELXL2014*/7 (Sheldrick 2014, Fc*=kFc[1+0.001xFc$^2\lambda^3$/sin(2θ)]$^{-1/4}$
Primary atom site location: structure-invariant direct methods	Extinction coefficient: 0.0069 (5)

(E)-*N*-(1,2-dichlorovinyl)-*N*-(4-methoxybenzyl)-4-methyl-benzenesulfonamide – SB851_HY – (2.52b)

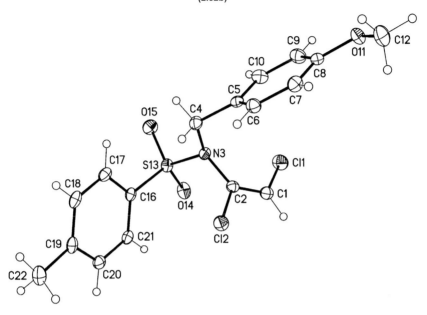

Crystal data

$C_{17}H_{17}Cl_2NO_3S$	$Z = 2$
$M_r = 386.27$	$F(000) = 400$
Triclinic, P-1 (no.2)	$D_x = 1.478$ Mg m^{-3}
$a = 8.0624$ (4) Å	Mo $K\alpha$ radiation, $\lambda = 0.71073$ Å
$b = 10.3401$ (5) Å	Cell parameters from 9974 reflections
$c = 10.8967$ (6) Å	$\theta = 2.6$–$27.5°$
$\alpha = 81.992$ (2)°	$\mu = 0.51$ mm^{-1}
$\beta = 86.286$ (2)°	$T = 123$ K
$\gamma = 74.817$ (2)°	Blocks, colourless
$V = 867.78$ (8) Å^3	$0.24 \times 0.16 \times 0.06$ mm

Data collection

Bruker D8 VENTURE diffractometer with Photon100 detector	3989 independent reflections
Radiation source: fine-focus sealed tube, IµS microfocus	3504 reflections with $I > 2\sigma(I)$
Detector resolution: 10.4167 pixels mm^{-1}	$R_{int} = 0.033$
rotation in ϕ and ω, 1°, shutterless scans	$\theta_{max} = 27.5°$, $\theta_{min} = 2.6°$
Absorption correction: multi-scan *SADABS* (Sheldrick, 2014)	$h = -10 \rightarrow 9$
$T_{min} = 0.887$, $T_{max} = 0.958$	$k = -13 \rightarrow 13$
21248 measured reflections	$l = -14 \rightarrow 14$

Refinement

Refinement on F^2	Secondary atom site location: difference Fourier map
Least-squares matrix: full	Hydrogen site location: difference Fourier map
$R[F^2 > 2\sigma(F^2)] = 0.029$	H-atom parameters constrained
$wR(F^2) = 0.071$	$w = 1/[\sigma^2(F_o^2) + (0.0278P)^2 + 0.5356P]$ where $P = (F_o^2 + 2F_c^2)/3$
$S = 1.04$	$(\Delta/\sigma)_{max} = 0.001$
3989 reflections	$\Delta\rangle_{max} = 0.38$ e Å^{-3}
220 parameters	$\Delta\rangle_{min} = -0.35$ e Å^{-3}
0 restraints	Extinction correction: *SHELXL2014/7* (Sheldrick 2014, Fc'=kFc[1+0.001xFc²λ³/sin(2θ)]$^{-1/4}$
Primary atom site location: structure-invariant direct methods	Extinction coefficient: 0.0096 (13)

(E)-N-(4-(tertbutyl)benzyl)-N-(1,2-dichlorovinyl)-4-methylbenzenesulfonamide – SB848_HY –
(2.52c)

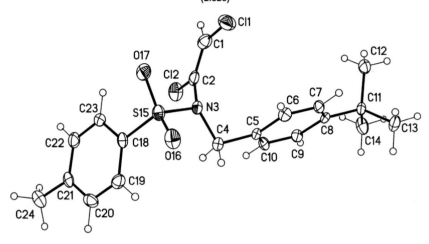

Crystal data

$C_{20}H_{23}Cl_2NO_2S$	$F(000) = 864$
$M_r = 412.35$	$D_x = 1.375$ Mg m^{-3}
Monoclinic, $P2_1/n$ (no.14)	Cu $K\alpha$ radiation, $\lambda = 1.54178$ Å
$a = 12.2988$ (6) Å	Cell parameters from 4123 reflections
$b = 11.0785$ (6) Å	$\theta = 4.1$–$67.9°$
$c = 15.3502$ (8) Å	$\mu = 4.02$ mm^{-1}
$\beta = 107.703$ (3)°	$T = 123$ K
$V = 1992.46$ (18) Å^3	Rods, colourless
$Z = 4$	$0.12 \times 0.02 \times 0.02$ mm

Data collection

Bruker D8 VENTURE diffractometer with Photon100 detector	3637 independent reflections
Radiation source: INCOATEC microfocus sealed tube	2638 reflections with $I > 2\sigma(I)$
Detector resolution: 10.4167 pixels mm^{-1}	$R_{int} = 0.073$
rotation in ϕ, 1°, shutterless scans	$\theta_{max} = 68.5°$, $\theta_{min} = 4.1°$
Absorption correction: multi-scan SADABS (Sheldrick, 2014)	$h = -14 \rightarrow 14$
$T_{min} = 0.784$, $T_{max} = 0.929$	$k = -13 \rightarrow 12$
12732 measured reflections	$l = -18 \rightarrow 14$

Refinement

Refinement on F^2	Primary atom site location: structure-invariant direct methods
Least-squares matrix: full	Secondary atom site location: difference Fourier map
$R[F^2 > 2\sigma(F^2)] = 0.056$	Hydrogen site location: difference Fourier map
$wR(F^2) = 0.134$	H-atom parameters constrained
$S = 1.02$	$w = 1/[\sigma^2(F_o^2) + (0.0565P)^2 + 2.2811P]$ where $P = (F_o^2 + 2F_c^2)/3$
3637 reflections	$(\Delta/\sigma)_{max} < 0.001$
236 parameters	$\Delta\rangle_{max} = 0.65$ e Å^{-3}
0 restraints	$\Delta\rangle_{min} = -0.35$ e Å^{-3}

N-ethynyl-N-(4-methoxybenzyl)-4-methylbenzenesulfonamide – SB842_HY – (2.56)

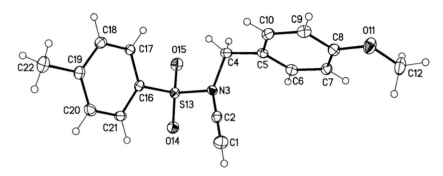

Crystal data

$C_{17}H_{17}NO_3S$	$D_x = 1.350$ Mg m^{-3}
$M_r = 315.37$	Cu $K\alpha$ radiation, $\lambda = 1.54178$ Å
Orthorhombic, $Pca2_1$ (no.29)	Cell parameters from 9882 reflections
$a = 15.5137$ (6) Å	$\theta = 5.5–72.0°$
$b = 6.2289$ (3) Å	$\mu = 1.96$ mm^{-1}
$c = 16.0592$ (6) Å	$T = 123$ K
$V = 1551.85$ (11) Å^3	Blocks, colourless
$Z = 4$	$0.34 \times 0.22 \times 0.12$ mm
$F(000) = 664$	

Data collection

Bruker D8 VENTURE diffractometer with Photon100 detector	3011 independent reflections
Radiation source: INCOATEC microfocus sealed tube	2985 reflections with $I > 2\sigma(I)$
Detector resolution: 10.4167 pixels mm^{-1}	$R_{int} = 0.021$
rotation in ϕ and ω, 1°, shutterless scans	$\theta_{max} = 72.1°$, $\theta_{min} = 5.5°$
Absorption correction: multi-scan *SADABS* (Sheldrick, 2014)	$h = -18{\rightarrow}19$
$T_{min} = 0.623$, $T_{max} = 0.795$	$k = -6{\rightarrow}7$
10488 measured reflections	$l = -19{\rightarrow}18$

Refinement

Refinement on F^2	Hydrogen site location: difference Fourier map
Least-squares matrix: full	H-atom parameters constrained
$R[F^2 > 2\sigma(F^2)] = 0.024$	$w = 1/[\sigma^2(F_o^2) + (0.0388P)^2 + 0.3139P]$ where $P = (F_o^2 + 2F_c^2)/3$
$wR(F^2) = 0.064$	$(\Delta/\sigma)_{max} = 0.001$
$S = 1.08$	$\Delta\rangle_{max} = 0.25$ e Å^{-3}
3011 reflections	$\Delta\rangle_{min} = -0.30$ e Å^{-3}
202 parameters	Extinction correction: *SHELXL2014*/7 (Sheldrick, 2014,
1 restraint	Extinction coefficient: 0.0022 (3)
Primary atom site location: structure-invariant direct methods	Absolute structure: Flack x determined using 1406 quotients [(I+)-(I-)]/[(I+)+(I-)] (Parsons, Flack and Wagner, Acta Cryst. B69 (2013) 249-259).
Secondary atom site location: difference Fourier map	Absolute structure parameter: -0.009 (6)

Fc*=kFc[1+0.001xFc^{2}λ^3/sin(2θ)]$^{-1/4}$

N-(4-(*tert*butyl)benzyl)-N-ethynyl-4-methylbenzenesulfonamide – SB857_HY – (2.57)

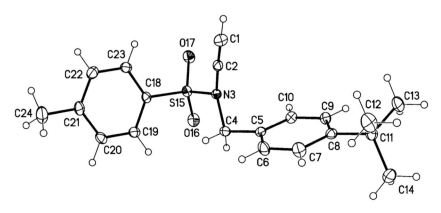

Crystal data

C₂₀H₂₃NO₂S	$F(000) = 728$
$M_r = 341.45$	$D_x = 1.233$ Mg m⁻³
Monoclinic, $P2_1/c$ *(no.14)*	Mo $K\alpha$ radiation, $\lambda = 0.71073$ Å
$a = 6.2821$ (5) Å	Cell parameters from 9902 reflections
$b = 15.6425$ (10) Å	$\theta = 2.5–28.7°$
$c = 18.7296$ (13) Å	$\mu = 0.19$ mm⁻¹
$\beta = 92.042$ (2)°	$T = 123$ K
$V = 1839.3$ (2) Å³	Blocks, colourless
$Z = 4$	$0.40 \times 0.20 \times 0.20$ mm

Data collection

Bruker D8 VENTURE diffractometer with Photon100 detector	4226 independent reflections
Radiation source: INCOATEC microfocus sealed tube	3809 reflections with $I > 2\sigma(I)$
Detector resolution: 10.4167 pixels mm⁻¹	$R_{int} = 0.030$
rotation in φ and ω, 1°, shutterless scans	$\theta_{max} = 27.6°$, $\theta_{min} = 2.5°$
Absorption correction: multi-scan *SADABS* (Sheldrick, 2014)	$h = -8\rightarrow8$
$T_{min} = 0.922$, $T_{max} = 0.971$	$k = -20\rightarrow20$
43163 measured reflections	$l = -24\rightarrow24$

Refinement

Refinement on F^2	Primary atom site location: structure-invariant direct methods
Least-squares matrix: full	Secondary atom site location: difference Fourier map
$R[F^2 > 2\sigma(F^2)] = 0.034$	Hydrogen site location: difference Fourier map
$wR(F^2) = 0.088$	H-atom parameters constrained
$S = 1.05$	$w = 1/[\sigma^2(F_o^2) + (0.0404P)^2 + 0.9808P]$ where $P = (F_o^2 + 2F_c^2)/3$
4226 reflections	$(\Delta/\sigma)_{max} = 0.001$
218 parameters	$\Delta\rangle_{max} = 0.32$ e Å⁻³
0 restraints	$\Delta\rangle_{min} = -0.50$ e Å⁻³

2.7 References

[1]   H. Zaugg, L. Swett, G. Stone, *J. Org. Chem.*, **1958**, *23* (9), 1389-1390, *Notes: An Unusual Reaction of Propargyl Bromide.*

[2]   H. G. Viehe, *Angew. Chem. Int. Ed.*, **1963**, *2* (8), 477-477, *Synthesis of Substituted Acetylenic Compounds.*

[3]   J. Ficini, *Tetrahedron*, **1976**, *32* (13), 1449-1486, *Ynamine: A versatile tool in organic synthesis.*

[4]   H. G. Viehe, *Angew. Chem. Int. Ed.*, **1967**, *6* (9), 767-778, *Synthesis and Reactions of the Alkynylamines.*

[5]   C. A. Zificsak, J. A. Mulder, R. P. Hsung, C. Rameshkumar, L.-L. Wei, *Tetrahedron*, **2001**, *57* (36), 7575-7606, *Recent advances in the chemistry of ynamines and ynamides.*

[6]   Z. Janousek, J. Collard, H. G. Viehe, *Angew. Chem. Int. Ed.*, **1972**, *11* (10), 917-918, *Reaction of Secondary Acetamides with N-Dichloromethylene-N,N-dimethylammonium Chloride.*

[7]   H. Y. Li, L. F. You, X. J. Zhang, W. L. Johnson, R. Figueroa, R. P. Hsung, *Heterocycles*, **2007**, *74*, 553-568, *Syntheses of amide-substituted isoxazoles and pyrazoles via regioselective [3+2] cycloadditions of terminally unsubstituted ynamides.*

[8]   M. Ijsselstijn, J.-C. Cintrat, *Tetrahedron*, **2006**, *62* (16), 3837-3842, *Click chemistry with ynamides.*

[9]   B. Witulski, T. Stengel, *Angew. Chem. Int. Ed.*, **1998**, *37* (4), 489-492, *N-Functionalized 1-Alkynylamides: New Building Blocks for Transition Metal Mediated Inter- and Intramolecular [2+2+1] Cycloadditions.*

[10]   X. Zhang, Y. Zhang, J. Huang, R. P. Hsung, K. C. M. Kurtz, J. Oppenheimer, M. E. Petersen, I. K. Sagamanova, L. Shen, M. R. Tracey, *J. Org. Chem.*, **2006**, *71* (11), 4170-4177, *Copper(II)-Catalyzed Amidations of Alkynyl Bromides as a General Synthesis of Ynamides and Z-Enamides. An Intramolecular Amidation for the Synthesis of Macrocyclic Ynamides.*

[11]   R. P. Hsung, C. A. Zificsak, L.-L. Wei, C. J. Douglas, H. Xiong, J. A. Mulder, *Org. Lett.*, **1999**, *1* (8), 1237-1240, *Lewis Acid Promoted Hetero [2 + 2] Cycloaddition Reactions of Aldehydes with 10-Propynyl-9(10H)-acridone. A Highly Stereoselective Synthesis of Acrylic Acid Derivatives and 1,3-Dienes Using an Electron Deficient Variant of Ynamine.*

[12]   R. Yamasaki, N. Terashima, I. Sotome, S. Komagawa, S. Saito, *J. Org. Chem.*, **2010**, *75* (2), 480-483, *Nickel-Catalyzed [3 + 2 + 2] Cycloaddition of Ethyl Cyclopropylideneacetate and Heteroatom-Substituted Alkynes: Application to Selective Three-Component Reaction with 1,3-Diynes.*

[13]   R. Tanaka, A. Yuza, Y. Watai, D. Suzuki, Y. Takayama, F. Sato, H. Urabe, *J. Am. Chem. Soc.*, **2005**, *127* (21), 7774-7780, *One-Pot Synthesis of Metalated Pyridines from Two Acetylenes, a Nitrile, and a Titanium(II) Alkoxide.*

[14]   M. Bendikov, H. M. Duong, E. Bolanos, F. Wudl, *Org. Lett.*, **2005**, *7* (5), 783-786, *An Unexpected Two-Group Migration Involving a Sulfonylnamide to Nitrile Rearrangement. Mechanistic Studies of a Thermal N → C Tosyl Rearrangement.*

[15]   J. Collard-Motte, Z. Janousek; *Synthesis of ynamines* in *Synthetic Organic Chemistry*; Springer Berlin Heidelberg: Berlin, Heidelberg, 1986.

[16]   J. A. Mulder, K. C. M. Kurtz, R. P. Hsung, *ChemInform*, **2003**, *34* (43), no-no, *In Search of an Atom-Economical Synthesis of Chiral Ynamides.*

[17]   G. Evano, A. Coste, K. Jouvin, *Angew. Chem. Int. Ed.*, **2010**, *49* (16), 2840-2859, *Ynamides: Versatile Tools in Organic Synthesis.*

[18]   K. A. DeKorver, H. Y. Li, A. G. Lohse, R. Hayashi, Z. J. Lu, Y. Zhang, R. P. Hsung, *Chem. Rev.*, **2010**, *110* (9), 5064-5106, *Ynamides: A Modern Functional Group for the New Millennium.*

[19]   X.-N. Wang, H.-S. Yeom, L.-C. Fang, S. He, Z.-X. Ma, B. L. Kedrowski, R. P. Hsung, *Acc. Chem. Res.*, **2014**, *47* (2), 560-578, *Ynamides in Ring Forming Transformations.*

[20]   A. M. Cook, C. Wolf, *Tetrahedron Lett.*, **2015**, *56* (19), 2377-2392, *Terminal ynamides: synthesis, coupling reactions, and additions to common electrophiles.*

[21]   R. P. Hsung, *Tetrahedron*, **2006**, *62* (16), 3781-3782, *Chemistry of electron-deficient ynamines and ynamides.*

[22]   J. Huang, H. Xiong, R. P. Hsung, C. Rameshkumar, J. A. Mulder, T. P. Grebe, *Org. Lett.*, **2002**, *4* (14), 2417-2420, *The First Successful Base-Promoted Isomerization of Propargyl Amides to Chiral Ynamides. Applications in Ring-Closing Metathesis of Ene−Ynamides and Tandem RCM of Diene−Ynamides.*

[23] L.-L. Wei, J. A. Mulder, H. Xiong, C. A. Zificsak, C. J. Douglas, R. P. Hsung, *Tetrahedron*, **2001**, *57* (3), 459-466, *Efficient preparations of novel ynamides and allenamides.*

[24] P. Murch, B. L. Williamson, P. J. Stang, *Synthesis*, **1994**, *1994* (12), 1255-1256, *Push-Pull Ynamines via Alkynyliodonium Chemistry.*

[25] F. Denonne, P. Seiler, F. Diederich, *Helv. Chim. Acta*, **2003**, *86* (9), 3096-3117, *Towards the Synthesis of Azoacetylenes.*

[26] A. Balsamo, B. Macchia, F. Macchia, A. Rossello, P. Domiano, *Tetrahedron Lett.*, **1985**, *26* (34), 4141-4144, *3-(4-iodomethyl-2-oxo-1-azetidinyl)propynoic acidt-butyl ester: a new β-lactam derivative, synthesis of an ynamide by reaction of 4-iodomethylazetidin-2-one with the t-butyl ester of propiolic acid in the presence of copper(i).*

[27] M. O. Frederick, J. A. Mulder, M. R. Tracey, R. P. Hsung, J. Huang, K. C. M. Kurtz, L. Shen, C. J. Douglas, *J. Am. Chem. Soc.*, **2003**, *125* (9), 2368-2369, *A Copper-Catalyzed C−N Bond Formation Involving sp-Hybridized Carbons. A Direct Entry to Chiral Ynamides via N-Alkynylation of Amides.*

[28] J. R. Dunetz, R. L. Danheiser, *Org. Lett.*, **2003**, *5* (21), 4011-4014, *Copper-Mediated N-Alkynylation of Carbamates, Ureas, and Sulfonamides. A General Method for the Synthesis of Ynamides.*

[29] K. Dooleweerdt, H. Birkedal, T. Ruhland, T. Skrydstrup, *J. Org. Chem.*, **2008**, *73* (23), 9447-9450, *Irregularities in the Effect of Potassium Phosphate in Ynamide Synthesis.*

[30] I. K. Sagamanova, K. C. M. Kurtz, R. P. Hsung; *Practical Synthesis of a Chiral Ynamide: (R)-4-Phenyl-3-(2-Triisopropylsilyl-Ethynyl)Oxazolidin-2-One in Org. Synth.*; John Wiley & Sons, Inc., 2003.

[31] X. Zhang, H. Li, Y. You, Y. Tang, R. P. Hsung, *Adv. Synth. Cat.*, **2006**, *348* (16-17), 2437-2442, *Copper Salt-Catalyzed Azide-[3 + 2] Cycloadditions of Ynamides and Bis-Ynamides.*

[32] Y. Zhang, R. P. Hsung, M. R. Tracey, K. C. M. Kurtz, E. L. Vera, *Org. Lett.*, **2004**, *6* (7), 1151-1154, *Copper Sulfate-Pentahydrate-1,10-Phenanthroline Catalyzed Amidations of Alkynyl Bromides. Synthesis of Heteroaromatic Amine Substituted Ynamides.*

[33] B. Yao, Z. Liang, T. Niu, Y. Zhang, *J. Org. Chem.*, **2009**, *74* (12), 4630-4633, *Iron-Catalyzed Amidation of Alkynyl Bromides: A Facile Route for the Preparation of Ynamides.*

[34] A. F. Abdel-Magid, K. G. Carson, B. D. Harris, C. A. Maryanoff, R. D. Shah, *J. Org. Chem.*, **1996**, *61* (11), 3849-3862, *Reductive Amination of Aldehydes and Ketones with Sodium Triacetoxyborohydride. Studies on Direct and Indirect Reductive Amination Procedures.*

[35] C.-W. Li, K. Pati, G.-Y. Lin, S. M. A. Sohel, H.-H. Hung, R.-S. Liu, *Angew. Chem. Int. Ed.*, **2010**, *49* (51), 9891-9894, *Gold-Catalyzed Oxidative Ring Expansions and Ring Cleavages of Alkynylcyclopropanes by Intermolecular Reactions Oxidized by Diphenylsulfoxide.*

[36] Y.-P. Wang, R. L. Danheiser, *Tetrahedron Lett.*, **2011**, *52* (17), 2111-2114, *Synthesis of 2-iodoynamides and regioselective [2+2] cycloadditions with ketene.*

[37] R. V. Joshi, Z.-Q. Xu, M. B. Ksebati, D. Kessel, T. H. Corbett, J. C. Drach, J. Zemlicka, *J. Chem. Soc. Perkin Trans. 1*, **1994**, (8), 1089-1098, *Synthesis, transformations and biological activity of chloro enamines and ynamines derived from chloroalkenyl- and alkynyl-N-substituted purine and pyrimidine bases of nucleic acids.*

[38] D. Brückner, *Synlett*, **2000**, *2000* (10), 1402-1404, *Synthesis of Ynamides and Ynol Ethers via Formamides and Formates.*

[39] D. Rodríguez, L. Castedo, C. Saá, *Synlett*, **2004**, *2004* (05), 0783-0786, *New Alkynyl Amides by Negishi Coupling.*

[40] D. Rodríguez, M. F. Martínez-Esperón, L. Castedo, C. Saá, *Synlett*, **2007**, *2007* (12), 1963-1965, *Synthesis of Disubstituted Ynamides from β,β-Dichloroenamides and Electrophiles.*

[41] S. J. Mansfield, C. D. Campbell, M. W. Jones, E. A. Anderson, *Chem. Commun.*, **2015**, *51* (16), 3316-3319, *A robust and modular synthesis of ynamides.*

[42] M. R. Tracey, Y. S. Zhang, M. O. Frederick, J. A. Mulder, R. P. Hsung, *Org. Lett.*, **2004**, *6* (13), 2209-2212, *Sonogashira cross-couplings of ynamides. Syntheses of urethane- and sulfonamide-terminated conjugated phenylacetylenic systems.*

[43] T. Wezeman, S. Zhong, M. Nieger, S. Bräse, *Angew. Chem. Int. Ed.*, **2016**, *55* (11), 3823-3827, *Synthesis of Highly Functionalized 4-Aminoquinolines.*

[44] R. B. Merrifield, *J. Am. Chem. Soc.*, **1963**, *85* (14), 2149-2154, *Solid Phase Peptide Synthesis. I. The Synthesis of a Tetrapeptide.*

[45]   G. A. Truran, K. S. Aiken, T. R. Fleming, P. J. Webb, J. H. Markgraf, *J. Chem. Ed.*, **2002**, *79* (1), 85, *Solid-Phase Organic Synthesis and Combinatorial Chemistry: A Laboratory Preparation of Oligopeptides.*

[46]   J. G. Parsons, C. S. Sheehan, Z. Wu, I. W. James, A. M. Bray; *A Review of Solid-Phase Organic Synthesis on SynPhase™ Lanterns and SynPhase™ Crowns* in *Methods Enzymol.*; Academic Press, 2003; Vol. Volume 369.

[47]   R. I. Storer, T. Takemoto, P. S. Jackson, S. V. Ley, *Angew. Chem. Int. Ed.*, **2003**, *115* (22), 2625-2629, *A Total Synthesis of Epothilones Using Solid-Supported Reagents and Scavengers.*

[48]   K. Knepper, S. Vanderheiden, S. Bräse, *Beilstein J. Org. Chem.*, **2012**, *8*, 1191-1199, *Synthesis of diverse indole libraries on polystyrene resin - Scope and limitations of an organometallic reaction on solid supports.*

[49]   S. Bräse, J. H. Kirchhoff, J. Kobberling, *Tetrahedron*, **2003**, *59* (7), 885-939, *Palladium-catalyzed reactions in solid phase organic synthesis.*

[50]   S. V. Ley, I. R. Baxendale, *Nat. Rev. Drug Discov.*, **2002**, *1* (8), 573-586, *New tools and concepts for modern organic synthesis.*

[51]   G. W. Kenner, J. R. McDermot, R. C. Sheppard, *J. Chem. Soc. Chem. Comm.*, **1971**, (12), 636-637, *Safety Catch Principle in Solid Phase Peptide Synthesis.*

[52]   S. Bräse, S. Dahmen, *Chem. Eur. J.*, **2000**, *6* (11), 1899-1905, *Traceless Linkers-Only Disappearing Links in Solid-Phase Organic Synthesis?*

[53]   K. Knepper, R. E. Ziegert; *T1 and T2 – Versatile Triazene Linker Groups* in *Linker Strategies in Solid-Phase Organic Synthesis*; John Wiley & Sons, Ltd, 2009.

[54]   P. J. H. Scott, *Linker Strategies in Solid-Phase Organic Synthesis*; John Wiley & Sons, Ltd, 2009.

[55]   N. Jung, M. Wiehn, S. Bräse; *Multifunctional Linkers for Combinatorial Solid Phase Synthesis* in *Combinatorial Chemistry on Solid Supports*; Springer Berlin Heidelberg, 2007; Vol. 278.

[56]   P. J. H. Scott; *Linker Strategies in Modern Solid-Phase Organic Synthesis* in *Solid-Phase Organic Synthesis*; John Wiley & Sons, Inc., 2011.

[57]   S. Bräse, *Combinatorial Chemistry on Solid Supports*; Springer-Verlag Berlin Heidelberg, 2007.

[58]   H.-M. Eggenweiler, *Drug Discov. Today*, **1998**, *3* (12), 552-560, *Linkers for solid-phase synthesis of small molecules: coupling and cleavage techniques.*

[59]   T. Muller, S. Bräse; *Combinatorial and solid-phase syntheses* in *The Mizoroki-Heck Reaction*; John Wiley & Sons Ltd., 2009.

[60]   E. Kaiser, R. L. Colescott, C. D. Bossinger, P. I. Cook, *Anal. Biochem.*, **1970**, *34* (2), 595-598, *Color test for detection of free terminal amino groups in the solid-phase synthesis of peptides.*

[61]   W. Huber, A. Bubendorf, A. Grieder, D. Obrecht, *Anal. Chim. Acta*, **1999**, *393* (1-3), 213-221, *Monitoring solid phase synthesis by infrared spectroscopic techniques.*

[62]   S. Bräse, *Acc. Chem. Res.*, **2004**, *37* (10), 805-816, *The Virtue of the Multifunctional Triazene Linkers in the Efficient Solid-Phase Synthesis of Heterocycle Libraries.*

[63]   L. Benati, G. Bencivenni, R. Leardini, D. Nanni, M. Minozzi, P. Spagnolo, R. Scialpi, G. Zanardi, *Org. Lett.*, **2006**, *8* (12), 2499-2502, *Reaction of Azides with Dichloroindium Hydride: Very Mild Production of Amines and Pyrrolidin-2-imines through Possible Indium–Aminyl Radicals.*

[64]   S. Bräse, D. Enders, J. Köbberling, F. Avemaria, *Angew. Chem. Int. Ed.*, **1998**, *37* (24), 3413-3415, *A Surprising Solid-Phase Effect: Development of a Recyclable "Traceless" Linker System for Reactions on Solid Support.*

[65]   H. E. Gottlieb, V. Kotlyar, A. Nudelman, *J. Org. Chem.*, **1997**, *62* (21), 7512-7515, *NMR Chemical Shifts of Common Laboratory Solvents as Trace Impurities.*

Chapter 3. 4-Aminoquinolines

The research presented in this Chapter has been conducted at the Institute of Organic Chemistry at the Karlsruhe Institute of Technology in Karlsruhe, Germany (with Prof. Dr. Stefan BRÄSE) and at BAYER CropScience GmbH in Frankfurt-Höchst, Germany (with Dr. Stephen D. LINDELL). Dr. Stephen D. LINDELL and Dr. Sabilla ZHONG are thanked for their help during the practical work on this Chapter. Parts of this work have been published as T. Wezeman, S. Zhong, M. Nieger and S. Bräse, *Angew. Chem. Int. Ed.* **2016**, *55*, 3823 – 3827.

3.0 Abstract & Graphical abstract

A diverse set of highly substituted 4-aminoquinolines has been synthesized in a mild one-step procedure employing the electrophilic activation of readily accessible amides with triflic anhydride and 2-chloropyridine and their subsequent reaction with a range of sulfonyl ynamides. The advantage of this approach is that − in stark contrast to pre-existing synthetic strategies towards 4-aminoquinolines − this method allows for full synthetic flexibility regarding substituents on the quinoline ring.

3.1 What have 4-aminoquinolines ever done for us?

Malaria

The most famous examples of 4-aminoquinolines known to the organic chemist are likely the antimalarial drugs CHLOROQUINE (**3.1**) and AMODIAQUINE (**3.2**). Although both have been discovered decades ago, to this date we still heavily rely on them to treat malaria. The sad truth is that because malaria is a drug that mainly affects poorer countries, it is less appealing for the pharmaceutical industry to invest millions into the development of new antimalarial drugs, although there certainly are exceptions to this "rule".[1]

Malaria is an infectious disease caused by protozoa of the *Plasmodium* genus, which are transmitted by mosquitoes. Symptoms include fever, fatigue, vomiting, headaches, yellow skin, seizures, coma, or death; as such it remains the most lethal human parasitic infection, killing roughly 2000 people per day in 2014.[2,3] Onset of symptoms typically occurs between 7-15 days after being bitten by an infected mosquito.

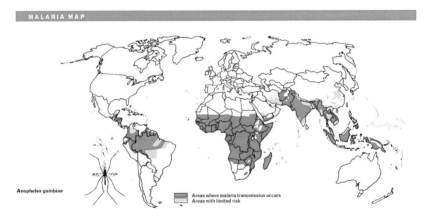

Figure 3.1 Worldwide Spreading and Malaria Risk Chart, as adapted from the World Malaria Risk Chart 2016, reproduced with permission from the International Association for Medical Assistance to Travellers.[4]

Chloroquine

CHLOROQUINE (**3.1**), after having been discovered in 1934 by ANDERSAG from BAYER AG, was firstly ignored due to human toxicity issues. It was eventually introduced for the prophylactic treatment of malaria in 1946. Since then CHLOROQUINE and several of its derivatives have been used extensively and are still being used today to help prevent and combat malaria. Recent studies show that similar 4-aminoquinolines could also serve as potent bacterial toxin inhibiters.[5]

Plasmodium parasites depend on both a female mosquito and a vertebrate, such as a human, for their reproduction. Detailed information on the lifecycle of the *plasmodium* parasites is widely available,[3] but goes beyond the scope of this Chapter. Important to note is that once a human is infected, the parasite eventually enters a red blood cell, where it must degrade hemoglobin to acquire essential nutrients such as amino acids. During this degradation process toxic heme is released, which is subsequently crystallized in the form of hemozoin. CHLOROQUINE inhibits this biocrystallization process and forms an even more toxic complex with the heme, which disrupts cell membrane function and eventually leads to parasite cell death.[6]

However, parasites that export chloroquine from their cells or that do not form the hemozoin crystals are resistant to chloroquine and in light of the emergence and spreading of resistant *Plasmodium* species,[7-9] development of new, more efficient anti-malarials is an ever present issue.[10-13] This is perhaps best exemplified by the 2015 Nobel Prize for Physiology or Medicine that was given to TU for her discovery of the Artemisinins (**3.3**), another famous anti-malarial that she was tasked with finding after the Chinese government wanted to find a way to combat chloroquine-resistant parasites.

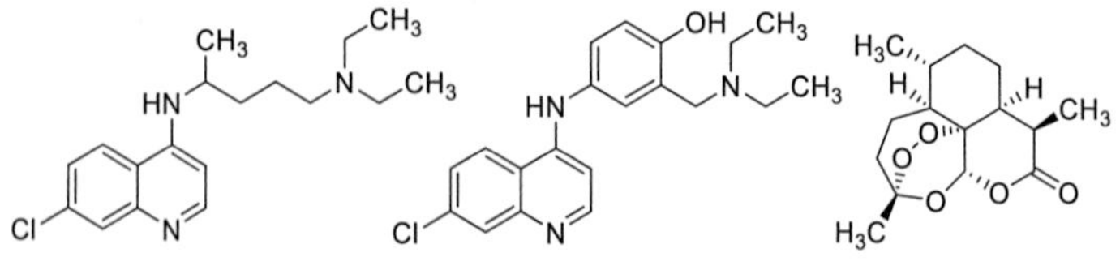

3.1 Chloroquine   **3.2 Amodiaquine**   **3.3 Artemisinin**

Figure 3.2 Three of the most prominent anti-malarials known to date.

Since quinolines are well-known for their antimalarial properties,[14,15] perhaps the development of novel routes to access novel 4-amino substituted quinolines − that appear of special pharmaceutical interest[10,16,17] − in a modular way can contribute to finally realizing a world where malaria is no longer a threat.

3.2 Synthetic approaches to quinolines

Quinolines can be made in several different ways and their synthesis has been of interest since the end of the 19th century. Several classic methods include the Skraup,[18] Doebner-von Miller,[19,20] Friedländer,[21,22] Pfitzinger,[23] and Combes[24,25] syntheses, all of whom consist mostly on a cyclo-condensation type reaction with an aniline and carbonyl compounds, as can be seen in Scheme **3.1**. Drawbacks of these still often used reactions are the usually harsh conditions needed and limitations in substrate scope.

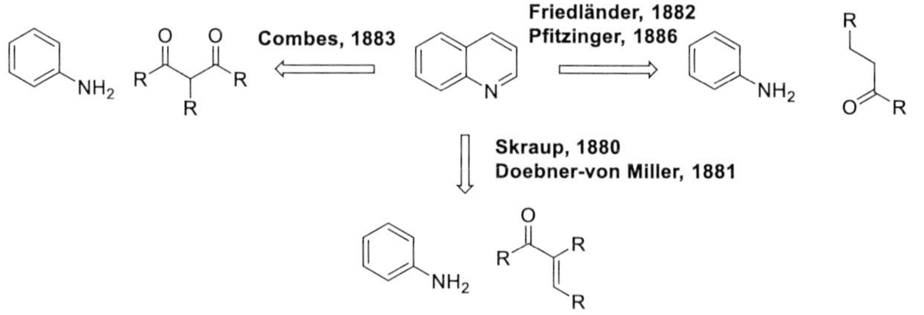

Scheme 3.1 Classical quinoline synthesis relying primarily on cyclo-condensations.

Besides these "old-school" methods, several more modern approaches to generate quinolines have been developed (see Scheme **3.2**).

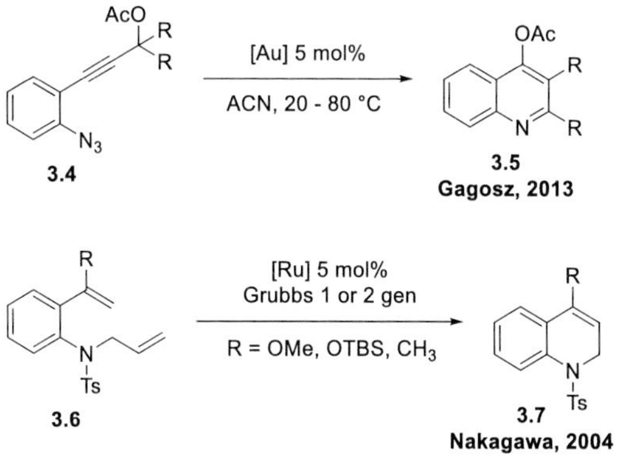

Scheme 3.2 A small selection of modern quinoline syntheses using transition metal catalysis.

These include employing gold catalysis to convert 2-alkynyl arylazide derivatives **3.4** into substituted 4-acetoxy quinolines **3.5**[26] and ring-closing metathesis to access 4-hydroxy or

4-methylquinolines **3.7**.[27] Other strategies involve solvent-free, microwave-assisted, so-called "green" InCl$_3$/SiO$_2$-catalyzed Skraup reactions (Scheme **3.3**).[28,29]

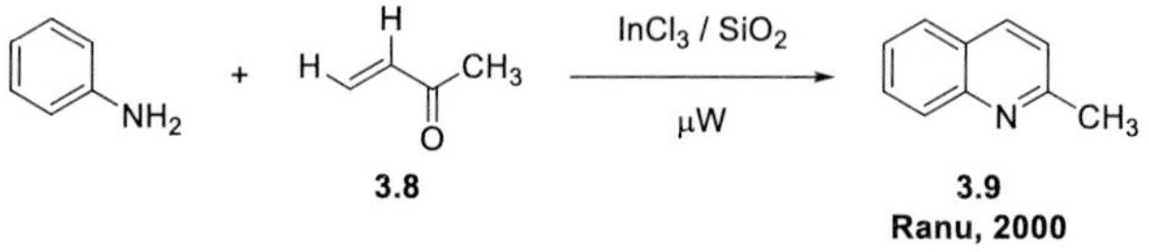

3.8   **3.9**
                                        Ranu, 2000

Scheme 3.3 Indium chloride impregnated silica gel-catalyzed Skraup reaction.

Although the quinoline core structure is relatively accessible, not all of these routes allow the implementation of a 4-amino group. Over the last century these 4-aminoquinolines have been prepared by reduction of quinoline-4-phenylhydrazine **3.10**[30,31] or via hypobromite oxidation of 2,3-dimethylquinoline-4-carboxyamides **3.12**.[32] 5-Aminotetrahydroacridine **3.16**, a tricyclic derivative, can be accessed via its iminodihydro-oxazine intermediate **3.14** by heating anthranilonitrile in cyclohexanone in the presence of a quaternary ammonium catalyst (Scheme **3.4**).[33]

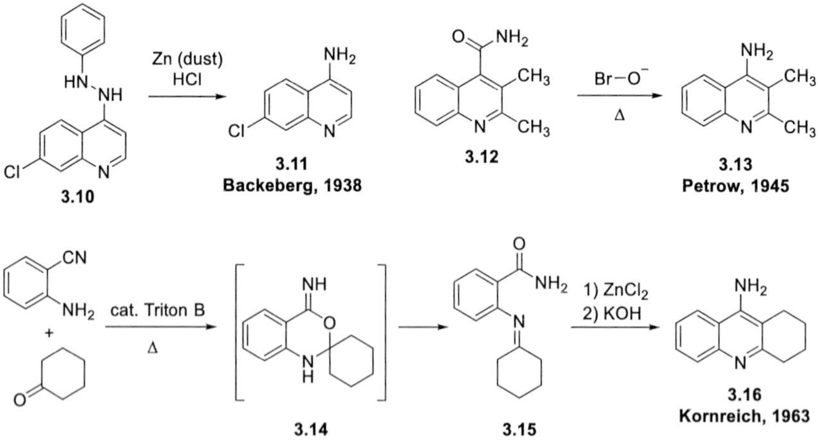

3.10   **3.11**   **3.12**   **3.13**
                Backeberg, 1938   **Petrow, 1945**

3.14   **3.15**   **3.16**
                                    Kornreich, 1963

Scheme 3.4 Selection of early 4-aminoquinoline syntheses.

A relatively popular route to 4-aminoquinolines, that was specifically designed to allow diversification at the 4-amino group, goes via a 4-chloroquinoline precursor **3.18**.[16] The benefit of the key step, a nucleophilic attack, is that it permits variation of the nucleophile, thus creating a small library of analogs to the anti-malarial CHLOROQUINE (**3.1**).

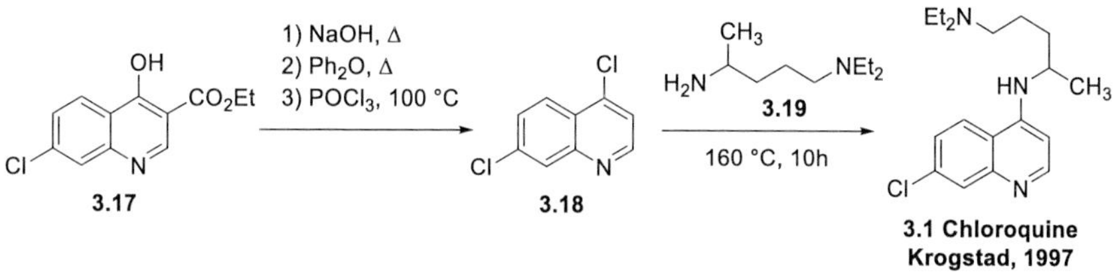

Scheme 3.5 Synthesis of Chloroquine and analogs via a 4-chloroquine precursor.

Alternatively the 4-aminoquinoline core can be obtained by rearrangement of pyrazolium-3-carboxylates **3.20** via pyrazol-3-ylidene intermediates **3.22**,[34] by reacting 2-(trifluoromethyl)-4H-3,1-benzoxazinones **3.24** with ynamines **3.25**[35] or with the use of an aerobic oxidative Pd-catalyzed imidoylative coupling with a double C−H activation, as in the case with the synthesis of 4-aminoquinoline **3.28**.[36]

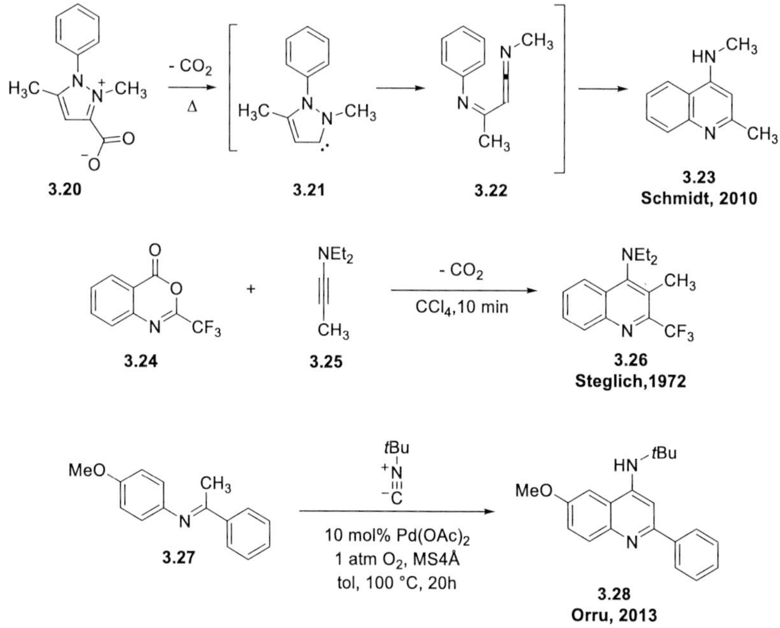

Scheme 3.6 4-Aminoquinoline synthesis with simultaneous installment of the 4-amino group and construction of the pyridine ring.

A need to overcome the current synthetic limitations

From all the routes discussed above the 4-chloro precursor strategy seems the most useful for library synthesis, as evidenced by the synthesis of over thirty chloroquine analogs and their subsequent screening in antimalarial assays.[37] However, when one carefully assesses the synthesized analogs one can see clearly that most modifications are being done by starting with different substitution patterns on the quinoline ring. This severely limits the possibilities and leads to labor intensive library constructions. Putting this knowledge in context of the other presented routes to 4-aminoquinolines, a clear synthetic shortcoming becomes obvious: There is no truly modular way to efficiently produce highly functionalized 4-aminoquinolines. The introduction of substituents at both the C-2, C-3 and benzylic C-5 to C-8 positions remained a challenge – *until now*.

3.3 Electrophilically-activated amides

The modular synthesis of highly functionalized 4-aminoquinoline, discussed in this Chapter, relies on the reaction of ynamides, discussed in Chapter 2, with electrophilically-activated amides. Over the last decade the MOVASSAGHI, MAULIDE and CHARETTE groups have reported extensively on the triflic anhydride[38] (Tf_2O) and (halo)pyridine-based amide activation protocol.

In the late 1990s the CHARETTE group found that the combination of triflic anhydride with pyridine can be very efficiently put to use in functional group modification, as shown in Scheme **3.7**.[39-42]

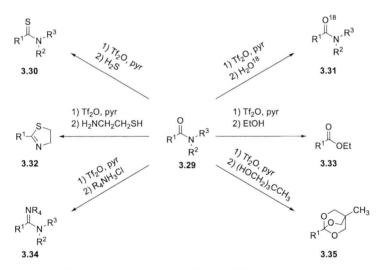

Scheme 3.7 CHARETTE's early work with triflic anhydride/pyridine for functional group transformations.

They followed up their research with a detailed NMR-spectroscopic study in 2001 that revealed the pyridinium intermediates responsible for the reactivity (shown in Scheme **3.8**).[43]

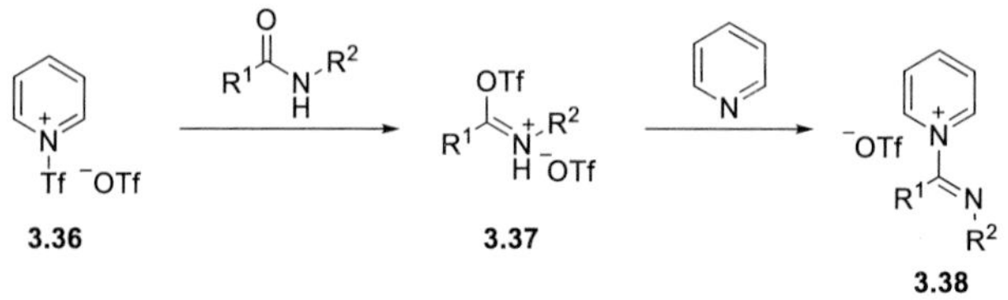

Scheme 3.8 Proposed mechanistic pathway for the formation of pyridinium intermediates **3.38** from an initially formed *N*-(trifluoromethylsulfonyl)pyridinium trifluoromethanesulfonate **3.37**.

Additionally they were able to apply the electrophilic activation procedure in the synthesis of several natural products, e.g. (±)-tetraponerine T4 (**3.41**)[44] and indolizidines **3.44**.[45]

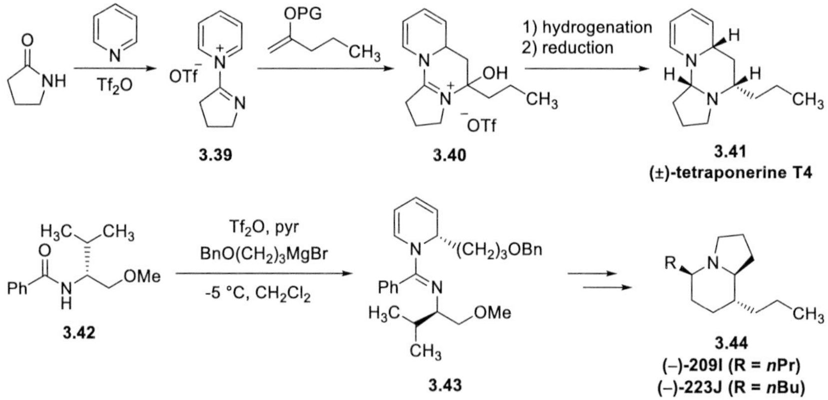

Scheme 3.9 Natural product syntheses relying on the amide activation procedure with triflic anhydride and pyridine.

In recent years the triflic anhydride/pyridine electrophilic activation of amides has been implemented in various ways to produce a broad spectrum of products, as depicted in Scheme **3.10**.[46-57]

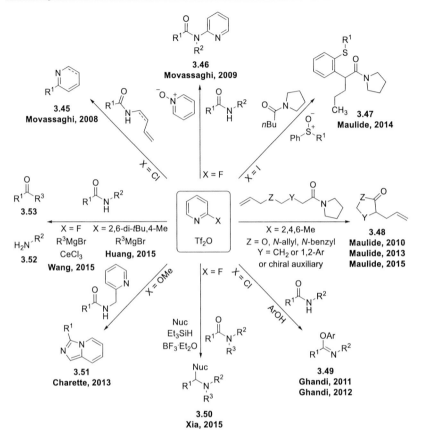

Scheme 3.10 Broad overview of implementation of the triflic anhydride/pyridine electrophilic activation of amides to produce a wide range of products.[46-57] Please note that **3.48** is formed only after a Claisen rearrangement.

The role of the pyridine

A keen observer might have at this point noted the use of differently substituted pyridines among the reactions shown in Scheme **3.10**. In order to increase the understanding of the influence of the substituent on the pyridine ring, detailed *in situ* infrared spectroscopic studies have been performed by MOVASSAGHI in 2015.[58] They observed that out of their eleven tested pyridine substrates 2-chloropyridine was the only one that did not form a *N*-(trifluoromethylsulfonyl)pyridinium trifluoromethanesulfonate **3.55**. Other electron-deficient pyridines, such as the often used 2-fluoropyridine, showed significant formation of **3.55** only after warming to room temperature.

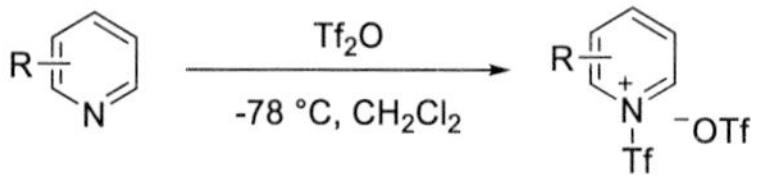

3.54   **3.55**

Scheme 3.11 Formation of *N*-(trifluoromethylsulfonyl)pyridinium trifluoromethanesulfonate.

The electron-rich pyridines included in the tests were observed to form complex **3.55** rapidly. Although the formation of the pyridinium complex **3.55** does not fully inhibit the desired amide activation, it can still form complexes like **3.38** and **3.57** following the mechanism shown in Scheme **3.8**, and as a competing reaction it will inherently slow the reaction down. Due to its electronic and steric properties 2-chloropyridine is observed *not* to form this complex. This allows the triflic anhydride to react directly with the amide in the desired fashion, as shown in Scheme **3.12**, without the competing formation of **3.55**. Their studies also showed that in some cases the formation of nitrilium ions **3.58** occurred. They concluded that although this might depend to some degree on the substitution pattern on the pyridine, a clear trend could be observed where *N*-alkyl amide substrates preferred the formation of nitrilium ions **3.58** to a greater extend as compared to *N*-aryl amides.[52]

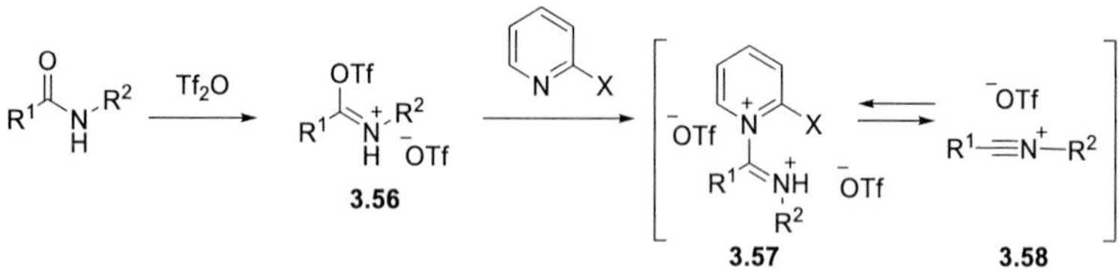

3.56   **3.57**   **3.58**

Scheme 3.12 Mechanistic pathway with direct addition of triflic anhydride to the amide, followed by substitution of the pyridine and formation of amidinium ion **3.57** and/or nitrilium ion **3.58** depending on the nature of the amide.

Lastly, they observed that amides showing higher propensity to form nitrilium ions, tended to react with isoquinoline *N*-oxides in better yields. This can be explained by the high reactivity of the nitrilium ions.

Adding a triple bond to the mix

In 2006 MOVASSAGHI and his coworkers started to explore what would happen if they introduced a σ-nucleophile such as a nitrile to an activated amide.[59] Over the course of the next few years they published several papers, including an overview article[60] and a few methodology papers[61-63] where they fully exploit the reactions of the amides and the nitriles to pyrimidines and quinazolines, as shown in Scheme **3.13**. They covered a wide range of substituted amides and nitriles.[64] Additionally they found that a similarly facile reaction to pyridines and quinolines can be employed by changing the σ-nucleophilic nitrile for an electron-rich π-nucleophile such as an alkyne or an enol ether (leading to products like quinoline **3.59b** after loss of the silyl ether).[65]

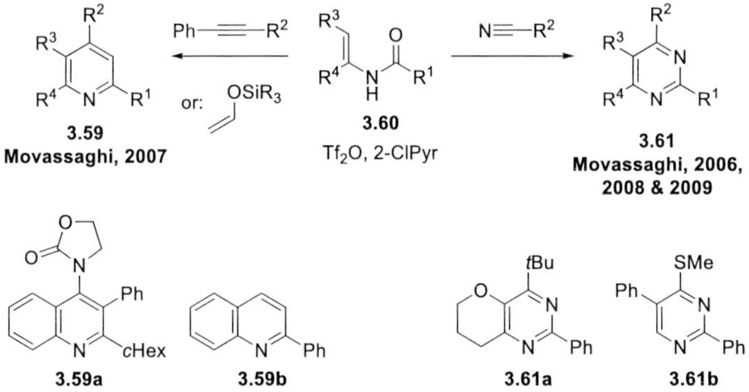

Scheme **3.13** MOVASSAGHI's pyrimidine and pyridine syntheses and some key examples from both routes.

Their rather extensive exploration of the scope of the pyridine/quinoline reaction also included the synthesis of four 4-aminoquinolines **3.59a**, based on an oxazolidin-2-one ynamide. This intrigued me. Keeping the current synthetic limitations of 4-aminoquinolines in mind, it was decided to explore whether my diverse routes towards a wide range of substituted sulfonyl ynamides (see Chapter 2) could be used to produce highly functionalized and diversified 4-aminoquinolines.

3.4 Synthesis of highly functionalized 4-aminoquinolines

Many of the currently used methods to synthesize 4-aminoquinolines have shown to be rather limiting in their tolerance towards functionalization on the quinoline ring, especially on the C-2 and C-3 positions. In order to have synthetic access to libraries of 4-aminoquinolines the organic chemist needs to be able to introduce diverse functional groups into the system. By employing the electrophilic amide activation procedure described in Chapter 3.3, in combination with the ynamides designed in Chapter 2, virtually all positions on the quinoline ring are accessible. A plausible mechanism, based on the similar work done by MOVASSAGHI described in Chapter 3.3 is depicted in Scheme **3.14**.

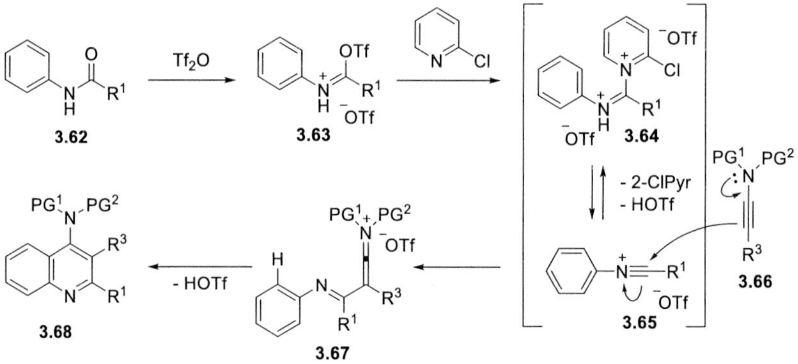

Scheme 3.14 Proposed mechanistic pathway with direct addition of triflic anhydride to the amide, followed by substitution of the pyridine and formation of amidinium ion **3.64**. Attack of the ynamide on the nitrilium ion **3.65** is followed by electrocyclization to the final quinoline.

A modular approach

Key to an effective library synthesis is the availability of a modular system. Scheme **3.15** shows how diversity is achieved in our 4-aminoquinoline system and where each substitution originates from. The amides **3.70**, which can easily be prepared from their corresponding acids or acid chlorides and anilines, provide the substitution at C-2 and C-5 to C-8 (e.g. Scheme **3.15**, R^4 and R^5). The ynamide **3.71** supplies the protected 4-amino and the C-3 functionality (e.g. Scheme **3.15**, R^3). The synthesis of the used sulfonyl ynamides **3.71** is discussed extensively in Chapter 2.

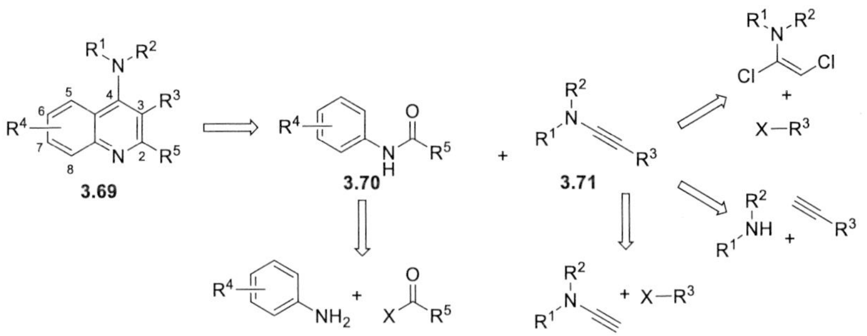

Scheme 3.15 Retrosynthetic analysis of the modular 4-aminoquinoline **3.69** design.

Reaction optimization study

In order to find the optimal conditions for the quinoline synthesis a series of experiments were performed (see Table **3.1**). For the activation procedure the amides were reacted with triflic anhydride at $-78\ ^{\circ}C$ in the presence of 2-chloropyridine. After 5 minutes at this temperature the reaction mixture was gently warmed to $0\ ^{\circ}C$ and after 5-10 minutes typically a reddish color appeared. At that point the solution of ynamides was added and the mixture stirred at $0\ ^{\circ}C$ until consumption of all ynamide was observed. It was quickly found that altering reaction times of the individual steps yielded very impure crude mixtures with lower amounts of product than desirable. Therefore only the ratios of reagents were varied in order to see what conditions would be best.

Entries 1-4 of Table **3.1** show clearly that increasing the ratio of activated amide, compared to ynamide, yielded poorer results. This was mainly due to the larger amounts of side-products formed, resulting in more challenging isolations. Increasing the amount of 2-chloropyridine had no clear effect and heating of the reaction mixture where the activated amide was in excess also did not improve the yield (entry 3, Table **3.1**).

Table 3.1 Selected entries for the optimization of reaction conditions for the activated amide protocol leading to the 4-aminoquinolines.

R² ... N(H) ... O ... **3.70** cyclohexyl

3.71 R¹ ... S(O₂)N≡CH ... H₃C

Tf₂O, 2-ClPyr / CH₂Cl₂

3.69 R¹ ... S(O₂)N ... H ... cyclohexyl ... N ... R²

Entry	R¹	R²	Amide (equiv.)	2-ClPyr (equiv.)	Tf₂O (equiv.)	Ynamide (equiv.)	T [°C][a]	Yield [%][b]
1	OMe	H	1.0	1.2	1.2	1.2	0	55
2	OMe	H	1.3	1.3	1.3	1.0	0	30
3	OMe	H	1.3	1.3	1.3	1.0	50	26
4	OMe	H	1.3	2.6	1.5	1.0	0	32
5	H	H	1.0	1.2	1.2	1.2	0	25
6	H	H	1.3	1.3	1.3	1.0	0	41
7	H	H	1.3	1.3	1.3	1.0	50	18
8	H	H	1.0, MS4Å	1.2	1.2	1.2	0	51
9	H	Cl	1.0	1.2	1.1	1.2	0	43
10	H	Cl	1.3	1.3	1.2	1.0	0	23
11	H	Cl	1.0, MS4Å	1.2	1.1	1.2	0	53

a) Reaction mixtures are prepared at –78 °C (5 min stirring) before they are stirred at shown temperature, heating is facilitated with aid of microwave irradiation; b) isolated yields. MS4Å = molecular sieves with 4 Å pores.

Although heating of the reaction at 50 °C (entries 3 and 7, Table **3.1**) showed poor results, heating of the reaction to 120 °C using microwave irradiation for 20 minutes was found to still deliver quinolines (e.g. **3.89** and **3.96**, Figure **3.6**) in reasonable yields.

Oddly enough, when the original conditions were applied to the same reaction with a different ynamide a much lower yield was observed. In this case it did seem to help to use the activated amide in excess, increasing the yield from 25% to 41% (compare entries 5 and 6, Table **3.1**) However, when the same was attempted with another amide substrate, the reverse was observed (entries 9 and 10, Table **3.1**). At this point it can be concluded that likely a more complex situation is at hand. In order to see if this seemingly odd response to the changes in equivalents could be prevented, an attempt was made to eliminate any other sources of yield fluctuations.

And indeed: The best results were obtained using the original conditions as proposed by MOVASSAGHI, but with the addition of activated 4Å molecular sieves to remove any traces of water. Not only is the terminal ynamide fairly sensitive to water, the presence of water during the activation of the amide is detrimental to the overall reaction. In the case of entry 5, the ynamide might still have contained traces of water, so when more activated amide was employed, more of the amide was still available to react. On the other hand, in entry 9 the amide used here may have still contained water, thus lowering the yield when more was added to the reaction.

The addition of activated 4Å molecular sieves to the reaction mixture prior to cooling to −78 °C and addition of the triflic anhydride, as well as the addition of the sieves to the solution of ynamide before it is added to the reaction, was found to drastically increase consistency in yields.

A diverse set of 4-aminoquinolines

With the optimized conditions in place, the exploration of the scope of possible 4-aminoquinolines began. For enhanced clarity of the drawings in this subchapter, the starting materials and conditions are omitted. All 4-aminoquinolines were synthesized according to general procedure **GP3.1** with the corresponding amide (1.00 equiv.), 2-chloropyridine (1.20 equiv.) and Tf$_2$O (1.10 equiv.) in CH$_2$Cl$_2$ at –78 °C by addition of the matching sulfonyl ynamide (1.10-1.20 equiv.) in CH$_2$Cl$_2$ at 0 °C. Several 4-aminoquinolines described in this chapter were synthesized and characterized in collaboration with Dr. Sabilla ZHONG and their synthetic and analytical data can be found online.[66]

Perhaps the easiest groups to vary are the substituents on the aromatic ring. Although not all reactions proceeded in great yields, in principle the C-5 to C-8 can be freely functionalized. In order to simplify the product isolation and prevent the formation of isomers, most substituents were chosen in such a manner that they would be *para* to the amide.

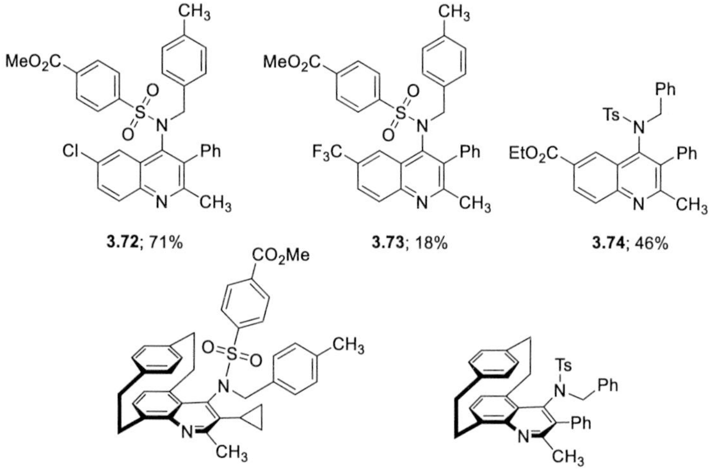

Figure 3.3 Quinolines with a variety C-5 to C-8 modifications, synthesized according to general procedure **GP3.1**.

In order to show the broad applicability of the methodology, it was found that the ynamides also readily react with paracyclophane-based amides, as shown for compounds **3.75** and **3.76**, creating very interesting planar chiral compounds, albeit in a racemic manner. However, issues with the purifications led to the isolation of a mixture of **3.75** with an inseparable side product and **3.76** in poor yield.

The C-3 functional groups on the quinoline core are inherited from the sulfonyl ynamide. The ynamides used for **3.46** and **3.40** were prepared using the copper-catalyzed pathway described in Chapter 2.2 from phenyl- and cyclopropyl-acetylene. A simple proton at C-3 can be obtained by using the water-sensitive terminal ynamide **2.43b** or by its triisopropylsilane (TIPS) precursor **2.40b**. Although it is unclear in what stage of the reaction the ynamide/quinoline loses the TIPS protection group, the amount of isolated product is rather low, making it worthwhile to deprotect beforehand in a separate step instead of a domino fashion.

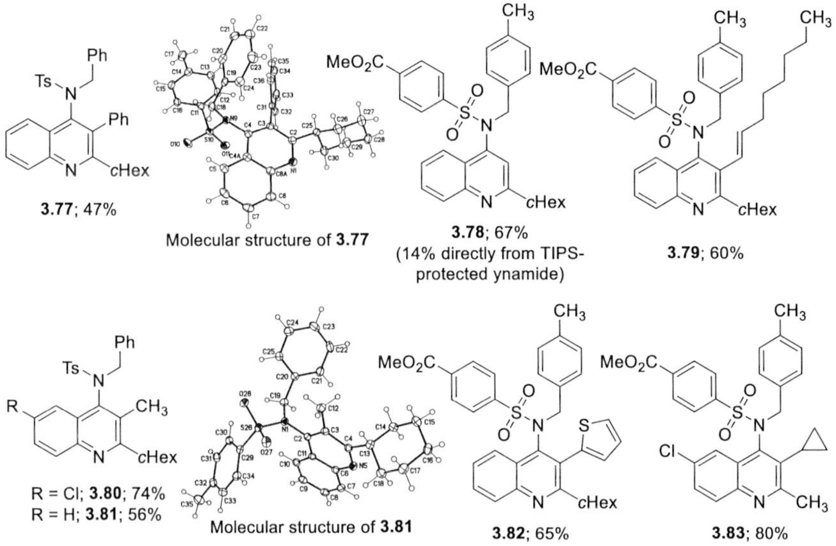

Figure 3.4 Quinolines with different C-3 modifications, synthesized according to general procedure

GP3.1.

Quinolines **3.79** and **3.82** were made using ynamides **2.61** and **2.62**, which were obtained after additional Sonogashira reactions on their terminal precursor. The diverse group of C-3 modified quinolines depicted in Figure **3.4** shows that the usage of the sulfonyl ynamides as synthon for the ring construction is extremely beneficial. The direct access to the ynamides shown in Chapter 2 could easily be exploited to prepare a vast amount of C-3 derivatives. Moreover it shows again how efficient and broadly applicable the amide activation protocol can be.

By selecting the acid or acid chloride that is used to prepare the amides, the C-2 substitution in the final quinolines can be chosen. Figure **3.5** shows quinolines with the three different C-2 amides that were at hand. From literature it was known that the implementation of a formamide is not compatible due to the facile and competitive formation of isonitriles.[64] Although only three different groups were explored from this category, in principle a wide selection should be accessible.

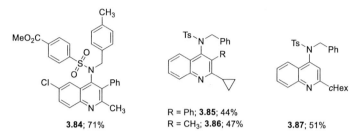

R = Ph; **3.85**; 44%
R = CH₃; **3.86**; 47%

3.87; 51%

Figure 3.5 Quinolines with different C-2 modifications, synthesized according to general procedure **GP3.1**.

The ynamide bears two protecting groups to ensure its stability, prevent side-reaction and to temper the reactivity of the ynamine to an ynamide. However, in respect of library synthesis, one might consider these two groups to offer additional points of diversity. Figure **3.6** depicts the different varieties of protecting groups used. Although the tosyl group is often used, it also offers no synthetic flexibility. The methyl ether analog however, can be converted easily into its benzoic acid with lithium hydroxide and used in subsequent peptide chemistry, for example to couple it to a solid support for follow-up chemistry or to create analogs.

The benzyl protecting group has been varied as well and although there is at times a rather large difference in yield between two nearly identical reactions – see Figure **3.6**; compare **3.90 - 3.92** – no clear trend has been observed. Instead differences in yield might rise from minor levels of impurities in the starting materials. Ynamide **2.43e**, that gave quinoline **3.92** in 90% yield, bears a *tert*-butyl group and was found to crystallize out very nicely in large blocks, thus simplifying its purification procedure and leading to very dry product. Its chlorinated analog ynamide **2.43d** was more difficult to purify and was isolated as slightly yellow solid, indicating the presence of possible impurities. Possibly traces of impurities or water could have diminished the yields. To test the influence of the benzyl group, it was replaced with a furfural and the reaction was found to still proceed. Overall one might conclude that although the nature of the protecting groups on the ynamide have some minor influence on the yields, other factors likely have bigger impacts.

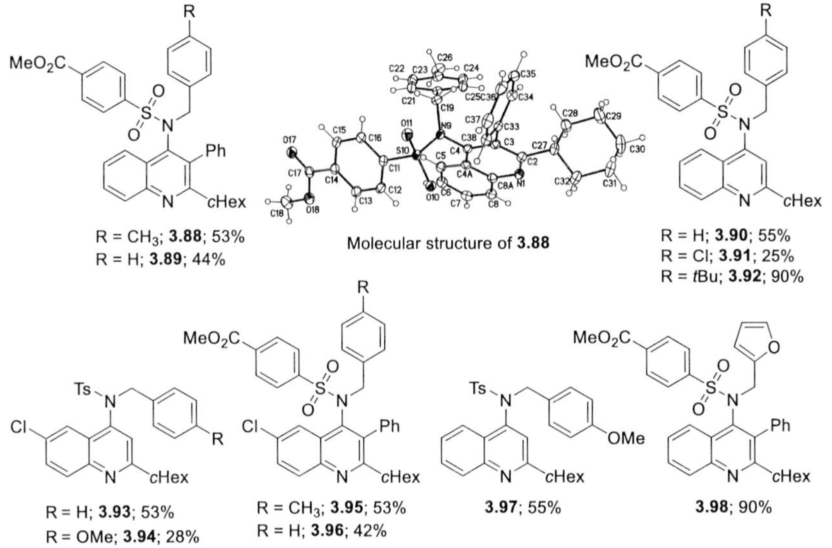

R = CH₃; **3.88**; 53%
R = H; **3.89**; 44%

Molecular structure of **3.88**

R = H; **3.90**; 55%
R = Cl; **3.91**; 25%
R = tBu; **3.92**; 90%

R = H; **3.93**; 53%
R = OMe; **3.94**; 28%

R = CH₃; **3.95**; 53%
R = H; **3.96**; 42%

3.97; 55%

3.98; 90%

Figure 3.6 Quinolines with different 4-amino protecting groups, synthesized according to general procedure **GP3.1**.

The installment of different groups on the amino moiety of the 4-aminoquinoline yields an even larger library of compounds for biological screens. However, doing this diversification by introducing these groups via the ynamide might not be the most convenient or least labor-intensive way. A free amine on the other hand could be easily derivatized by formation of amide bonds. These type of reactions could even be done using modern robotics, thus speeding up the process. To do this one would first need to get rid of the protecting groups the ynamide brings to the table.

3.5 Access to deprotected 4-aminoquinolines

In order for the prepared 4-aminoquinolines to be more useful from a library synthesis and pharmaceutical point of view, it is highly desirable to be able to deprotect the tosylated and benzylated amine under simple and straightforward conditions. The free amines allow access to a much larger and diverse library – via for example amide bonds with acids and acid chlorides. Conventional detosylation reactions usually employ inelegant and harsh conditions, such as using sodium or lithium metal with naphthalene at cold temperatures,[67,68] using strong reductive agents such as sodium bis(2-methoxyethoxy)aluminium hydride in refluxing benzene,[69] pyridine/HF mixtures[70] or HBr in acetic acid,[71] which are unsuitable for more delicate compounds (Scheme **3.16**).

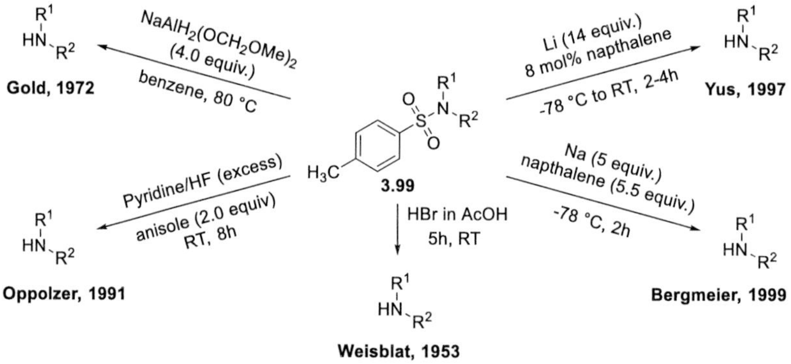

Scheme 3.16 Deprotection of tosylamides under typically harsh conditions.

Fortunately more mild conditions have been designed in more recent years to overcome this synthetic hurdle. Some of these options are the sonication of the tosylamide in magnesium-methanol mixtures,[72] treatment with low-valent titanium, generated *in situ* from Ti(O*i*Pr)$_4$, trimethylsilyl chloride and magnesium powder[73] and forcing a condensation with a bismesylate under basic conditions, followed by the addition of sodium amalgam.[74] A very interesting method to remove the tosyl group under mild conditions is the reaction with samarium iodide[75] – in the presence of an amide such as triethylamine or pyrrolidine and water.[76] Although initial tests proved this method to work on debenzylated tosyl-4-aminoquinolines and to remove the tosyl instantly, issues in reproducibility led to the exploration of alternatives that would be more robust.

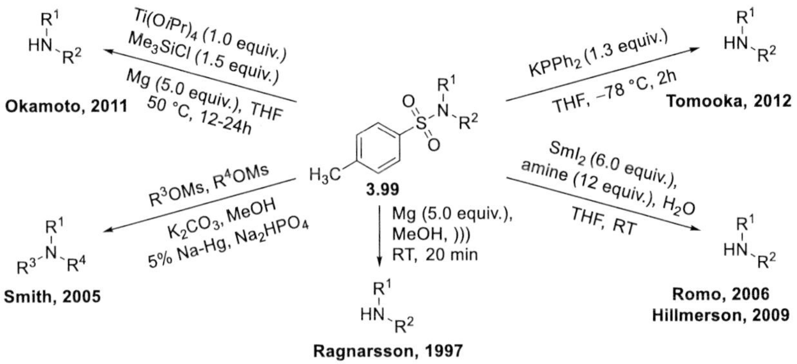

Okamoto, 2011
Ti(OiPr)$_4$ (1.0 equiv.)
Me$_3$SiCl (1.5 equiv.)
Mg (5.0 equiv.), THF
50 °C, 12-24h

Tomooka, 2012
KPPh$_2$ (1.3 equiv.)
THF, –78 °C, 2h

Smith, 2005
R^{3}OMs, R^{4}OMs
K$_2$CO$_3$, MeOH
5% Na-Hg, Na$_2$HPO$_4$

Romo, 2006
Hillmerson, 2009
SmI$_2$ (6.0 equiv.),
amine (12 equiv.), H$_2$O
THF, RT

Ragnarsson, 1997
Mg (5.0 equiv.),
MeOH,)))
RT, 20 min

3.99

Scheme 3.17 Deprotection of tosylamides under mild conditions.

Employing a procedure recently reported by TOMOOKA *et al.* using potassium diphenylphosphide was found to be very effective.[77] A short reaction optimization showed that two equivalents of the phosphide led to the best results and 4-aminoquinoline **3.101** was isolated in 79% yield.

TOMOOKA and coworkers proposed several different plausible mechanisms for this deprotection. One of the possible pathways constitutes of a direct nucleophilic attack of the phosphor atom on the nitrogen and is shown in Scheme **3.18**. An alternative stepwise mechanism involving initial attack of the phosphor on the sulfur atom could be excluded due to lack of evidence regarding certain intermediates, but the authors note that their third proposed mechanism, employing a SET reaction, could only partly be disproved. However, with the aid of DFT calculations they were able to determine that the direct attack would be most probable.

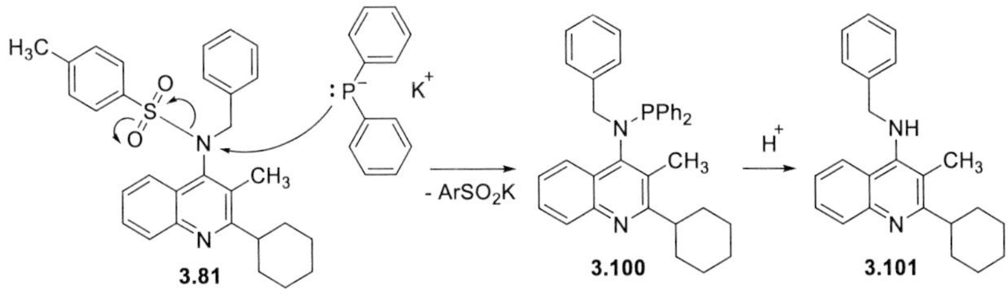

3.81   3.100   3.101

Scheme 3.18 Proposed mechanism for the nucleophilic tosyl deprotection using potassium diphenylphosphide.

Besides the tosyl group, the benzyl group should be removed as well. This could be easily accomplished by conventional hydrogenation using 0.2 equivalents of wetted Pd/C and a balloon of hydrogen gas at room temperature for several days.

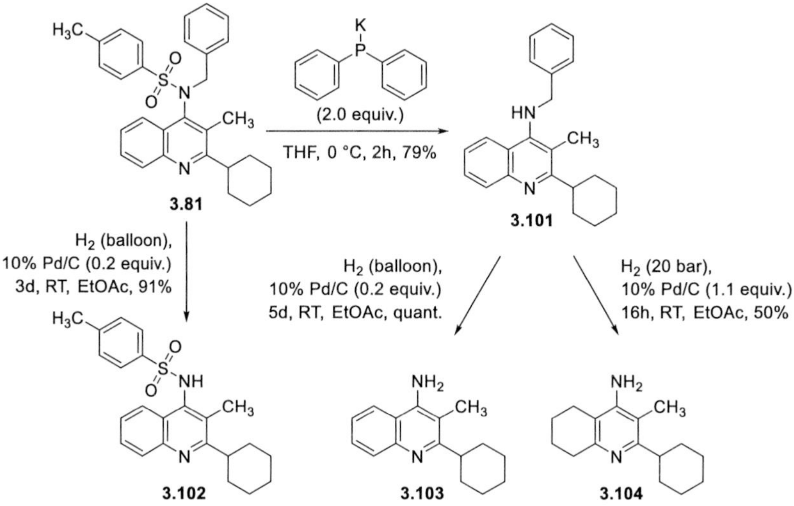

Scheme 3.19 Step-wise detosylation and debenzylation of 4-aminoquinolines.

In an attempt to find faster conditions *N*-benzyl 4-aminoquinoline **3.101** was submitted to 20 bar hydrogen with 0.2 equivalents of dry Pd/C in a pressure reactor. When after 3 hours no promising conversion was observed, the catalyst loading was increased to 1.1 equivalent and stirred overnight in the pressure reactor. Surprisingly not only the *N*-benzyl was removed under these conditions, but also the benzene ring of the quinoline **3.104** was reduced. Removal of the benzyl group from quinoline **3.81** without prior removal of the tosyl group is also possible, using the same conditions at atmospheric pressure, with similar yields (91%).

Since stirring the reaction for 5 days with a balloon of hydrogen is not the most convenient method, the use of the H-Cube® Continuous-flow Hydrogenation Reactor by THALESNANO was explored. In this flow device the dissolved starting material is heated in a coil and then pumped through a catalyst cartridge while *in situ* generated hydrogen is continuously added (see Scheme **3.20**). Although several different catalyst cartridges are available a simple 10% Pd/C at 60 °C yielded full conversion after several iterations. Depending on the concentration of the quinoline up to 3-5 iterations were needed to consume all the starting material (by HPLC).

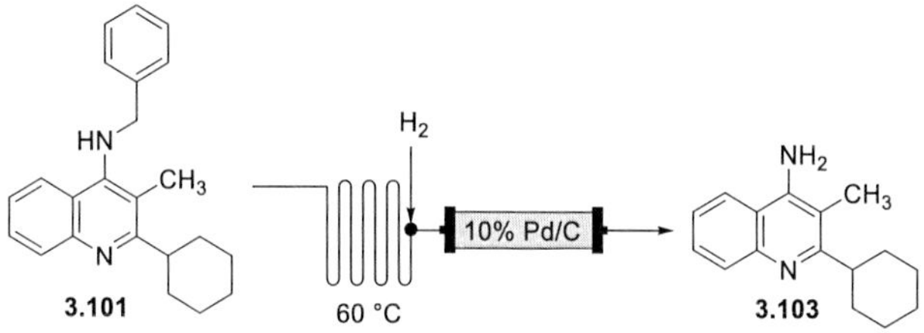

Scheme 3.20 Schematic illustration of the H-Cube® Continuous-flow Hydrogenation Reactor used to hydrogenate the benzyl group.

A quick solvent screen revealed that the reaction did not proceed notably faster in ethanol or acetic acid. An interesting next step could be to find conditions that can deprotect both the benzyl and tosyl groups in one step, as most conditions for the removal of the tosyl group are reductive in nature as well.

3.6 Conclusion and outlook

Since 4-aminoquinolines are medicinally relevant compounds, especially considering the ongoing global battle against malaria, novel synthetic routes that allow straightforward library construction are an ever present need. By using a mild one-step procedure that employs readily available electrophilically-activated amides that react with sulfonyl ynamides, a new and modular access to a diverse set of 4-aminoquinolines has been explored. By varying the amides and ynamides all quinoline positions can be easily functionalized. The choice of amide influences the groups on the C-2 and C-5 to C-8 positions, and even complex amides, derived from paracyclophanes, have been shown to be well tolerated. The C-3 position of the quinoline, that normally is rather difficult to modify, can be easily substituted by use of different ynamides, that can be made in a modular and straightforward way, as described in Chapter 2. Additionally, the tosyl and benzyl-protected 4-aminoquinolines can be deprotected using nucleophilic phosphide and hydrogenation in excellent yields.

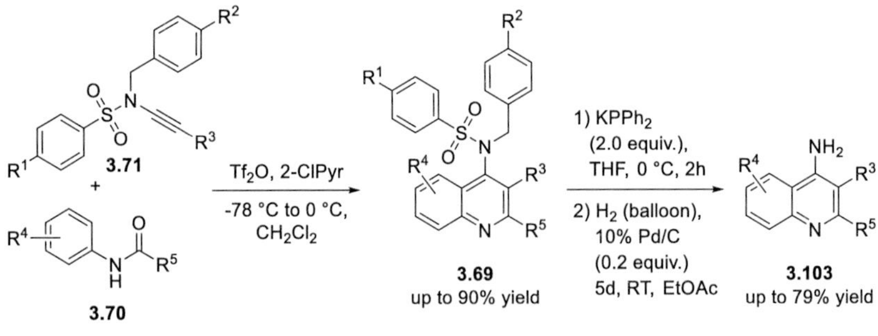

Scheme 3.21 4-Aminoquinoline synthesis and subsequent deprotection steps.

By combining the modularly accessible sulfonyl ynamides with electrophilically-activated amides a truly flexible system towards 4-aminoquinoline has been developed. The subsequent deprotection of the amine allows for the preparation of screening libraries in a straightforward manner.

Although the methodology has now been confirmed to provide easy and direct access to diverse and unprotected 4-aminoquinolines, this has yet to be done on a larger scale. With sufficient quantities of quinoline **3.103** in hand the free amine can be used to conveniently provide a library of screening compounds, for example via amide bond formation with acid chlorides.

Furthermore it would be very interesting to see if the activated amide methodology to form novel heterocyclic structures is compatible with other functional groups besides alkynes. The MOVASSAGHI group has extensively explored the use of nitriles, but to my best knowledge isonitriles and azides have not been attempted. If this would work in a similar manner, pyrroles **3.109** and imidazole **3.107** could be accessible.

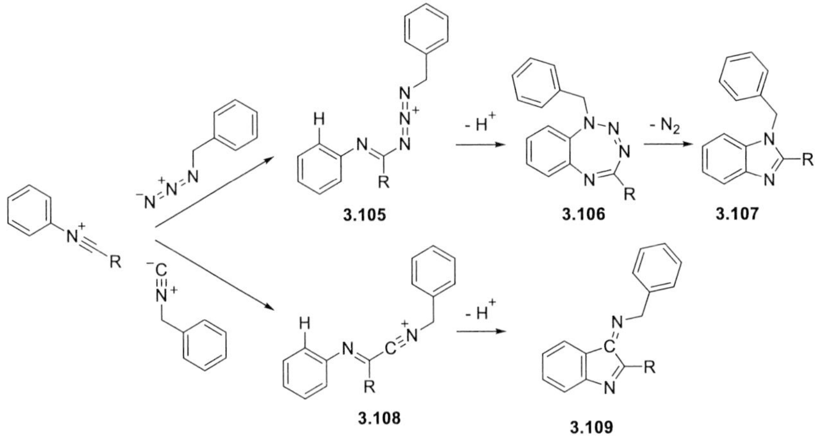

Scheme 3.22 Other reactions to be investigated.

3.7 Experimental

General remarks

Nuclear magnetic resonance spectroscopy (NMR): ^{1}H-NMR spectra were recorded using the following devices: *Bruker* Avance 300 (300 MHz), *Bruker* Avance 400 (400 MHz) or a *Bruker* Avance DRX 500 (500 MHz); ^{13}C-NMR spectra were recorded using the following devices: *Bruker* Avance 300 (75 MHz), *Bruker* Avance 400 (100 MHz) or a *Bruker* Avance DRX 500 (125 MHz). All measurements were carried out at room temperature. The following solvents from *Eurisotop* were used: chloroform-d_1, methanol-d_4, acetone-d_6, DMSO-d_6. Chemical shifts δ were expressed in parts per million (ppm) and referenced to chloroform (^{1}H: δ = 7.26 ppm, ^{13}C: δ = 77.00 ppm), methanol (^{1}H: δ = 3.31 ppm, ^{13}C: δ = 49.00 ppm), acetone (^{1}H: δ = 2.05 ppm, ^{13}C: δ = 30.83 ppm) or DMSO (^{1}H: δ = 2.50 ppm, ^{13}C: δ = 39.43 ppm).[78] The signal structure is described as follows: s = singlet, d = doublet, t = triplet, q = quartet, quin = quintet, bs = broad singlet, m = multiplet, dt = doublet of triplets, dd = doublet of doublets, td = triplet of doublets, ddd = doublet of doublet of doublets. The spectra were analyzed according to first order and all coupling constants are absolute values and expressed in Hertz (Hz). The multiplicities of the signals of ^{13}C-NMR spectra were determined using characteristic chemical shifts and DEPT (Distortionless Enhancement by Polarization Transfer) and are described as follows: "+" = primary or tertiary (positive DEPT-135 signal), "–" = secondary (negative DEPT-135 signal), "C_q" = quarternary carbon atoms (no DEPT signal).

Infrared spectroscopy (IR): IR spectra were recorded on a *Bruker* IFS 88 using ATR (Attenuated Total Reflection). The intensities of the peaks are given as follows: vs = very strong (0 - 10% transmission), s = strong (11 - 40% transmission), m = medium (41 - 70% transmission), w = weak (71 - 90% transmission), vw = very weak (91 - 100% transmission).

Mass spectrometry (EI-MS, ESI-MS, FAB-MS): Mass spectra were recorded on a *Finnigan* MAT 95 using EI-MS (Electron ionization mass spectrometry) at 70 eV or FAB-MS (Fast Atom Bombardment Mass Spectroscopy) with 3-NBA as matrix. In special cases an *Agilent Technologies* 6230 TOF-LC/MS was used to record ESI-TOF-MS spectra in the positive mode. The molecular fragments are stated as the ratio of mass over charge *m/z*. The intensities of the

signals are given in percent relative to the intensity of the base signal (100%). The molecular ion is abbreviated $[M]^+$ for EI-MS and ESI-MS, the protonated molecular ion is abbreviated $[M+H]^+$ for FAB-MS.

Thin layer chromatography (TLC): Analytical thin layer chromatography was carried out using silica coated aluminium plates (silica 60, F_{254}, layer thickness: 0.25 mm) with fluorescence indicator by *Merck*. Detection proceeded under UV light at $\lambda = 254$ nm. For development phosphomolybdic acid solution (5% phosphomolybdic acid in ethanol, dip solution); potassium permanganate solution (1.00 g potassium permanganate, 2.00 g acetic acid, 5.00 g sodium bicarbonate in 100 mL water, dip solution) was used followed by heating in a hot air stream. Preparative thin layer chromatography was carried out using either the silica coated aluminium plates (silica 60, F_{254}, layer thickness: 0.25 mm) or the silica coated PSC glass plates (silica 60, F_{254}, layer thickness: 2.0 mm) by *Merck*.

Analytical balance: Used device: *Sartorius* Basic, model LA310S, range from 0.1 mg to 310.0 g.

Microwave: Reactions heated using microwave irradiation were carried out in a single mode *CEM* Discover LabMate microwave operated with *CEM*'s Synergy software. This instrument works with a constantly focused power source (0 - 300W). Irradiation can be adjusted via power- or temperature control. The temperature was monitored with an infrared sensor.

Solvents and reagents: Solvents of technical quality were distilled prior to use. Solvents of p.a. (*per analysi*) quality were commercially purchased (*Acros, Fisher Scientific, Sigma Aldrich*) and used without further purification. Absolute solvents were either commercially purchased as absolute solvents stored over molecular sieves under argon atmosphere or freshly prepared by distillation of p.a. quality over a drying agent (dichloromethane from calcium hydride, THF from sodium using benzophenone as indicator, and diethyl ether from sodium) and stored under argon atmosphere. Reagents were commercially purchased (*ABCR, Acros, Alfa Aesar, Fluka, J&K, Sigma Aldrich, TCI, VWR*) and used without further purification if not stated otherwise.

Reactions: For reactions with air- or moisture-sensitive reagents, the glassware with a PTFE coated magnetic stir bar was heated under high vacuum with a heat gun, filled with argon and closed with a rubber septum. All solution phase reactions were performed using the typical

Schlenk procedures with argon as inert gas. Liquids were transferred with V2A-steel cannulas. Solids were used as powders if not indicated otherwise. Reactions at low temperatures were cooled in flat Dewar flasks from *Isotherm*. The following cooling mixtures were used: 0 °C (Ice/Water), 0 to −10 °C (Ice/Water/NaCl), −10 to −78 °C (acetone/dry ice *or* isopropanol/dry ice). In general, solvents were removed at preferably low temperatures (40 °C) under reduced pressure. If not stated otherwise, solutions of ammonium chloride, sodium chloride and sodium hydrogen carbonate are aqueous, saturated solutions. Reaction progress of liquid phase reaction was checked with thin-layer chromatography (TLC). Crude products were purified according to literature procedures by preparative TLC or flash column chromatography using Silica gel 60 (0.063×0.200 mm, 70-230 mesh ASTM) (*Merck*), Geduran® Silica gel 60 (0.040×0.063 mm, 230-400 mesh ASTM) (*Merck*) or Celite® (*Fluka*) and sea sand (calcined, purified with hydrochloric acid, *Riedel-de Haën*) as stationary phase. Eluents (mobile phase) were p.a. and volumetrically measured.

General remark: Procedures and spectroscopic data on compounds mentioned in this chapter that were prepared and analyzed by Dr. Sabilla ZHONG can be found in the electronic supplementary information of "T. Wezeman, S. Zhong, M. Nieger and S. Bräse, *Angew. Chem. Int. Ed.* **2016**, *55*, 3823 – 3827".

General procedure GP3.1: Synthesis of quinolines from ynamides and amides

A solution of the amide (1.00 equiv.) and 2-chloropyridine (1.20 equiv.) in CH$_2$Cl$_2$ (3.0 mL) was cooled to −78 °C. Tf$_2$O (1.10 equiv.) was added, it was stirred for 5 min at −78 °C and then warmed to 0 °C. A solution of the ynamide (1.10-1.20 equiv.) in CH$_2$Cl$_2$ (2.0 mL) was added dropwise. The mixture was stirred at 0 °C until consumption of the starting material was observed. It was then quenched with 20% aqueous NaOH solution (3.0 mL). The organic layer was dried over Na$_2$SO$_4$ and the solvent was removed under reduced pressure. The crude product was purified by column chromatography.

N-phenylcyclohexanecarboxamide (3.70a)

N,N′-Diisopropylcarbodiimide (2.17 g, 17.2 mmol, 1.10 equiv.) was added slowly to a solution of cyclohexanecarboxylic acid (2.00 g, 15.6 mmol), aniline (1.45 g, 15.6 mmol, 1.00 equiv.), and 1-hydroxybenzotriazole (478 mg, 3.12 mmol, 0.20 equiv.) in THF (40 mL). The reaction was stirred for 4d at RT under argon atmosphere. The mixture was then quenched with brine, extracted with ethyl acetate, and dried with anhydrous sodium sulfate. The solvent was removed under reduced pressure and the resulting crude was purified by flash chromatography (*c*Hex/EtOAc = 9:1) to yield **3.70a** as a white solid (2.57 g, 13.9 mmol, 81%). **^{1}H-NMR** (300 MHz, CDCl$_3$): δ = 7.65 (s, 1H, N*H*), 7.53 (d, *J* = 7.8 Hz, 2H, C*H*$_{Ar}$), 7.26 (t, *J* = 7.7 Hz, 2H, C*H*$_{Ar}$), 7.05 (t, *J* = 7.3 Hz, 1H, C*H*$_{Ar}$), 2.22 (tt, *J* = 0.1 Hz, 1H, C*H*), 2.00 - 1.41 (m, 7H, C*H*$_2$), 1.37 - 1.10 (m, 3H, C*H*$_2$) ppm. – **^{13}C-NMR** (75 MHz, CDCl$_3$): δ = 174.6 (*C*$_q$, *C*=O), 138.1 (*C*$_q$), 128.8 (3 × *C*H$_{Ar}$), 123.9 (+, *C*H$_{Ar}$), 119.9 (+, *C*H$_{Ar}$), 46.4 (+, *C*H), 29.6 (2 × *C*H$_2$), 25.6 (3 × *C*H$_2$) ppm. – **IR** (ATR): ṽ = 3238 (w), 2930 (m), 2849 (w), 1656 (w), 1596 (s), 1545 (s), 1489 (s), 1440 (s), 1250 (m) cm^{-1}. – **MS**: (70 eV,

EI), m/z (%) 93.0 (100), 203.1 (37) [M]$^+$. – **HRMS** (EI, $C_{13}H_{17}NO$): calc. 203.1305; found 203.1304.

N-(4-Chlorophenyl)cyclohexanecarboxamide (3.70b)

A flask was charged with cyclohexane carboxylic acid (1.94 mL, 2.00 g, 15.6 mmol, 1.00 equiv.) and 1-hydroxybenzotriazole (422 mg, 3.12 mmol, 0.20 equiv.). Next, dry THF (40 mL), N,N'-di-isopropylcarbodiimide (2.66 mL, 2.17 g, 17.2 mmol, 1.10 equiv.) and 4-chloroaniline (1.98 g, 15.6 mmol, 1.00 equiv.) were added. The mixture was stirred at RT for three days before being quenched with brine, extracted with EtOAc and dried over anhydrous sodium sulfate. The solvent was removed under reduced pressure and the crude product was recrystallized from ethanol to yield **3.70b** as colorless to white crystals (1.72 g, 7.18 mmol, 46%). The analytical data was in accordance with the literature.[79]

N-Benzyl-N-(2-cyclohexyl-3-phenylquinolin-4-yl)-4-methylbenzenesulfonamide (3.77)

This compound was synthesized following **GP3.1** from amide **3.70a** (36.4 mg, 0.179 mmol, 1.00 equiv.) and the corresponding ynamide (71.3 mg, 0.197 mmol, 1.10 equiv.); reaction time: 1h. The crude product was purified by column chromatography (cHex/EtOAc = 10:1) and the quinoline was obtained as a colorless solid (46.0 mg, 84.1 μmol, 47%). R_f (cHex/EtOAc = 5:1) = 0.47. – **m.p.** = 214 - 217 °C. – **^{1}H-NMR** (500 MHz, CDCl$_3$): δ = 8.14 (d, ^3J = 8.5 Hz, 1H, CH_{Ar}), 7.99 (d, ^3J = 8.5 Hz, 1H, CH_{Ar}), 7.68 (td, ^3J = 7.6, ^4J = 1.1 Hz, 1H, CH_{Ar}), 7.41 - 7.46 (m, 2H, 2 × CH_{Ar}), 7.34 - 7.40 (m, 2H, 2 × CH_{Ar}), 7.17 - 7.23 (m, 3H, 3 × CH_{Ar}), 7.13 - 7.16 (m, 2H, 2 × CH_{Ar}), 7.08 - 7.12 (m, 1H, CH_{Ar}), 7.04 (t, ^3J = 7.7 Hz, 2H, 2 × CH_{Ar}), 6.44 (d, ^3J = 7.8 Hz, 2H, 2 × CH_{Ar}), 5.89 (d, ^3J = 7.2 Hz, 1H, CH_{Ar}), 4.62 (d, ^2J = 14.1 Hz, 1H, NCH_2), 4.43 (d, ^2J = 14.1 Hz, 1H, NCH_2), 2.44 (s, 3H, CH_3), 2.31 - 2.38 (m, 1H, CH), 1.65 - 1.80 (m, 5H, CH_2), 1.55 - 1.62 (m, 2H, CH_2), 1.22 - 1.33 (m, 1H, CH_2), 0.93 - 1.04 (m, 2H, CH_2) ppm. – **^{13}C-NMR** (125 MHz, CDCl$_3$): δ = 166.7 (C_q), 149.3 (C_q), 143.8 (C_q), 140.9 (C_q), 137.0 (C_q),

136.9 (C_q), 136.4 (C_q), 134.4 (C_q), 131.1 (+, CH_{Ar}), 130.2 (+, $2 \times CH_{Ar}$), 129.8 (+, CH_{Ar}), 129.4 (+, $2 \times CH_{Ar}$), 129.10 (+, CH_{Ar}), 129.08 (+, CH_{Ar}), 128.6 (+, $2 \times CH_{Ar}$), 128.3 (+, CH_{Ar}), 128.2 (+, $2 \times CH_{Ar}$), 128.0 (+, CH_{Ar}), 127.5 (+, CH_{Ar}), 127.4 (+, CH_{Ar}), 126.4 (+, CH_{Ar}), 125.2 (C_q), 125.0 (+, CH_{Ar}), 54.9 (−, NCH_2), 43.4 (+, CH), 32.4 (−, CH_2), 32.3 (−, CH_2), 26.5 (−, CH_2), 26.4 (−, CH_2), 25.9 (−, CH_2), 21.6 (+, CH_3) ppm. − **IR** (ATR): $\tilde{\nu}$ = 2912 (vw), 2848 (vw), 1578 (vw), 1486 (vw), 1443 (vw), 1358 (vw), 1319 (w), 1147 (w), 1086 (w), 1021 (w) cm^{-1}. − **MS** (EI), m/z (%): 546 (64) [M]$^+$, 491 (69), 391 (100) [M–$C_7H_7O_2S$]$^+$, 337 (48), 335 (43), 245 (23). − **HRMS** (EI, $C_{35}H_{34}N_2O_2{}^{32}S$): calc. 546.2336; found 546.2337.

N-benzyl-N-(6-chloro-2-cyclohexyl-3-methylquinolin-4-yl)-4-methylbenzenesulfonamide (3.80)

This compound was synthesized following **GP3.1** from amide **3.70b** (102.8, 0.343 mmol, 1.20 equiv.) and the corresponding ynamide (68.0 mg, 0.286 mmol, 1.00 equiv.); reaction time: 2h. The crude product was purified by column chromatography (Hept/EtOAc = 10:1) and quinoline **3.80** was obtained as a colorless crystal (110.0 mg, 0.212 mmol, 74%). **^{1}H-NMR** (400 MHz, CDCl$_3$): δ = 7.92 (d, ^3J = 8.8 Hz, 1H, CH_{Ar}), 7.7 (d, ^3J = 8.1 Hz, 2H, CH_{Ar}), 7.42 (dd, ^3J = 9.1, 2.3 Hz, 1H, CH_{Ar}), 7.31 (d, ^3J = 8.3 Hz, 2H, CH_{Ar}), 7.21 - 7.26 (m, 1H, CH_{Ar}), 7.10 - 7.19 (m, 2H, CH_{Ar}), 6.99 (d, ^3J = 7.6 Hz, 2H, CH_{Ar}), 6.73 (d, ^3J = 2.3 Hz, 1H, CH_{Ar}), 4.74 (d, ^2J = 297.3, 13.9 Hz, 2H, CH_2), 2.77 - 2.94 (m, 1H, CH), 2.48 (s, 3H, CH_3), 1.94 (s, 3H, CH_3), 1.79 - 1.90 (m, 3H, $3 \times CH_2$), 1.76 (br. s., 2H, $2 \times CH_2$), 1.51 - 1.66 (m, 2H, $2 \times CH_2$), 1.29 - 1.44 (m, 3H, $3 \times CH_2$) ppm. − **^{13}C-NMR** (101 MHz, CDCl$_3$): δ = 167.4 (C_q), 146.1 (C_q), 144.2 (C_q), 139.6 (C_q), 137.6 (C_q), 134.7 (C_q), 133.9 (C_q), 131.5 (+, CH_{Ar}), 131.2 (+, $2 \times CH_{Ar}$), 129.9 (+, $2 \times CH_{Ar}$), 129.9 (+, CH_{Ar}), 128.8 (+, $2 \times CH_{Ar}$), 128.3 (+, CH_{Ar}), 128.2 (+, $2 \times CH_{Ar}$), 127.4 (+, CH_{Ar}), 125.2 (+, CH_{Ar}), 122.3 (+, CH_{Ar}), 55.1 (−, CH_2), 43.1 (+, CH), 32.0 (−, CH_2), 31.3 (−, CH_2), 26.7 (−, CH_2), 26.5 (−, CH_2), 26.0 (−, CH_2), 21.5 (+, CH_3), 14.9 (+, CH_3) ppm. − **IR** (ATR): $\tilde{\nu}$ = 2922 (m), 2850 (w), 1584 (w), 1478 (m), 1448 (w), 1350 (m), 1160 (s), 1090 (m), 1054 (m) cm^{-1}. − **MS** (EI) m/z (%): 518.4 (9) [M]$^+$, 450.3 (27), 220.1 (43), 91.1 (100). − **HRMS** (EI, $C_{30}H_{31}N_2O_2{}^{32}S{}^{35}Cl$): calc. 518.1789; found 518.1790.

N-benzyl-*N*-(2-cyclohexyl-3-methylquinolin-4-yl)-4-methylbenzenesulfonamide (3.81)

This compound was synthesized following **GP3.1** from amide **3.70a** (1320 mg, 4.41 mmol, 1.20 equiv.) and the corresponding ynamide (747.0 mg, 3.675 mmol, 1.00 equiv.); reaction time: 12h. The crude product was purified by column chromatography (Hept/EtOAc = 10:1) and quinoline **3.81** was obtained as colorless crystals (1005 mg, 2.07 mmol, 56%). **¹H-NMR** (400 MHz, CDCl₃): δ = 8.02 (d, ^3J = 8.3 Hz, 1H, CH_{Ar}), 7.70 (d, ^3J = 8.1 Hz, 2H, CH_{Ar}), 7.52 (ddd, ^3J = 8.2, 6.7, 1.5 Hz, 1H, CH_{Ar}), 7.09 - 7.34 (m, 4H, CH_{Ar}), 6.97 (d, ^3J = 7.6 Hz, 2H, CH_{Ar}), 5.08 (d, ^3J = 241.0, 13.9 Hz, 2H, CH_2), 2.81 - 2.92 (m, 1H, CH), 2.47 (s, 3H, CH_3), 1.86 (s, 4H, CH_3), 1.64 (br. s., 2H, CH_2), 1.26 - 1.46 (m, 4H, CH_2) ppm. – **¹³C-NMR** (101 MHz, CDCl₃): δ = 166.8 (C_q), 147.9 (C_q), 143.7 (C_q), 140.5 (C_q), 137.9 (C_q), 134.9 (C_q), 132.3 (C_q), 129.9 (+, 2 × CH_{Ar}), 129.7 (+, 2 × CH_{Ar}), 129.7 (+, 2 × CH_{Ar}), 128.1 (+, 2 × CH_{Ar}), 128.0 (+, 2 × CH_{Ar}), 127.7 (+, CH_{Ar}), 125.5 (+, CH_{Ar}), 124.7 (C_q), 123.3 (+, CH_{Ar}), 55.0 (−, CH_2), 43.1 (+, CH), 32.0 (−, CH_2), 31.5 (−, CH_2), 26.8 (−, CH_2), 26.6 (−, CH_2), 26.1 (−, CH_2), 21.5 (+, CH_3), 14.9 (+, CH_3) ppm. – **IR** (ATR): ṽ = 2923 (w), 2854 (w), 1588 (w), 1488 (w), 1446 (w), 1401 (w), 1372 (m), 1205 (m), 1089 (m), 1027 (m) cm⁻¹. – **MS** (FAB) *m/z* (%): 485.2 (100) [M+H]⁺, 328.2 (31). – **HRMS** (FAB, $C_{30}H_{33}N_2O_2{}^{32}S$): calc. 485.2257; found 485.2259.

N-benzyl-*N*-(2-cyclohexylquinolin-4-yl)-4-methylbenzenesulfonamide (3.87)

This compound was synthesized following **GP3.1** from the corresponding ynamide (109.5 mg, 0.384 mmol, 1.20 equiv.) and amide **3.70a** (65.0 mg, 0.320 mmol, 1.00 equiv.) in the presence of activated 4Å molecular sieves; reaction time: 2h. The crude product was purified by column chromatography (Hept/EtOAc = 10:1) and quinoline **3.87** was obtained as a white solid (76.0 mg, 0.162 mmol, 51%). **¹H-NMR** (400 MHz, CDCl₃): δ = 7.95 (dd, ^3J = 8.3, 3.3 Hz, 2H, CH_{Ar}), 7.56 - 7.65 (m, 3H, CH_{Ar}), 7.38 (t, ^3J = 7.6 Hz, 1H, CH_{Ar}), 7.30 (d, ^3J = 8.1 Hz, 2H, CH_{Ar}), 7.11 (s, 5H, CH_{Ar}), 6.56 (s, 1H, CH_{Ar}), 4.81 (br. d, ^2J = 158.4 Hz, 2H, CH_2), 2.75 (t, ^3J = 11.7 Hz, 1H, CH), 2.46 (s, 3H, CH_3), 1.83 (d, ^3J = 10.6 Hz, 4H, CH_2), 1.75 (d, ^3J = 12.9

Hz, 1H, CH_2), 1.17 - 1.47 (m, 5H, CH_2) ppm. – ^{13}**C-NMR** (101 MHz, CDCl$_3$): δ = 166.3 (C_q), 149.3 (C_q), 144.7 (C_q), 144.0 (C_q), 135.0 (C_q), 135.0 (C_q), 129.7 (+, 2 × CH_{Ar}), 129.6 (+, CH_{Ar}), 129.0 (+, CH_{Ar}), 128.7 (+, CH_{Ar}), 128.3 (+, CH_{Ar}), 128.0 (+, 2 × CH_{Ar}), 127.9 (+, CH_{Ar}), 127.9 (+, CH_{Ar}), 127.7 (+, CH_{Ar}), 126.3 (+, CH_{Ar}), 126.0 (+, CH_{Ar}), 123.9 (+, CH_{Ar}), 119.4 (+, CH_{Ar}), 55.9 (−, CH_2), 47.1 (+, CH), 32.5 (−, 2 × CH_2), 26.3 (−, 2 × CH_2), 25.9 (−, CH_2), 21.5 (+, CH_3) ppm. – **IR** (ATR): ṽ = 2920 (w), 2847 (w), 1701 (w), 1597 (m), 1554 (w), 1495 (w), 1445 (m), 1411 (s), 1157 (s), 1120 (w), 1092 (m), 977 (m) cm^{-1}. – **MS** (FAB) m/z (%): 471.2 (100) [M+H]$^+$, 315.2 (33). – **HRMS** (FAB, C$_{29}$H$_{31}$N$_2$O$_2$^{32}S): calc. 471.2101; found 471.2100.

Methyl 4-(*N*-benzyl-*N*-(2-cyclohexylquinolin-4-yl)sulfamoyl)benzoate (3.90)

This compound was synthesized following **GP3.1** from amide **3.70a** (26.6 mg, 0.131 mmol, 1.00 equiv.) and the corresponding ynamide (51.7 mg, 0.157 mmol, 1.20 equiv.); reaction time: 2h. The crude product was purified by column chromatography (*c*Hex/EtOAc = 5:1) and quinoline **3.90** was obtained as a white solid (37.0 mg, 0.072 mmol, 55%). R_f (*c*Hex/

EtOAc = 3:1) = 0.39 – ^{1}**H-NMR** (300 MHz, CDCl$_3$): δ = 8.16 (d, ^3J = 8.6 Hz, 2H, CH_{Ar}), 7.97 (d, ^3J = 8.4 Hz, 1H, CH_{Ar}), 7.79 (d, ^3J = 8.4 Hz, 3H, CH_{Ar}), 7.62 (ddd, $^{3,4}J$ = 8.4, 6.9, 1.4 Hz, 1H, CH_{Ar}), 7.37 (ddd, $^{3,4}J$ = 8.2, 6.9, 1.1 Hz, 1H, CH_{Ar}), 7.17 - 7.06 (m, 5H, CH_{Ar}), 6.56 (s, 1H, CH_{Ar}), 4.84 (d, ^3J = 41.4 Hz, 2H, NCH_2), 3.98 (s, 3H, OCH_3), 2.81 - 2.67 (m, 1H, CH_2), 1.90 - 1.66 (m, 5H, CH_2), 1.46 - 1.16 (m, 5H, CH_2) ppm. – ^{13}**C-NMR** (75 MHz, CDCl$_3$): δ = 166.5 (C_q), 165.5 (C_q), 149.4 (C_q), 143.9 (C_q), 142.3 (C_q), 134.7 (C_q), 134.1 (+, 2 × CH_{Ar}), 130.7 (+, CH_{Ar}), 130.1 (+, CH_{Ar}), 129.8 (+, CH_{Ar}), 129.0 (+, CH_{Ar}), 128.3 (+, CH_{Ar}), 128.1 (+, CH_{Ar}), 127.8 (2 × CH_{Ar}), 126.2 (+, CH_{Ar}), 125.8 (+, CH_{Ar}), 123.3 (+, CH_{Ar}), 120.1 (+, CH_{Ar}), 119.9 (C_q), 56.1 (+, NCH_2), 53.2 (+, CH), 52.6 (+, CH_2), 52.2 (+, CH_2), 47.0 (−, OCH_3), 32.5 (+, CH_2), 26.3 (+, CH_2), 25.9 (+, CH_2) ppm. – **IR** (ATR): ṽ = 3279 (w), 2922 (s), 2850 (s), 1726 (w), 1701 (s), 1635 (s), 1598 (vw), 1521 (w), 1438 (w), 1399 (w), 1344 (w), 1277 (w), 1227 (s), 1159 (w), 1105 (w), 1087 (w), 1045 (vw) cm^{-1}. – **MS** (FAB), m/z (%): 515 (47) [M+H]$^+$, 279 (59), 255 (35), 91 (100). – **HRMS** (FAB, C$_{30}$H$_{31}$N$_2$O$_4$^{32}S): calc. 515.2005; found 515.1969.

Methyl 4-(*N*-(4-chlorobenzyl)-*N*-(2-cyclohexylquinolin-4-yl)sulfamoyl)benzoate (3.91)

This compound was synthesized following **GP3.1** from amide **3.70a** (23.4 mg, 0.115 mmol, 1.00 equiv.) and the corresponding ynamide (50.2 mg, 0.138 mmol, 1.20 equiv.); reaction time: 2h. The crude product was purified by column chromatography (*c*Hex/EtOAc = 5:1) and quinoline **3.91** was obtained as a white solid (16.0 mg, 0.029 mmol, 25%). R_f (*c*Hex/EtOAc = 3:1) = 0.44 – **^{1}H-NMR** (300 MHz, CDCl$_3$): δ = 8.16 (d, ^3J = 8.6 Hz, 2H, CH_{Ar}), 7.99 (d, ^3J = 8.5 Hz, 1H, CH_{Ar}), 7.78 (d, ^3J = 8.6 Hz, 2H, CH_{Ar}), 7.74 (d, ^3J = 9.0 Hz, 1H, CH_{Ar}), 7.64 (ddd, J = 8.4, 7.0, 1.3 Hz, 1H, CH_{Ar}), 7.38 (ddd, J = 8.2, 7.0, 1.1 Hz, 1H, CH_{Ar}), 7.08 (dd, ^3J = 25.6, 8.5 Hz, 4H, CH_{Ar}), 6.56 (s, 1H, CH_{Ar}), 4.80 (d, ^3J = 40.5 Hz, 2H, NCH_2), 3.99 (s, 3H, CH_3), 2.82 - 2.68 (m, 1H, CH), 1.93 - 1.55 (m, 5H, CH_2), 1.45 - 1.16 (m, 5H, CH_2) ppm. – **^{13}C-NMR** (75 MHz, CDCl$_3$): δ = 166.6 (C$_q$), 165.5 (C$_q$), 149.5 (C$_q$), 143.7 (2 × C$_q$), 142.1 (2 × C$_q$), 134.3 (2 × CH_{Ar}), 133.3 (2 × CH_{Ar}), 130.4 (+, CH_{Ar}), 129.9 (+, CH_{Ar}), 128.7 (+, CH_{Ar}), 128.6 (+, CH_{Ar}), 127.8 (+, CH_{Ar}), 127.6 (+, CH_{Ar}), 125.6 (+, CH_{Ar}), 120.0 (+, CH_{Ar}), 119.9 (C$_q$), 114 (+, CH_{Ar}), 64.5 (+, CH), 55.3 (+, NCH_2), 47.0 (−, OCH_3), 32.5 (+, 2 × CH_2), 26.2 (+, 2 × CH_2), 25.8 (+, CH_2) ppm. – **IR** (ATR): ṽ = 2921 (w), 2848 (vw), 1723 (s), 1594 (vw), 1489 (w), 1433 (w), 1401 (vw), 1335 (w), 1272 (s), 1242 (w), 1159 (w), 1013 (w) cm^{-1}. – **MS** (EI), *m/z* (%): 549 (15) [M]$^+$, 81 (100). – **HRMS** (EI, C$_{30}$H$_{30}$N$_2$O$_4$^{35}Cl^{32}S): calc. 549.1615; found 549.1599.

Methyl 4-(*N*-(4-(*t*-butyl)benzyl)-*N*-(2-cyclohexylquinolin-4-yl)sulfamoyl)benzoate (3.92)

This compound was synthesized following **GP3.1** from amide **3.70a** (22.9 mg, 0.113 mmol, 1.00 equiv.) and the corresponding ynamide (52.1 mg, 0.135 mmol, 1.20 equiv.); reaction time: 2h. The crude product was purified by column chromatography (*c*Hex/EtOAc = 5:1) and quinoline **3.92** was obtained as a white solid (58.1 mg, 0.102 mmol, 90%). R_f (*c*Hex/EtOAc = 3:1) = 0.48 – **^{1}H-NMR** (300 MHz, CDCl$_3$): δ = 8.14 (d, ^3J = 8.5 Hz, 2H, CH_{Ar}), 7.99 (d, ^3J = 8.0 Hz, 1H, CH_{Ar}), 7.80 (d, ^3J = 8.4 Hz, 2H, CH_{Ar}),

7.75 - 7.70 (m, 1H, CH_{Ar}), 7.62 (ddd, $^{3,4}J$ = 8.4, 7.0, 1.3 Hz, 1H, CH_{Ar}), 7.39 - 7.32 (m, 1H, CH_{Ar}), 7.15 (d, ^3J = 8.3 Hz, 2H, CH_{Ar}), 6.95 (d, ^3J = 8.3 Hz, 2H, CH_{Ar}), 6.54 (s, 1H, CH_{Ar}), 4.82 (d, ^3J = 17.4 Hz, 2H, NCH_2), 3.98 (s, 3H, OCH_3), 2.81 - 2.67 (m, 1H, CH_2), 1.89 - 1.68 (m, 5H, CH_2), 1.47 - 1.25 (m, 5H, CH_2), 1.20 (s, 9H, 3 × CH_3) ppm. – **^{13}C-NMR** (75 MHz, CDCl$_3$): δ = 166.6 (C_q), 165.5 (C_q), 151.3 (C_q), 149.4 (C_q), 143.6 (C_q), 142.8 (C_q), 134.0 (C_q), 131.8 (+, 2 × CH_{Ar}), 130.7 (+, CH_{Ar}), 129.7 (+, CH_{Ar}), 128.9 (+, 2 × CH_{Ar}), 128.0 (+, CH_{Ar}), 127.6 (+, CH_{Ar}), 126.2 (+, CH_{Ar}), 125.7 (+, CH_{Ar}), 125.4 (+, CH_{Ar}), 123.4 (+, CH_{Ar}), 120.9 (+, CH_{Ar}), 120.8 (C_q), 55.6 (+, NCH_2), 53.2 (+, CH), 52.7 (+, CH_2), 52.2 (+, CH_2), 47.1 (OCH_3), 34.4 (+, CH_3), 32.6 (+, CH_2), 32.4 (+, CH), 31.2 (+, CH_3), 31.1 (+, CH_3), 26.2 (+, CH_2), 25.8 (+, CH_2) ppm. – **IR** (ATR): ṽ = 3310 (w), 2921 (s), 2850 (s), 1730 (s), 1660 (w), 1599 (w), 1347 (w), 1272 (s), 1160 (s), 1102 (w), 1086 (w), 1054 (w), 1015 (w) cm^{-1}. – **MS** (EI), m/z (%): 571 (62) [M]$^+$, 371 (20), 147 (100). – **HRMS** (EI, C$_{34}$H$_{39}$N$_2$O$_4$^{32}S): calc. 571.2631; found 571.2606.

N-Benzyl-*N*-(6-chloro-2-cyclohexylquinolin-4-yl)-4-methylbenzenesulfonamide (3.93)

This compound was synthesized following **GP3.1** from the corresponding ynamide (70.0 mg, 0.294 mmol, 1.00 equiv.) and amide **3.70b** (100.8 mg, 0.353 mmol, 1.20 equiv.) in the presence of activated 4Å molecular sieves; reaction time: 1h. The crude product was purified by column chromatography (*c*Hex/ EtOAc = 10:1) and quinoline **3.93** was obtained as a colorless solid (79.0 mg, 156 μmol, 53%). R_f (*c*Hex/EtOAc = 10:1) = 0.13 – **^{1}H-NMR** (400 MHz, CDCl$_3$): δ = 7.86 (d, ^3J = 9.1 Hz, 1H, CH_{Ar}), 7.70 (d, ^4J = 2.3 Hz, 1H, CH_{Ar}), 7.60 (d, ^3J = 8.3 Hz, 2H, 2 × CH_{Ar}), 7.51 (dd, ^3J = 8.8, ^4J = 2.3 Hz, 1H, CH_{Ar}), 7.32 (d, ^3J = 8.3 Hz, 2H, 2 × CH_{Ar}), 7.08 - 7.16 (m, 5H, 5 × CH_{Ar}), 6.60 (s, 1H, CH_{Ar}), 4.93 (bs, 1H, NCH_2), 4.65 (bs, 1H, NCH_2), 2.72 (tt, ^3J = 11.7, ^3J = 3.1 Hz, 1H, CH), 2.47 (s, 3H, CH_3), 1.73 - 1.86 (m, 5H, CH_2), 1.19 - 1.45 (m, 5H, CH_2) ppm. – **^{13}C-NMR** (100 MHz, CDCl$_3$): δ = 166.8 (C_q), 147.7 (C_q), 144.3 (C_q), 143.7 (C_q), 135.0 (C_q), 134.8 (C_q), 132.1 (C_q), 130.5 (+, CH_{Ar}), 130.5 (+, CH_{Ar}), 129.8 (+, 2 × CH_{Ar}), 129.1 (+, 2 × CH_{Ar}), 128.4 (+, 2 × CH_{Ar}), 128.1 (+, CH_{Ar}), 128.0 (+, 2 × CH_{Ar}), 126.9 (C_q), 122.7 (+, CH_{Ar}), 120.9 (+, CH_{Ar}), 55.9 (–, NCH_2), 47.0 (+, CH), 32.4 (–, CH_2), 26.3 (–, CH_2), 25.9 (–, CH_2), 21.6 (+, CH_3) ppm. – **IR** (ATR): ṽ = 3277 (vw), 2924 (w), 2851 (w), 1650 (w), 1594 (w), 1519 (w), 1490 (w), 1446 (w), 1398 (w), 1351 (m), 1162 (m), 1111 (w), 1089 (w), 1051 (w) cm^{-1}. – **MS** (EI), *m/z* (%): 506/504 (3/8) [M]$^+$, 451/449 (6/14), 438/436 (6/14), 351/349 (31/99) [M–C$_7$H$_7$O$_2$S]$^+$, 239/237 (19/58), 129/127 (28/92), 91 (19) [C$_7$H$_7$]$^+$, 83 (57), 55 (100). – **HRMS** (EI, C$_{29}$H$_{29}$N$_2$O$_2$^{35}Cl^{32}S): calc. 504.1633; found 504.1632.

N-(6-chloro-2-cyclohexylquinolin-4-yl)-*N*-(4-methoxybenzyl)-4-methylbenzenesulfon-amide (3.94)

This compound was synthesized following **GP3.1** from amide **3.70b** (102.0 mg, 0.323 mmol, 1.00 equiv.) and the corresponding ynamide (99.9 mg, 0.485 mmol, 1.30 equiv.) with 1.50 equivalents of 2-chloropyridine and 1.50 equivalents of triflic anhydride; reaction time: 1h. The crude product was purified by column chromatography (Hept/EtOAc = 10:1) and quinoline **3.94** was obtained as a colorless crystal (48.0 mg, 0.090 mmol, 28%). **^{1}H-NMR** (400 MHz, CDCl$_3$): δ = 7.87 (d, ^3J = 8.8 Hz, 1H, C*H*$_{Ar}$), 7.72 (d, ^3J = 2.2 Hz, 1H, C*H*$_{Ar}$), 7.59 (d, ^3J = 8.0 Hz, 2H, C*H*$_{Ar}$), 7.51 (dd, ^3J = 8.8, 2.2 Hz, 1H, C*H*$_{Ar}$), 7.32 (d, ^3J = 8.5 Hz, 2H, C*H*$_{Ar}$), 6.99 (d, ^3J = 8.5 Hz, 2H, C*H*$_{Ar}$), 6.61 - 6.68 (m, 2H, C*H*$_{Ar}$), 6.57 (s, 1H, C*H*$_{Ar}$), 4.49 - 4.95 (m, 2H, C*H*$_2$), 3.68 (s, 3H, OC*H*$_3$), 2.67 - 2.78 (m, 1H, C*H*), 2.47 (s, 3H, C*H*$_2$), 1.67 - 1.89 (m, 5H, C*H*$_2$), 1.16 - 1.46 (m, 5H, C*H*$_2$) ppm. – **^{13}C-NMR** (101 MHz, CDCl$_3$): δ = 166.7 (*C*$_q$), 159.4 (*C*$_q$), 147.7 (*C*$_q$), 144.2 (*C*$_q$), 143.8 (*C*$_q$), 135.1 (*C*$_q$), 132.1 (*C*$_q$), 130.5 (+, 3 × C*H*$_{Ar}$), 130.4 (+, 2 × C*H*$_{Ar}$), 129.7 (+, 2 × C*H*$_{Ar}$), 127.9 (+, C*H*$_{Ar}$), 127.1 (+, C*H*$_{Ar}$), 126.8 (+, C*H*$_{Ar}$), 122.8 (+, C*H*$_{Ar}$), 120.8 (+, C*H*$_{Ar}$), 113.8 (+, 2 × C*H*$_{Ar}$), 55.4 (+, C*H*$_3$), 55.1 (–, C*H*$_2$), 47.0 (+, C*H*), 32.4 (–, C*H*$_2$), 26.3 (–, C*H*$_2$), 25.9 (–, C*H*$_2$), 21.6 (+, C*H*$_3$) ppm. – **IR** (ATR): ṽ = 2915 (w), 2846 (w), 1594 (w), 1553 (w), 1511 (w), 1445 (w), 1345 (m), 1242 (m), 1160 (m), 1087 (w) cm^{-1}. – **MS** (EI) *m/z* (%): 534.4 (4) [M]$^+$, 379.3 (21), 121.1 (100). – **HRMS** (EI, C$_{30}$H$_{31}$N$_2$O$_3$^{32}S^{35}Cl): calc. 534.1738; found 534.1740.

Methyl 4-(*N*-benzyl-*N*-(6-chloro-2-cyclohexyl-3-phenylquinolin-4-yl)sulfamoyl)benzoate (3.96)

This compound was synthesized following **GP3.1** from amide **3.70b** (29.2 mg, 0.123 mmol, 1.00 equiv.) and the corresponding ynamide (54.8 mg, 0.135 mmol, 1.10 equiv.), reaction time: 1h, followed by an additional 10 minutes heated at 100 °C (microwave irradiation). The crude product was purified by column chromatography (*c*Hex/EtOAc = 9:1) and the product was obtained as a lightly brownish solid (32.4 mg, 0.052 mmol, 42%). **^{1}H-NMR** (400 MHz, CDCl$_3$): δ = 8.05 - 8.00 (m, 3H, CH_{Ar}), 7.56 (dd, ^3J = 8.9, 2.3 Hz, 1H, CH_{Ar}), 7.52 (d, ^4J = 2.0 Hz, 1H, CH_{Ar}), 7.49 - 7.36 (m, 5H, CH_{Ar}), 7.24 - 7.16 (m, 2H, CH_{Ar}), 7.06 (t, ^3J = 7.7 Hz, 2H, CH_{Ar}), 6.57 (d, ^3J = 7.1 Hz, 2H, CH_{Ar}), 6.25 (d, ^3J = 7.6 Hz, 1H, CH_{Ar}), 4.49 (dd, J = 95.2, 14.2 Hz, 2H, NCH_2), 4.00 (s, 3H, OCH_3), 2.43 - 2.29 (m, 1H, CH), 1.81 - 1.51 (m, 6H, CH_2), 1.33 - 1.19 (m, 2H, CH_2), 1.08 - 0.92 (m, 2H, CH_2) ppm. – **^{13}C-NMR** (100 MHz, CDCl$_3$): δ = 167.2 (*C*O$_2$CH$_3$), 165.6 (C_{Ar}), 147.4 (C_{Ar}), 143.9 (C_{Ar}), 140.1 (C_{Ar}), 137.5 (C_{Ar}), 134.0 (C_{Ar}), 133.8 (C_{Ar}), 131.3 (C_{Ar}), 130.8 (C_{Ar}), 130.0 (3 × C_{Ar}), 130.0 (2 × C_{Ar}), 129.7 (C_{Ar}), 129.2 (C_{Ar}), 128.9 (C_{Ar}), 128.6 (C_{Ar}), 128.4 (2 × C_{Ar}), 128.3 (C_{Ar}), 128.2 (C_{Ar}), 128.1 (C_{Ar}), 127.9 (C_{Ar}), 127.7 (C_{Ar}), 123.7 (C_{Ar}), 55.4 (N*C*H$_2$), 52.7 (*C*H), 43.4 (O*C*H$_3$), 32.4 (*C*H$_2$), 32.3 (*C*H$_2$), 26.9 (*C*H$_2$), 26.3 (*C*H$_2$), 25.8 (*C*H$_2$) ppm.– **IR** (ATR): ṽ = 2920 (w), 2853 (w), 1718 (m), 1571 (w), 1490 (w), 1432 (w), 1397 (w), 1350 (m), 1276 (m), 1158 (m), 1106 (m), 1085 (w), 1032 (w), 1014 (w) cm^{-1}. – **MS** (EI), *m/z* (%): 625 (20) [M+H]$^+$, 507 (19), 374 (67), 372 (83), 91 (100). – **HRMS** (EI, C$_{36}$H$_{34}$N$_2$O$_4$^{35}Cl^{32}S): calc. 624.1922; found 624.1922.

N-(2-cyclohexylquinolin-4-yl)-N-(4-methoxybenzyl)-4-methylbenzenesulfonamide (3.97)

This compound was synthesized following **GP3.1** from the corresponding ynamide (106.1 mg, 0.336 mmol, 1.20 equiv.) and amide **3.70a** (57.0 mg, 0.280 mmol, 1.00 equiv.); reaction time: 2h. The crude product was purified by column chromatography (Hept/EtOAc = 10:1) and quinoline **3.97** was obtained as a white solid (77.0 mg,

0.154 mmol, 55%). **¹H-NMR** (400 MHz, CDCl₃): δ = 7.95 (dd, ^3J = 8.0, 4.8 Hz, 2H, CH_{Ar}), 7.57 - 7.67 (m, 3H, CH_{Ar}), 7.39 (t, ^3J = 7.7 Hz, 1H, CH_{Ar}), 7.30 (d, ^3J = 8.0 Hz, 2H, CH_{Ar}), 7.00 (d, ^3J = 8.3 Hz, 2H, CH_{Ar}), 6.62 (d, ^3J = 8.5 Hz, 2H, CH_{Ar}), 6.53 (s, 1H, CH_{Ar}), 4.49 - 5.00 (m, 2H, CH_2), 3.66 (s, 3H, CH_3), 2.69 - 2.81 (m, 1H, CH), 2.46 (s, 3H, CH_3), 1.71 - 1.89 (m, 5H, CH_2), 1.17 - 1.48 (m, 5H) ppm. – **¹³C-NMR** (101 MHz, CDCl₃): δ = 166.3 (C_q), 159.3 (C_q), 149.4 (C_q), 144.7 (C_q), 143.9 (C_q), 135.3 (C_q), 130.4 (+, 2 × CH_{Ar}), 129.6 (+, 2 × CH_{Ar}), 129.6 (+, CH_{Ar}), 128.8 (+, 2 × CH_{Ar}), 128.1 (C_q), 127.1 (C_q), 126.5 (+, CH_{Ar}), 126.0 (+, CH_{Ar}), 124.0 (+, CH_{Ar}), 119.6 (+, CH_{Ar}), 113.7 (+, 2 × CH_{Ar}), 55.4 (−, CH_2), 55.1 (+, CH_3), 47.1 (+, CH), 32.5 (−, 2 × CH_2), 26.3 (−, CH_2), 26.0 (−, CH_2), 21.5 (+, CH_3) ppm. – **IR** (ATR): ṽ = 2918 (w), 2845 (w), 1599 (w), 1556 (m), 1509 (w), 1443 (w), 1339 (m), 1241 (m), 1156 (m), 1089 (m) cm⁻¹. – **MS** (FAB) *m/z* (%): 501.2 (74) [M+H]⁺, 345.2 (21), 121.1 (100). – **HRMS** (FAB, C₃₀H₃₃N₂O₃³²S): calc. 501.2206; found 501.2205.

N-benzyl-2-cyclohexyl-3-methylquinolin-4-amine (3.101)

In a 25 mL round-bottomed flask was added *N*-benzyl-N-(2-cyclohexyl-3-methylquinolin-4-yl)-4-methylbenzenesulfonamide **3.81** (48.5 mg, 0.10 mmol, 1.00 equiv.) and stirred at 0 °C for 5 min before potassium diphenylphosphide (0.5M solution in THF, 400 µl, 0.200 mmol, 2.00 equiv.) was added. After 2h stirring at 0 °C the reaction was complete (TLC). The reaction mixture was diluted with a 1M HCl solution and stirred for 1h. The reaction mixture was then quenched with sat. aq. NaHCO₃, The aqueous layer was extracted with EtOAc (1 × 10 mL) and CH₂Cl₂ (3 × 20 mL). The combined organic layers were dried with magnesium sulfate, filtered and concentrated under reduced pressure. The crude was purified with preparative TLC (*c*Hex/EtOAc/Et₃N = 90:5:1). Quinoline **3.101** was obtained as colorless oil (26.0 mg, 0.079 mmol, 79%). **¹H-NMR** (300 MHz, CDCl₃): δ = 8.00 (d, J = 8.1 Hz, 1H, CH_{Ar}), 7.92 (d, J = 8.5 Hz, 1H, CH_{Ar}), 7.52 - 7.61 (m, 1H, CH_{Ar}), 7.29 - 7.42 (m, 6H, CH_{Ar}), 4.48 (s, 2H, CH_2), 2.96 - 3.07 (m, 1H, CH), 2.32 (s, 3H, CH_3), 1.83 - 1.92 (m, 5H, CH_2), 1.77 - 1.82 (m, 1H, CH_2), 1.39 - 1.46 (m, 2H, CH_2), 1.27 (s, 3H, CH_2) ppm. – **¹³C-NMR** (101 MHz, CDCl₃): δ = 165.7 (C_q), 150.2 (C_q), 147.5 (C_q), 139.8 (C_q), 129.7 (+, CH_{Ar}), 128.8 (2 × CH_{Ar}), 127.9 (+, CH_{Ar}), 127.7 (2 × CH_{Ar}), 127.6 (+, CH_{Ar}), 124.4 (+, CH_{Ar}), 122.0 (+, CH_{Ar}), 121.3 (C_q), 116.5 (C_q), 53.9 (−, CH_2), 43.3 (+, CH), 31.9 (−, CH_2),

29.7 (−, CH_2), 26.9 (−, CH_2), 26.2 (−, CH_2), 13.2 (+, CH_3) ppm – **IR** (ATR): ṽ = 3060 (vw), 2919 (m), 2848 (m), 1614 (vw), 1583 (w), 1559 (w), 1494 (m), 1448 (m), 1372 (m), 1346 (m) cm⁻¹. – **MS** (EI) *m/z* (%): 330.4 (15) [M]⁺, 275.3 (21), 183.1 (32), 91.1 (100). – **HRMS** (EI, $C_{23}H_{26}N_2$): calc. 330.2090; found 330.2091.

2-Cyclohexyl-3-methylquinolin-4-amine (3.103)

In a 25 mL pear shaped flask was added *N*-benzyl-2-cyclohexyl-3-methylquinolin-4-amine **3.101** (15.0 mg, 0.045 mmol, 1.00 equiv.) and wetted palladium on carbon (10 wt% Pd, 9.7 mg, 0.20 equiv.) in EtOAc (2 mL). The argon atmosphere was replaced with hydrogen gas by gently bubbling the gas through the solution. After stirring for 5 days at room temperature with a balloon of hydrogen gas on the reaction, the reaction was complete (as judged by LCMS). The Pd/C was filtered off through a short plug of Celite® in a Pasteur pipet and the remaining solution was evaporated to yield the pure product **3.103** (10.8 mg, 0.045 mmol, quant.).
¹H-NMR (400 MHz, CDCl₃): δ = 7.96 (d, *J* = 8.3 Hz, 1H), 7.69 (dd, *J* = 8.8, 1.0 Hz, 1H), 7.55 (ddd, *J* = 8.3, 6.9, 1.4 Hz, 1H), 7.37 (ddd, *J* = 8.2, 7.0, 1.1 Hz, 1H), 4.64 (br. s., 2H, N*H*₂), 2.97 - 3.07 (m, 1H, C*H*), 2.27 (s, 3H, C*H*₃), 1.74 - 1.95 (m, 7H, C*H*₂), 1.38 - 1.49 (m, 3H, C*H*₂) ppm. – **¹³C-NMR** (101 MHz, CDCl₃): δ = 164.7 (C_q), 146.5 (C_q), 129.4 (CH$_{Ar}$), 128.1 (CH$_{Ar}$), 124.1 (CH$_{Ar}$), 119.5 (CH$_{Ar}$), 117.2 (C_q), 108.4 (C_q), 43.0 (+, CH), 31.7 (−, CH_2), 29.7 (−, CH_2), 26.9 (−, 2 × CH_2), 26.1 (−, CH_2), 12.0 (+, CH_3) ppm.– **IR** (ATR): ṽ = 3369 (w), 3231 (w), 2921 (m), 2849 (m), 1628 (m), 1559 (m), 1498 (m), 1445 (m), 1424 (m), 1372 (m) cm⁻¹. – **MS** (EI) *m/z* (%): 240.2 (100) [M]⁺, 244.3 (47), 185.1 (95). – **HRMS** (EI, $C_{16}H_{20}N_2$): calc. 240.1620; found 240.1621.

2-cyclohexyl-3-methyl-5,6,7,8-tetrahydroquinolin-4-amine (3.104)

In a 20 mL vial was added *N*-benzyl-2-cyclohexyl-3-methylquinolin-4-amine **3.101** (22.6 mg, 0.068 mmol, 1.00 equiv.) and dry palladium on carbon (10 wt% Pd, 20.0 mg, 0.20 equiv.) in EtOAc (3 mL) and placed inside the pressure reactor. The atmosphere was replaced with hydrogen gas by three iterations of increasing hydrogen pressure to 20 bar and releasing it gently. After stirring for 3h at room temperature the conversion was checked with LCMS. Since mostly starting material was observed the catalyst loading was increased to a total of 80 mg, (1.10 equiv.). After a total of 16h the 20 Bar hydrogen pressure was released slowly and ESI-LCMS showed full conversion. The Pd/C was filtered off through a short plug of Celite® in a Pasteur pipet and the remaining solution was evaporated and purified by preparative TLC to yield the pure product **3.104** (8.2 mg, 0.034 mmol, 50%). **^{1}H-NMR** (300 MHz, Acetone-d_6): δ = 4.72 (br. s., 2H), 2.75 - 2.84 (m, 1H), 2.64 (d, J = 5.8 Hz, 2H), 2.39 (d, J = 6.0 Hz, 2H), 2.04 (d, J = 2.3 Hz, 4H), 1.57 - 1.84 (m, 12H), 1.33 - 1.43 (m, 2H) ppm – **^{13}C-NMR** (100 MHz, Acetone-d_6): δ = 160.0 (C_q), 153.1 (C_q), 150.9 (C_q), 113.3 (C_q), 110.5 (C_q), 42.6 (+, CH), 33.7 (−, 2 × CH_2), 32.9 (−, CH_2), 27.6 (−, 2 × CH_2), 27.2 (−, CH_2), 24.1 (−, CH_2), 24.0 (−, CH_2), 23.8 (−, CH_2), 11.6 (+, CH_3) ppm. – **IR** (ATR): $\tilde{v}$ = 3488 (w), 3348 (w), 3231 (w), 2921 (m), 2846 (m), 1630 (m), 1568 (m), 1424 (m), 1372 (m) cm^{-1}. – **MS** (EI) m/z (%): 244.3 (56) [M]$^+$, 243.3 (21), 189.2 (100). – **HRMS** (EI, $C_{16}H_{24}N_2$): calc. 244.1934; found 244.1933.

X-ray Crystallographic data

All X-ray crystallographic work in this chapter has been performed by Dr. Martin NIEGER, the Laboratory of Inorganic Chemistry, University of Helsinki (Finland). All X-ray images shown in this chapter are depicted with displacement parameters drawn at 50 % probability level.

Methyl 4-(N-(2-cyclohexyl-3-phenylquinolin-4-yl)-N-(4-methylbenzyl)-sulfamoyl)benzoate – SB795_HY – CCDC 1402829 – (3.88)

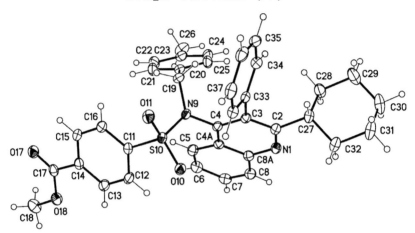

Crystal data

$C_{37}H_{36}N_2O_4S$	$F(000) = 1280$
$M_r = 604.74$	$D_x = 1.301$ Mg m^{-3}
Monoclinic, $P2_1/n$ (no.14)	Cu $K\alpha$ radiation, $\lambda = 1.54178$ Å
$a = 8.5361$ (3) Å	Cell parameters from 9956 reflections
$b = 19.0560$ (7) Å	$\theta = 4.6–72.0°$
$c = 19.3655$ (7) Å	$\mu = 1.28$ mm^{-1}
$\beta = 101.473$ (1)°	$T = 123$ K
$V = 3087.12$ (19) Å^3	Blocks, colourless
$Z = 4$	$0.20 \times 0.14 \times 0.06$ mm

Data collection

Bruker D8 Venture diffractometer with PHOTON 100 detector	5608 reflections with $I > 2\sigma(I)$
Radiation source: IμS microfocus	$R_{int} = 0.023$
rotation in φ and ω, 0.5°, shutterless scans	$\theta_{max} = 72.2°$, $\theta_{min} = 3.3°$
Absorption correction: multi-scan $SADABS$ (SHeldrick, 2014)	$h = -10 \rightarrow 10$
$T_{min} = 0.799$, $T_{max} = 0.889$	$k = -23 \rightarrow 23$
23932 measured reflections	$l = -19 \rightarrow 23$
6061 independent reflections	

Refinement

Refinement on F^2	Primary atom site location: structure-invariant direct methods
Least-squares matrix: full	Secondary atom site location: difference Fourier map
$R[F^2 > 2\sigma(F^2)] = 0.033$	Hydrogen site location: difference Fourier map
$wR(F^2) = 0.088$	H-atom parameters constrained
$S = 1.03$	$w = 1/[\sigma^2(F_o^2) + (0.0454P)^2 + 1.2746P]$ where $P = (F_o^2 + 2F_c^2)/3$
6061 reflections	$(\Delta/\sigma)_{max} = 0.002$
399 parameters	$\Delta\rangle_{max} = 0.29$ e Å^{-3}
0 restraints	$\Delta\rangle_{min} = -0.42$ e Å^{-3}

N-benzyl-N-(2-cyclohexyl-3-phenylquinolin-4-yl)-4-methylbenzene-sulfonate − SB797_HY −
CCDC 1402827 − (3.77)

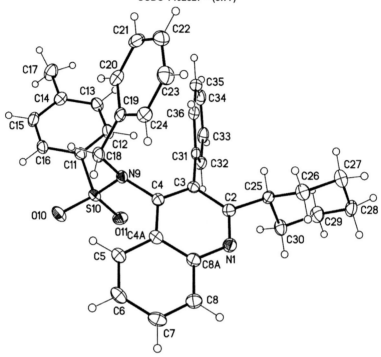

Crystal data

C₃₅H₃₄N₂O₂S	$F(000) = 1160$
$M_r = 546.70$	$D_x = 1.288$ Mg m⁻³
Monoclinic, $P2_1/c$ (no.14)	Cu $K\alpha$ radiation, $\lambda = 1.54178$ Å
$a = 9.8026$ (2) Å	Cell parameters from 9585 reflections
$b = 13.1470$ (3) Å	$\theta = 3.9$–72.1°
$c = 22.3358$ (5) Å	$\mu = 1.29$ mm⁻¹
$\beta = 101.644$ (1)°	$T = 123$ K
$V = 2819.28$ (11) Å³	Rods, colourless
$Z = 4$	0.16 × 0.06 × 0.03 mm

Data collection

Bruker D8 Venture diffractometer with Photon100 detector	4810 reflections with $I > 2\sigma(I)$
Radiation source: IμS microfocus	$R_{int} = 0.024$
rotation in ϕ and ω, 0.5°, shutterless scans	$\theta_{max} = 72.1°$, $\theta_{min} = 3.9°$
Absorption correction: multi-scan $SADABS$ (Sheldrick, 2015)	$h = -11{\rightarrow}12$
$T_{min} = 0.831$, $T_{max} = 0.929$	$k = -15{\rightarrow}16$
14022 measured reflections	$l = -27{\rightarrow}21$
5422 independent reflections	

Refinement

Refinement on F^2	Primary atom site location: structure-invariant direct methods
Least-squares matrix: full	Secondary atom site location: difference Fourier map
$R[F^2 > 2\sigma(F^2)] = 0.036$	Hydrogen site location: difference Fourier map
$wR(F^2) = 0.098$	H-atom parameters constrained
$S = 1.02$	$w = 1/[\sigma^2(F_o^2) + (0.0535P)^2 + 1.141P]$ where $P = (F_o^2 + 2F_c^2)/3$
5422 reflections	$(\Delta/\sigma)_{max} = 0.001$
362 parameters	$\Delta\rangle_{max} = 0.31$ e Å⁻³
0 restraints	$\Delta\rangle_{min} = -0.42$ e Å⁻³

N-benzyl-*N*-(2-cyclohexyl-3-methylquinolin-4-yl)-4-methylbenzenesulfonamide – SB852_HY –
(3.81)

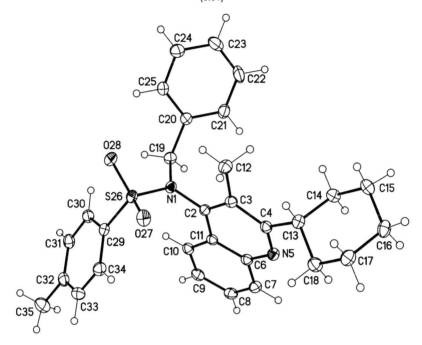

Crystal data

$C_{30}H_{32}N_2O_2S$	$F(000) = 1032$
$M_r = 484.63$	$D_x = 1.275$ Mg m^{-3}
Monoclinic, $P2_1/c$ *(no.14)*	Mo $K\alpha$ radiation, $\lambda = 0.71073$ Å
$a = 8.9871$ (4) Å	Cell parameters from 9878 reflections
$b = 17.7788$ (9) Å	$\theta = 2.3–27.5°$
$c = 16.3694$ (8) Å	$\mu = 0.16$ mm^{-1}
$\beta = 105.152$ (2)°	$T = 123$ K
$V = 2524.6$ (2) Å^3	Blocks, colourless
$Z = 4$	$0.30 \times 0.20 \times 0.10$ mm

Data collection

Bruker D8 VENTURE diffractometer with Photon100 detector	5820 independent reflections
Radiation source: INCOATEC microfocus sealed tube	4852 reflections with $I > 2\sigma(I)$
Detector resolution: 10.4167 pixels mm^{-1}	$R_{int} = 0.038$
rotation in ω, 1°, shutterless scans	$\theta_{max} = 27.6°$, $\theta_{min} = 2.3°$
Absorption correction: multi-scan *SADABS* (Sheldrick, 2014)	$h = -11 \rightarrow 11$
$T_{min} = 0.953$, $T_{max} = 0.987$	$k = -23 \rightarrow 23$
35569 measured reflections	$l = -21 \rightarrow 21$

Refinement

Refinement on F^2	Secondary atom site location: difference Fourier map
Least-squares matrix: full	Hydrogen site location: difference Fourier map
$R[F^2 > 2\sigma(F^2)] = 0.039$	H-atom parameters constrained
$wR(F^2) = 0.101$	$w = 1/[\sigma^2(F_o^2) + (0.0453P)^2 + 1.535P]$ where $P = (F_o^2 + 2F_c^2)/3$
$S = 1.02$	$(\Delta/\sigma)_{max} < 0.001$
5820 reflections	$\Delta\rangle_{max} = 0.57$ e Å^{-3}
319 parameters	$\Delta\rangle_{min} = -0.39$ e Å^{-3}
0 restraints	Extinction correction: *SHELXL2014*/7 (Sheldrick 2014, Fc*=kFc[1+0.001xFc²λ³/sin(2θ)]$^{-1/4}$
Primary atom site location: structure-invariant direct methods	Extinction coefficient: 0.0035 (6)

3.7 References

[1]   A. M. Thayer, *C&EN*, **2005**, *83* (43), 69-82, *Fighting Malaria: New antimalarial drugs are needed to ensure that effective and affordable treatments continue to be available and are not lost to parasite resistance.*

[2]   W.H.O., *World Malaria Report*; World Health Organization: http://www.who.int/malaria/publications/world-malaria-report-2015/en/, Geneva, Switzerland, 2015.

[3]   N. J. White, S. Pukrittayakamee, T. T. Hien, M. A. Faiz, O. A. Mokuolu, A. M. Dondorp, *Lancet,* **2014**, *383* (9918), 723-735, *Malaria.*

[4]   IAMAT; International association for medical assistance to travellers; *World Malaria Risk Chart 2016*: https://www.iamat.org/elibrary/view/id/1376, accessed 25/4/2016, last updated April 2016.

[5]   A.-M. Kreidler, R. Benz, H. Barth, *Arch. Tox.,* **2016**, 1-15, *Chloroquine derivatives block the translocation pores and inhibit cellular entry of Clostridium botulinum C2 toxin and Bacillus anthracis lethal toxin.*

[6]   A. F. G. Slater, *Pharmacol. Therap.,* **1993**, *57* (2), 203-235, *Chloroquine: Mechanism of drug action and resistance in plasmodium falciparum.*

[7]   R. N. Price, L. von Seidlein, N. Valecha, F. Nosten, J. K. Baird, N. J. White, *Lancet Inf. Dis.,* *14* (10), 982-991, *Global extent of chloroquine-resistant Plasmodium vivax: a systematic review and meta-analysis.*

[8]   B. Greenwood, T. Mutabingwa, *Nature,* **2002**, *415* (6872), 670-672, *Malaria in 2002.*

[9]   D. Payne, *Parasitology Today,* **1987**, *3* (8), 241-246, *Spread of chloroquine resistance in Plasmodium falciparum.*

[10]   A. Kumar, D. Paliwal, D. Saini, A. Thakur, S. Aggarwal, D. Kaushik, *Eur. J. Med. Chem.,* **2014**, *85* (0), 147-178, *A comprehensive review on synthetic approach for antimalarial agents.*

[11]   P. R. Graves, J. J. Kwiek, P. Fadden, R. Ray, K. Hardeman, A. M. Coley, M. Foley, T. A. J. Haystead, *Mol. Pharmacol.,* **2002**, *62* (6), 1364-1372, *Discovery of Novel Targets of Quinoline Drugs in the Human Purine Binding Proteome.*

[12]   H. R. Bhat, U. P. Singh, P. Gahtori, S. K. Ghosh, K. Gogoi, A. Prakash, R. K. Singh, *New J. Chem.,* **2013**, *37* (9), 2654-2662, *4-Aminoquinoline-1,3,5-triazine: Design, synthesis, in vitro antimalarial activity and docking studies.*

[13]   A. Kumar, K. Srivastava, S. Raja Kumar, S. K. Puri, P. M. S. Chauhan, *Bioorg. Med. Chem. Lett.,* **2010**, *20* (23), 7059-7063, *Synthesis of new 4-aminoquinolines and quinoline–acridine hybrids as antimalarial agents.*

[14]   M. Foley, L. Tilley, *Pharmacol. Therap.,* **1998**, *79* (1), 55-87, *Quinoline Antimalarials: Mechanisms of Action and Resistance and Prospects for New Agents.*

[15]   A. Encinas López; *Quinolines: Privileged Scaffolds in Medicinal Chemistry* in *Privileged Scaffolds in Medicinal Chemistry*; S. Bräse, Ed.; Royal Society of Chemistry, 2015; Vol. 1.

[16]   D. De, L. D. Byers, D. J. Krogstad, *J. Het. Chem.,* **1997**, *34* (1), 315-320, *Antimalarials: Synthesis of 4-aminoquinolines that circumvent drug resistance in malaria parasites.*

[17]   G. M. Steinberg, M. L. Mednick, J. Maddox, R. Rice, J. Cramer, *J. Med. Chem.,* **1975**, *18* (11), 1056-1061, *Hydrophobic binding site in acetylcholinesterase.*

[18]   Z. H. Skraup, *Chem. Ber.,* **1880**, *1* (1), 316-318, *Eine synthese des chinolins.*

[19]   O. Doebner, W. v. Miller, *Chem. Ber.,* **1881**, *14* (2), 2812-2817, *Ueber eine dem Chinolin homologe Base.*

[20]   O. Doebner, W. v. Miller, *Chem. Ber.,* **1883**, *16* (2), 2464-2472, *Ueber Chinaldinbasen.*

[21]   P. Friedländer, *Chem. Ber.,* **1882**, *15* (2), 2572-2575, *Ueber o-Amidobenzaldehyd.*

[22]   P. Friedländer, C. F. Gohring, *Chem. Ber.,* **1883**, *16* (2), 1833-1839, *Ueber eine Darstellungsmethode im Pyridinkern substituirter Chinolinderivate.*

[23]   W. Pfitzinger, *J. Prak. Chem.,* **1886**, *33* (1), 100-100, *Chinolinderivate aus Isatinsäure.*

[24]   F. W. Bergstrom, *Chem. Rev.,* **1944**, *35* (2), 77-277, *Heterocyclic Nitrogen Compounds. Part IIA. Hexacyclic Compounds: Pyridine, Quinoline, and Isoquinoline.*

[25]   A. Combes, *Bull. Soc. Chim. Fr.,* **1883**, *49*, 89, *Synthesis of quinoline derivatives from acetyl acetone.*

[26] C. Gronnier, G. Boissonnat, F. Gagosz, *Org. Lett.*, **2013**, *15* (16), 4234-4237, *Au-Catalyzed Formation of Functionalized Quinolines from 2-Alkynyl Arylazide Derivatives.*

[27] C. Theeraladanon, M. Arisawa, A. Nishida, M. Nakagawa, *Tetrahedron*, **2004**, *60* (13), 3017-3035, *A novel synthesis of substituted quinolines using ring-closing metathesis (RCM): its application to the synthesis of key intermediates for anti-malarial agents.*

[28] B. C. Ranu, A. Hajra, U. Jana, *Tetrahedron Lett.*, **2000**, *41* (4), 531-533, *Microwave-assisted simple synthesis of quinolines from anilines and alkyl vinyl ketones on the surface of silica gel in the presence of indium(III) chloride.*

[29] B. C. Ranu, A. Hajra, S. S. Dey, U. Jana, *Tetrahedron*, **2003**, *59* (6), 813-819, *Efficient microwave-assisted synthesis of quinolines and dihydroquinolines under solvent-free conditions.*

[30] R. C. Elderfield, W. J. Gensler, O. Birstein, F. J. Kreysa, J. T. Maynard, J. Galbreath, *J. Am. Chem. Soc.*, **1946**, *68* (7), 1250-1251, *Synthesis of Certain Simple 4-Aminoquinoline Derivatives.*

[31] O. G. Backeberg, *J. Chem. Soc. (Resumed)*, **1938**, (0), 1083-1087, *208. The reaction between phenylhydrazine and 4-chloroquinoline derivatives, and the preparation of the corresponding 4-benzeneazo- and 4-amino-compounds.*

[32] V. A. Petrow, *J. Chem. Soc. (Resumed)*, **1945**, (0), 18-22, *6. Novel types of styrylquinolinium compounds.*

[33] J. A. Moore, L. D. Kornreich, *Tetrahedron Lett.*, **1963**, *4* (20), 1277-1281, *A direct synthesis of 4-aminoquinolines.*

[34] A. Schmidt, N. Münster, A. Dreger, *Angew. Chem. Int. Ed.*, **2010**, *49* (15), 2790-2793, *Functionalized 4-Aminoquinolines by Rearrangement of Pyrazole N-Heterocyclic Carbenes.*

[35] G. Höfle, O. Hollitzer, W. Steglich, *Angew. Chem. Int. Ed.*, **1972**, *11* (8), 720-722, *4-(Dialkylamino)quinolines and 3,3-Disubstituted 4(3H)-Quinolones from 4H-3,1-Benzoxazinones and Ynamines.*

[36] T. Vlaar, B. U. W. Maes, E. Ruijter, R. V. A. Orru, *Chem. Het. Comp.*, **2013**, *49* (6), 902-908, *Synthesis of 4-aminoquinolines by aerobic oxidative palladium-catalyzed double C–H activation and isocyanide insertion.*

[37] P. B. Madrid, J. Sherrill, A. P. Liou, J. L. Weisman, J. L. DeRisi, R. K. Guy, *Bioorg. Med. Chem. Lett.*, **2005**, *15* (4), 1015-1018, *Synthesis of ring-substituted 4-aminoquinolines and evaluation of their antimalarial activities.*

[38] I. L. Baraznenok, V. G. Nenajdenko, E. S. Balenkova, *Tetrahedron*, **2000**, *56* (20), 3077-3119, *Chemical Transformations Induced by Triflic Anhydride.*

[39] A. B. Charette, P. Chua, *Tetrahedron Lett.*, **1997**, *38* (49), 8499-8502, *A new method for the conversion of secondary and tertiary amides to bridged orthoesters.*

[40] A. B. Charette, P. Chua, *Tetrahedron Lett.*, **1998**, *39* (3-4), 245-248, *Thiolysis and hydrolysis of imino and iminium triflates: Synthesis of secondary and tertiary thioamides and O-18-labeled amides.*

[41] A. B. Charette, P. Chua, *J. Org. Chem.*, **1998**, *63* (4), 908-909, *Mild method for the synthesis of thiazolines from secondary and tertiary amides.*

[42] A. B. Charette, M. Grenon, *Tetrahedron Lett.*, **2000**, *41* (11), 1677-1680, *Mild method for the synthesis of amidines by the electrophilic activation of amides.*

[43] A. B. Charette, M. Grenon, *Can. J. Chem.*, **2001**, *79* (11), 1694-1703, *Spectroscopic studies of the electrophilic activation of amides with triflic anhydride and pyridine.*

[44] A. B. Charette, S. Mathieu, J. Martel, *Org. Lett.*, **2005**, *7* (24), 5401-5404, *Electrophilic Activation of Lactams with Tf2O and Pyridine: Expedient Synthesis of (±)-Tetraponerine T4.*

[45] G. Lemonnier, A. B. Charette, *J. Org. Chem.*, **2010**, *75* (21), 7465-7467, *Stereoselective Synthesis of 2,3,6-Trisubstituted Tetrahydropyridines via Tf2O-Mediated Grob Fragmentation: Access to Indolizidines (-)-209I and (-)-223J.*

[46] G. Pelletier, A. B. Charette, *Org. Lett.*, **2013**, *15* (9), 2290-2293, *Triflic Anhydride Mediated Synthesis of Imidazo[1,5-a]azines.*

[47] M. Ghandi, M. Hasani, S. Salahi, *Monatsh. Chem.*, **2012**, *143* (3), 455-460, *Expedient one-pot synthesis of N-aryliminoethers via mild electrophilic activation of secondary amides.*

[48] M. Ghandi, S. Salahi, M. Hasani, *Tetrahedron Lett.*, **2011**, *52* (2), 270-273, *A mild, expedient, one-pot trifluoromethanesulfonic anhydride mediated synthesis of N-arylimidates.*

[49] P.-Q. Huang, Y.-H. Huang, K.-J. Xiao, Y. Wang, X.-E. Xia, *J. Org. Chem.*, **2015**, *80* (5), 2861-2868, *A General Method for the One-Pot Reductive Functionalization of Secondary Amides.*

[50]   P.-Q. Huang, Y. Wang, K.-J. Xiao, Y.-H. Huang, *Tetrahedron*, **2015**, *71* (24), 4248-4254, *A general method for the direct transformation of common tertiary amides into ketones and amines by addition of Grignard reagents.*

[51]   C. Madelaine, V. Valerio, N. Maulide, *Angew. Chem. Int. Ed.*, **2010**, *49* (9), 1583-1586, *Unexpected Electrophilic Rearrangements of Amides: A Stereoselective Entry to Challenging Substituted Lactones.*

[52]   J. W. Medley, M. Movassaghi, *J. Org. Chem.*, **2009**, *74* (3), 1341-1344, *Direct Dehydrative N-Pyridinylation of Amides.*

[53]   M. Movassaghi, M. D. Hill, *Org. Lett.*, **2008**, *10* (16), 3485-3488, *A Versatile Cyclodehydration Reaction for the Synthesis of Isoquinoline and β-Carboline Derivatives.*

[54]   M. Padmanaban, L. C. R. Carvalho, D. Petkova, J.-W. Lee, A. S. Santos, M. M. B. Marques, N. Maulide, *Tetrahedron*, **2015**, *71* (35), 5994-6005, *Investigation of cationic Claisen-type electrophilic rearrangements of amides.*

[55]   B. Peng, D. Geerdink, C. Farès, N. Maulide, *Angew. Chem. Int. Ed.*, **2014**, *53* (21), 5462-5466, *Chemoselective Intermolecular α-Arylation of Amides.*

[56]   B. Peng, D. Geerdink, N. Maulide, *J. Am. Chem. Soc.*, **2013**, *135* (40), 14968-14971, *Electrophilic Rearrangements of Chiral Amides: A Traceless Asymmetric α-Allylation.*

[57]   A.-E. Wang, Z. Chang, Y.-P. Liu, P.-Q. Huang, *Chin. Chem. Lett.*, **2015**, *26* (9), 1055-1058, *Mild N-deacylation of secondary amides by alkylation with organocerium reagents.*

[58]   K. L. White, M. Mewald, M. Movassaghi, *J. Org. Chem.*, **2015**, *80* (15), 7403-7411, *Direct Observation of Intermediates Involved in the Interruption of the Bischler-Napieralski Reaction.*

[59]   M. Movassaghi, M. D. Hill, *J. Am. Chem. Soc.*, **2006**, *128* (44), 14254-14255, *Single-Step Synthesis of Pyrimidine Derivatives.*

[60]   M. D. Hill, M. Movassaghi, *Chem. Eur. J.*, **2008**, *14* (23), 6836-6844, *New strategies for the synthesis of pyrimidine derivatives.*

[61]   O. K. Ahmad, J. W. Medley, A. Coste, M. Movassaghi, *Org. Synth.*, **2012**, *89*, 549-561, *Direct synthesis of azaheterocycles from N-aryl/vinyl amides. Synthesis of 4-(methylthio)-2-phenylquinazoline and 4-(4-methoxyphenyl)-2-phenylquinoline.*

[62]   M. D. Hill, M. Movassaghi, *Synthesis*, **2008**, (5), 823-827, *Direct synthesis of substituted pyrimidines and quinazolines.*

[63]   M. Movassaghi, M. D. Hill, *Nat. Protoc.*, **2007**, *2* (8), 2018-2023, *Synthesis of pyrimidines by direct condensation of amides and nitriles.*

[64]   O. K. Ahmad, M. D. Hill, M. Movassaghi, *J. Org. Chem.*, **2009**, *74* (21), 8460-8463, *Synthesis of Densely Substituted Pyrimidine Derivatives.*

[65]   M. Movassaghi, M. D. Hill, O. K. Ahmad, *J. Am. Chem. Soc.*, **2007**, *129* (33), 10096-10097, *Direct Synthesis of Pyridine Derivatives.*

[66]   T. Wezeman, S. Zhong, M. Nieger, S. Bräse, *Angew. Chem., Int. Ed.*, **2016**, *55* (11), 3823-3827, *Synthesis of Highly Functionalized 4-Aminoquinolines.*

[67]   S. C. Bergmeier, P. P. Seth, *Tetrahedron Lett.*, **1999**, *40* (34), 6181-6184, *A general method for deprotection of N-toluenesulfonyl aziridines using sodium naphthalenide.*

[68]   E. Alonso, D. J. Ramón, M. Yus, *Tetrahedron*, **1997**, *53* (42), 14355-14368, *Reductive deprotection of allyl, benzyl and sulfonyl substituted alcohols, amines and amides using a naphthalene-catalysed lithiation.*

[69]   E. H. Gold, E. Babad, *J. Org. Chem.*, **1972**, *37* (13), 2208-2210, *Reductive cleavage of sulfonamides with sodium bis(2-methoxyethoxy)aluminum hydride.*

[70]   W. Oppolzer, H. Bienayme, A. Genevois-Borella, *J. Am. Chem. Soc.*, **1991**, *113* (25), 9660-9661, *Enantioselective synthesis of (+)-3-isorauniticine via a catalytic tandem palladium-ene/carbonylation reaction.*

[71]   D. I. Weisblat, B. J. Magerlein, D. R. Myers, *J. Am. Chem. Soc.*, **1953**, *75* (15), 3630-3632, *The Cleavage of Sulfonamides.*

[72]   B. Nyasse, L. Grehn, U. Ragnarsson, *Chem. Commun.*, **1997**, (11), 1017-1018, *Mild, efficient cleavage of arenesulfonamides by magnesium reduction.*

[73]   N. Shohji, T. Kawaji, S. Okamoto, *Org. Lett.*, **2011**, *13* (10), 2626-2629, *Ti(O-i-Pr)₄/Me₃SiCl/Mg-Mediated Reductive Cleavage of Sulfonamides and Sulfonates to Amines and Alcohols.*

[74]   A. B. Smith, D.-S. Kim, *Org. Lett.,* **2005**, *7* (15), 3247-3250, *Total Synthesis of the Neotropical Poison-Frog Alkaloid (−)-205B.*

[75]   Z. Moussa, D. Romo, *Synlett,* **2006**, *2006* (19), 3294-3298, *Mild Deprotection of Primary N-(p-Toluenesulfonyl) Amides with SmI$_2$ -Following Trifluoroacetylation.*

[76]   T. Ankner, G. Hilmersson, *Org. Lett.,* **2009**, *11* (3), 503-506, *Instantaneous Deprotection of Tosylamides and Esters with SmI$_2$/Amine/Water.*

[77]   S. Yoshida, K. Igawa, K. Tomooka, *J. Am. Chem. Soc.,* **2012**, *134* (47), 19358-19361, *Nucleophilic Substitution Reaction at the Nitrogen of Arylsulfonamides with Phosphide Anion.*

[78]   H. E. Gottlieb, V. Kotlyar, A. Nudelman, *J. Org. Chem.,* **1997**, *62* (21), 7512-7515, *NMR Chemical Shifts of Common Laboratory Solvents as Trace Impurities.*

[79]   W. Li, X.-F. Wu, *J. Org. Chem.,* **2014**, *79* (21), 10410-10416, *Palladium-Catalyzed Carbonylative Synthesis of Benzoxazinones from N-(o-Bromoaryl)amides Using Paraformaldehyde as the Carbonyl Source.*

Chapter 4. Pyrazoles

The research presented in this Chapter has been conducted at the Institute of Organic Chemistry at the Karlsruhe Institute of Technology in Karlsruhe, Germany (with Prof. Dr. Stefan BRÄSE).

4.0 Abstract & Graphical abstract

By expanding the scope of the sydnone-alkyne cycloaddition to the more electron-rich sulfonyl ynamides a novel access to 3- and 4-aminopyrazoles has been described. Using terminal sulfonyl ynamides, C-4 unsubstituted sydnones and a copper sulfate, 1,10-phenanthroline and sodium ascorbate approach 1,4-pyrazoles are accessible in reasonable yields. Using other copper additives resulted in the facile conversion of the ynamide into an amide. A copper-free alternative using *in situ* prepared strained cyclic ynamides allowed for diversely C-4 substituted sydnones to undergo the desired cycloaddition, but always yielded a mixture of regio-isomeric products. Synthesis of an 8-membered 3-azacyclooctyne has been investigated and its presence confirmed by direct trapping with a sydnone.

4.1 Sydnones

Sydnones are considered to be one of the most popular members of the family of meso-ionic heteroaromatic compounds. The first synthesis of a sydnone was reported by EARL and MACKNEY in 1935 (Scheme **4.1**). While working on acetylations of nitrosoglycines they discovered what they thought looked like compound **4.2**.[1]

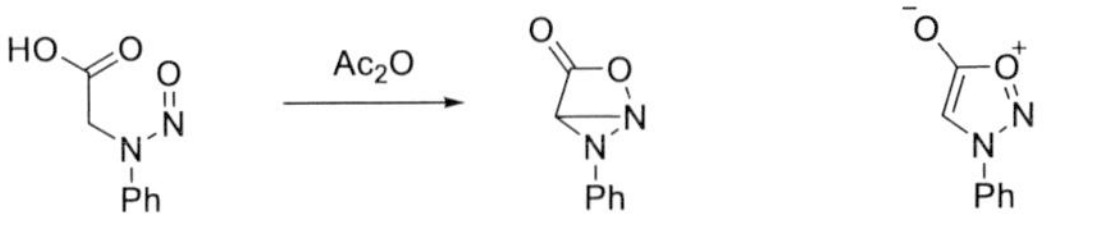

4.1   **4.2**   **4.3**
                              Initially assigned structure   Correct structure
                                 Earl & Mackney, 1935

Scheme 4.1 The first synthesis of a sydnone.

However, later work by BAKER, OLLIS and POOLE determined the correct structure of the sydnone – named after Sydney, where it was first synthesized – to be a meso-ionic compound like **4.3**.[2] Scheme **4.2** shows the common mesomeric structures of sydnones and the atom-numbering convention. As no uncharged species can be drawn, sydnones are typically represented with a positively charged oxygen in the ring and with an enolate-type exocyclic oxygen **4.4**.

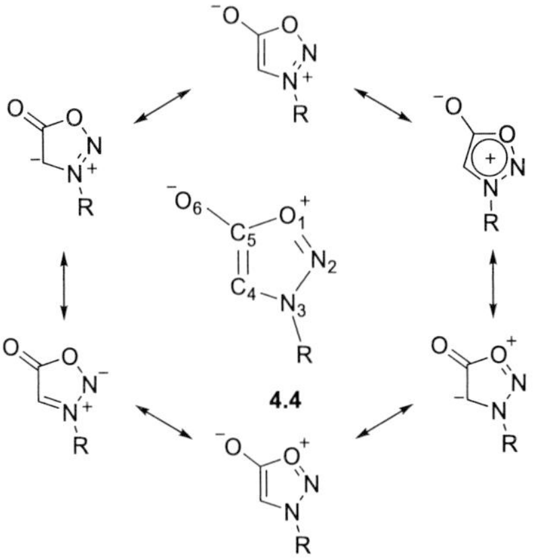

Scheme 4.2 Sydnones are meso-ionic compounds.

The synthesis of sydnones is typically rather straightforward, as the isolation and purification of the often highly crystalline products can usually be achieved by recrystallization from ethanol. Synthesis of the nitroso compounds **4.6** can be done classically by nitrosation of *N*-arylglycines **4.5** with sodium nitrite in concentrated acid or using amyl nitrite. Subsequent cyclodehydration via either acetic acid or trifluoroacetic anhydride yields the desired sydnones **4.4**.

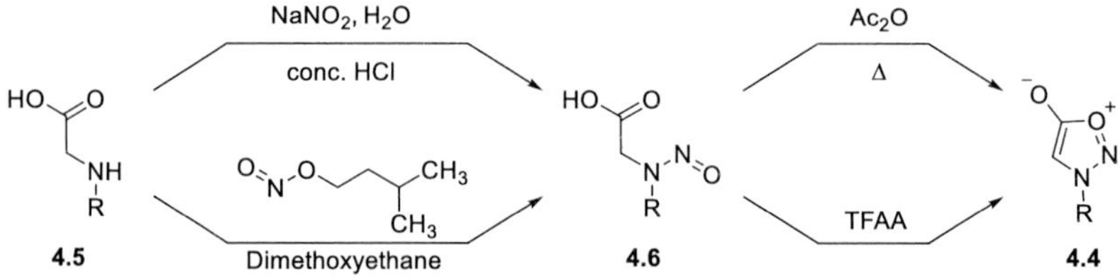

Scheme 4.3 Classical (top) and recent (lower) syntheses[3] of sydnones.

The sydnones **4.4** can be further derivatized by modification on the C-4 position (Scheme **4.4**). As this position is both acidic and nucleophilic two possible paths can be taken: electrophilic aromatic substitution or deprotonation and electrophile addition. Direct acylation can be done using sonication, perchloric acid and acetic anhydride.[4] Halogenated sydnones are easily accessible[5,6] and can be further modified using Sonogashira[7] or Suzuki couplings.[8] Alternatively the sydnones can be lithiated[9] and reacted with a wide range of electrophiles.

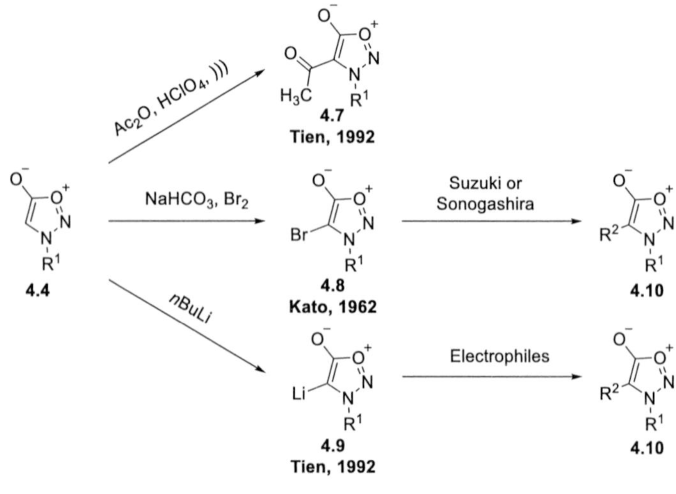

Scheme 4.4 Functionalization of C-4 using different methods.

In order to explore the chemical reactivity of sydnones to ynamides a small library of sydnones was prepared by Júlia COMAS-BARCELÓ from the Organic Chemistry department of the University of Sheffield. The sydnones were prepared via a two-step procedure consisting of the nitrosation and cyclodehydration of N-arylglycines **4.11** (Scheme **4.5**).[10] Further derivatization of the C-4 position of the sydnone was achieved by Pd-catalyzed direct arylation[11] or by lithiation and subsequent quenching with electrophiles.

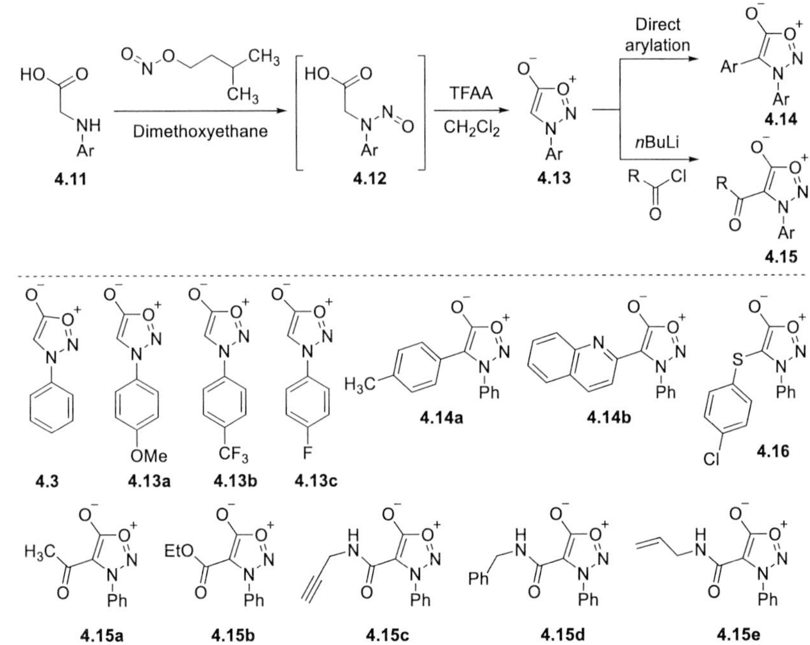

Scheme 4.5 Library of sydnones prepared by Júlia COMAS-BARCELÓ, Univ. Sheffield.

Interestingly, sydnones can be used as bio-orthogonal chemical agents,[12-14] and some have even been found to be biologically active.[15,16] More detailed information on sydnones and similar compounds can be found in overview articles published in the last decade.[10,17-19]

Sydnones are best known for their participation as 1,3-dipoles in cycloaddition reactions with alkynes, as pioneered by HUISGEN in the 1960s (Scheme **4.6**).[20,21] As these cycloadditions were uncatalyzed they typically required elevated temperatures, making them rather unpractical.

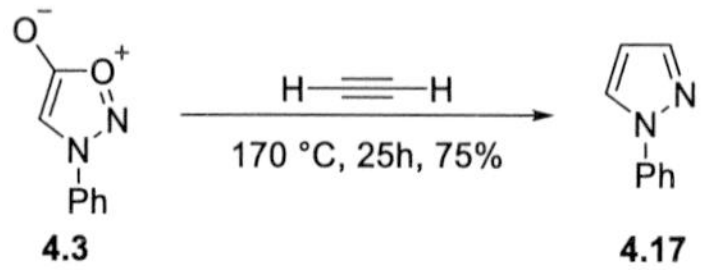

Scheme 4.6 HUISGEN's thermal cycloadditions with alkynes.

Nowadays more sophisticated catalyzed systems have been developed and sydnones serve as a convenient synthetic handle to prepare pyrazoles.[22,23]

4.2 Pyrazoles

Pyrazoles are a common heterocyclic motif found in many biologically active compounds. Ever since the classical pyrrole synthesis was developed by KNORR[24,25] in 1883 (see Scheme **4.7**) chemists have been developing new synthetic routes to access this widely used scaffold.

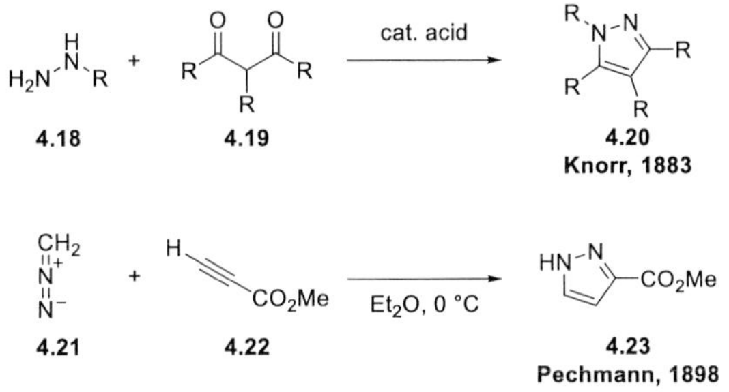

Scheme 4.7 The classical Knorr and Pechmann[26,27] pyrazole syntheses.

Due to the high interest for new routes to fully functionalized pyrazoles, which are known to possess biological activity[28-34] – the widely used CELECOXIB (**4.24**, Figure **4.1**) being a prime example – several investigations into pyrazoles syntheses have been published.[12,35-42]

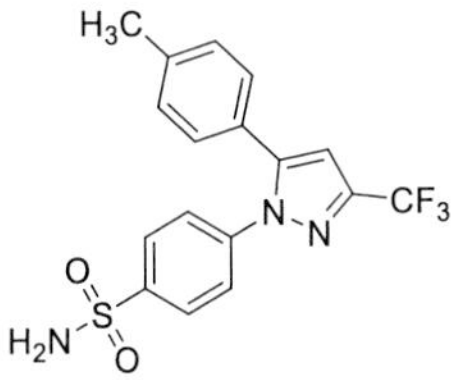

4.24
Celecoxib

Figure 4.1 Famous example of a biologically active pyrazole is CELECOXIB, a COX-2 inhibiting selective nonsteroidal anti-inflammatory drug.

Since the condensation of hydrazines with diketones, α,β-unsaturated nitriles[43] and similar compounds works so efficiently for the synthesis of pyrazoles, not many novel methods are being developed to avoid the use of toxic hydrazines.[38,44] And indeed, for the synthesis of aminopyrazoles few great advances have been made (Scheme **4.7**) – even though they show promising biological activity.[45] These compounds are often either made by reduction of functional groups on previously prepared pyrazole rings[46] (e.g. **4.26** → **4.27**) or by using the remains of azido (after loss of N_2)[47] (e.g. **4.31** → **4.32**) or nitroso[48,49] groups after condensation with hydrazines (e.g. **4.34** → **4.35**).

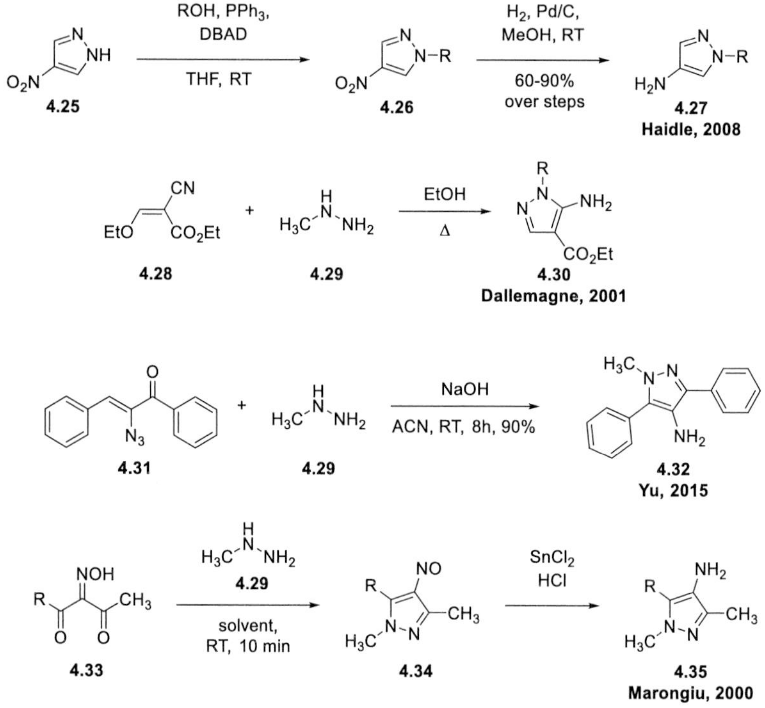

Scheme 4.7 Aminopyrazole syntheses, often using hydrazines as main reagents.

A procedure that avoids the use of toxic hydrazines, by using hydrazine synthons instead, employs sydnones that undergo cycloadditions with alkynes to form pyrazoles. These 1,3-dipolar cycloadditions tend to prefer the use of electron-deficient dienophiles and typically require high reaction temperatures and long reaction times. The regioselectivity of the obtained pyrazole products is often substrate-dependent.[10] Attempts to address this issue, such as the use of alkynylboronates[8,50] or copper additives to reverse the regioselectivity of the sydnone-alkyne cycloaddition reactions have been previously reported in the literature.[12,22,51]

Recent work by the HARRITY group[23] shows that the nature of the copper source can dramatically influence the regioselectivity of the pyrazoles. The copper-catalyzed sydnone alkyne cycloaddition (CuSAC) with copper triflate was found to proceed mainly Lewis acid-catalyzed resulting in the 1,3-substituted regio-isomer **4.39**. When copper acetate is used a copper(I) acetylide is formed and the 1,4-substituted pyrazole **4.38** is obtained instead.

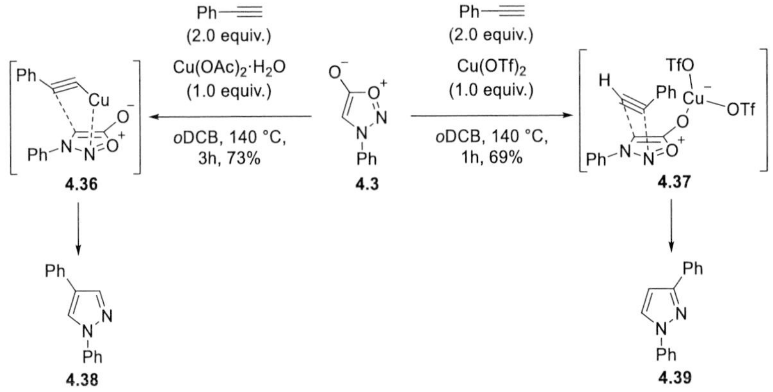

Scheme 4.8 Tunable regioselectivity by choice of copper additive. Transition states calculated by Béla FISER, Universidad del País Vasco, San Sebastián, Spain.[23]

However, the synthesis of sydnone-alkyne cycloaddition has not yet been used to prepare 3- or 4-aminopyrazoles. These are still made with hydrazines, as shown in Scheme **4.7** or − in rare cases − via diazonium salts **4.40**[52] or tetrazines **4.43** as in Scheme **4.9**.[53]

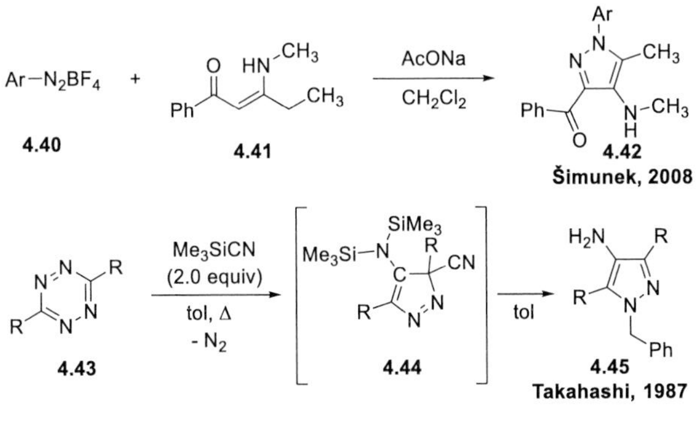

Scheme 4.9 Recent aminopyrazole syntheses avoiding the use of using hydrazines as main reagents.

This leads to conclude that innovation regarding the synthesis of 3- and 4-aminopyrazoles is still pretty much at a stand-still. In an effort to change this, and in meanwhile explore options to increase the tolerance of the CuSAC reaction for activated and electron-rich alkynes, the synthesis of aminopyrazoles was envisioned by reacting sydnones with sulfonyl ynamides, which to date have not been reported yet.[38,46,47,54]

4.3 Syntheses of 4-aminopyrazoles via copper catalysis

Since the sydnone-alkyne cycloaddition typically requires the use of electron-deficient dienophiles its synthetic usability is limited. By investigating the reaction between sydnones and the relatively electron-rich sulfonyl ynamides discussed in Chapter 2, these synthetic limitations could be lifted. If reactions with these reactive ynamides are successful they would allow for the construction of 3- and/or 4-aminopyrazoles via a novel pathway.

The first system tested was the uncatalyzed, thermal reaction. However, the purely thermal reaction between phenylsydnone **4.3** and sulfonyl ynamide **2.44** resulted in the slow degradation of the starting materials and no product formation was observed, therefore a more detailed investigation into possible additives was needed.

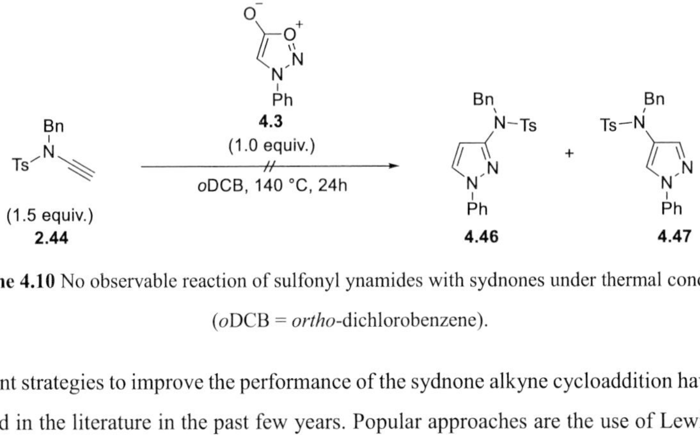

Scheme 4.10 No observable reaction of sulfonyl ynamides with sydnones under thermal conditions (*o*DCB = *ortho*-dichlorobenzene).

Different strategies to improve the performance of the sydnone alkyne cycloaddition have been reported in the literature in the past few years. Popular approaches are the use of Lewis acids or the addition of certain copper catalysts. The Lewis acids coordinate to the sydnones and typically favor the formation of the 1,3-substituted pyrazoles and the use of catalysts such as $Cu(OAc)_2 \cdot H_2O$ promote the formation of reactive copper(I) acetylides, leading to regioselective formation of 1,4-substituted pyrazoles.[22,23]

Catalyst screen

Based on these reports a concise screening of readily available copper catalysts was undertaken and their effects on the cycloaddition studied. Although terminal sulfonyl ynamides are known to be water-sensitive, this usually does not pose a significant problem since the desired reaction vastly outpaces any side-reactions. However, during the copper catalyst screen, the hydrolysis of the terminal sulfonyl ynamide **2.44** to its sulfonyl amide **4.47** was found to be a rather invasive problem.

Worse results were obtained when Cu(OTf)$_2$ (Table **4.1**, entry 1) was used, since full conversion of the ynamide to the amide was observed after a few minutes at room temperature instead of the expected Lewis acid-catalyzed 3-aminopyrazole product **4.46a**. When moving to Cu(OAc)$_2$ (Table **4.1**, entry 2 and 3), traces of product were detected by ESI-MS, and unreacted sydnone and large quantities of hydrolyzed ynamide **4.47** were isolated.

Table 4.1 Catalyst screening for the CuSAC reaction with sulfonyl ynamides.

Entry	[Cu]	Time	Temp [°C]	Result
1	Cu(OTf)$_2$[a]	15 min	RT	Only **4.48**
2	Cu(OAc)$_2$[b]	3-16h	100 or 140	Traces **4.46/4.47**, mostly **4.48**
3	Cu(OAc)$_2$·H$_2$O[b]	3-16h	100 or 140	Traces **4.46/4.47**, mostly **4.48**
4	CuI[c]	3d	60, 80 or 100	Traces **4.46/4.47**, mostly **4.48**
5	CuSO$_4$·5H$_2$O[d]	16h	80	**4.47/4.48** = 2:1

Conditions: a) Cu(OTf)$_2$ (1.0 equiv.), 0.2M in oDCB; b) Cu(OAc)$_2$ (1.0 equiv), 0.2M in oDCB, regio-isomeric ratio not determined; c) CuI (0.20 equiv.), 1,10-phenanthroline (0.20 equiv.), Et$_3$N (1.0 equiv.), 0.1M in DMF, regio-isomeric ratio not determined; d) CuSO$_4$·5H$_2$O (0.20 equiv.), Na-ascorbate (2.0 equiv.), 1,10-phenanthroline (0.20 equiv.), Et$_3$N (1.0 equiv.), 0.1M in tBuOH/H$_2$O = 1:1.

In an attempt to prevent hydrolysis the reaction was performed under anhydrous conditions including the addition of activated molecular sieves. When using CuI (Table **4.1**, entry 4) under these conditions, the amide formation was drastically reduced, but no significant product formation was observed, not even when using higher temperatures. Finally, when the "click chemistry"-like conditions reported by T$_{ARAN}$[12,22] (Table **4.1**, entry 5) were attempted, a 2:1

mixture of pyrazole and sulfonyl amide was generated. Isolation of the desired pyrazole product could be achieved by recrystallization from methanol.

Attempts to expand the scope

Due to the high sensitivity of terminal sulfonyl ynamides toward hydrolysis, the use of two internal ynamides **2.55** and **2.58** was tested. Unfortunately, although no side reaction towards the acetamide was observed, these substrates only yielded traces of the desired pyrazole products in the presence of anhydrous $Cu(OAc)_2$ – likely because it cannot form the required copper(I) acetylides. For the same reason these ynamides were found unsuitable for the $CuSO_4$-catalyzed reaction. Since the addition of Lewis acids, such as $Cu(OTf)_2$, mainly act by activating the sydnone this seemed the most promising lead. However, conditions using $Cu(OTf)_2$ also typically employ elevated temperatures and no product was isolated, as the internal sulfonyl ynamides might just not be thermally stable enough.

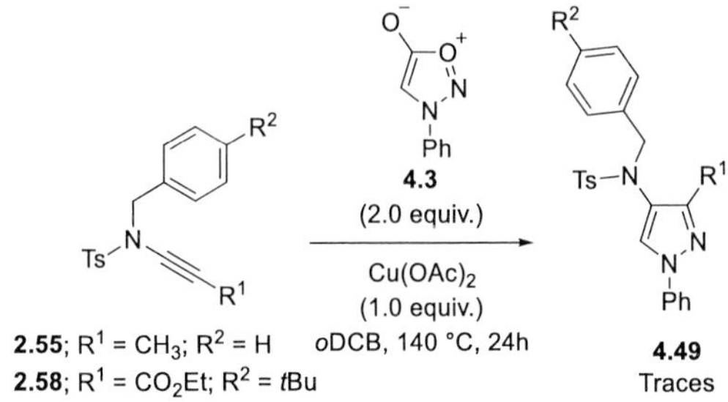

Scheme 4.11 Non-terminal sulfonyl ynamides do not undergo unwanted side-reactions to acetamides, but also do not undergo the CuSAC reaction in a facile manner. Only single regio-isomeric product shown; ratios were not determined due to insufficient quantities of product.

Instead of investigating and optimizing the CuSAC reaction with non-terminal sulfonyl ynamides it was decided to explore the scope for the cycloaddition reaction with the terminal ynamide. Several C-4 unsubstituted 3-arylsydnones were observed to deliver the corresponding 4-aminopyrazoles in a regioselective manner (see Scheme **4.12**), as confirmed by ^{1}H- and ^{13}C-NMR as well as with X-ray crystallographic analysis (see Figure **4.2**). However, all attempts to use C-4 substituted sydnones failed using the $CuSO_4$ protocol. It appears that the CuSAC reaction with ynamides is limited to C-4 unsubstituted 3-arylsydnones.

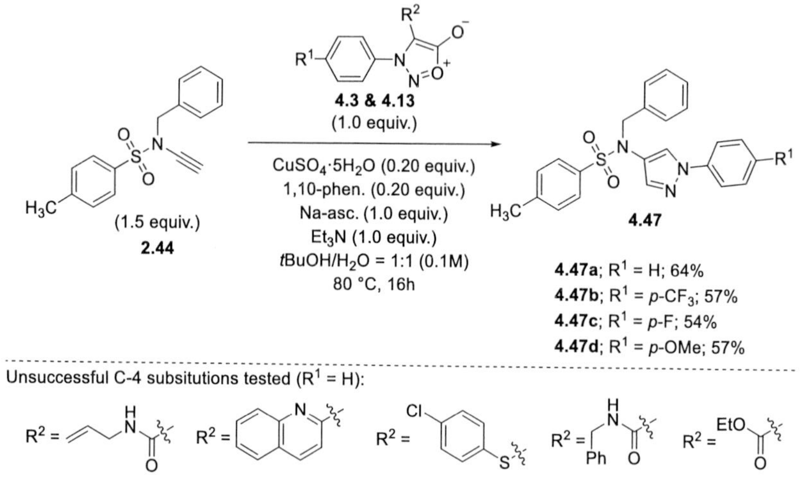

Scheme 4.12 Copper-catalyzed synthesis of 4-aminopyrazoles using C-4 unsubstituted sydnones.

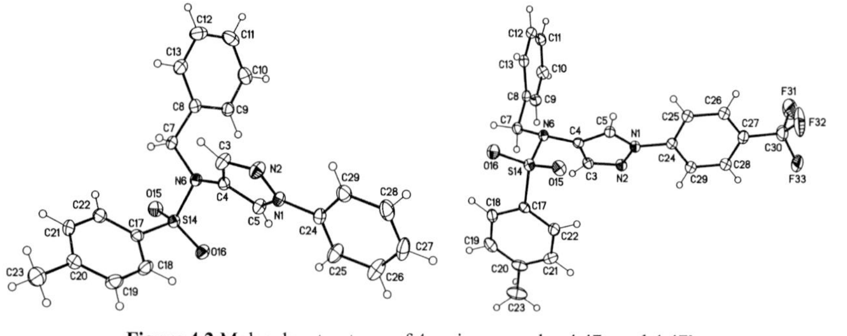

Figure 4.2 Molecular structures of 4-aminopyrazoles **4.47a** and **4.47b**.

Deprotection of the 4-aminopyrazole

In order to increase the synthetic versatility for these medically intriguing structures, the deprotection of the tosyl and benzyl groups were investigated. It was found that the tosyl group could be removed in moderate yield using potassium diphenylphosphide,[55] in a similar manner as the 4-aminoquinolines in Chapter 3 could be deprotected.

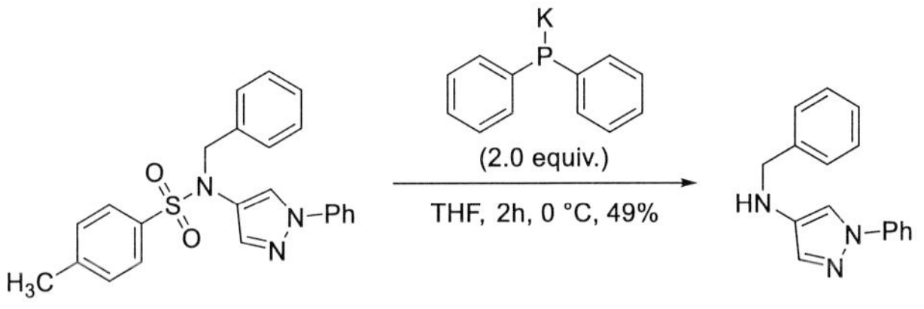

Scheme 4.13 Deprotection of the *N*-tosyl using potassium diphenylphosphide in moderate yield.

The benzyl group was found to be surprisingly resistant to hydrogenation. Even when a pressure of 10 bar hydrogen was applied, no conversion to secondary amine **4.50** was observed. Attempts to remove the benzyl group using oxidative conditions with bromo radicals, also provided unreacted starting material.[56]

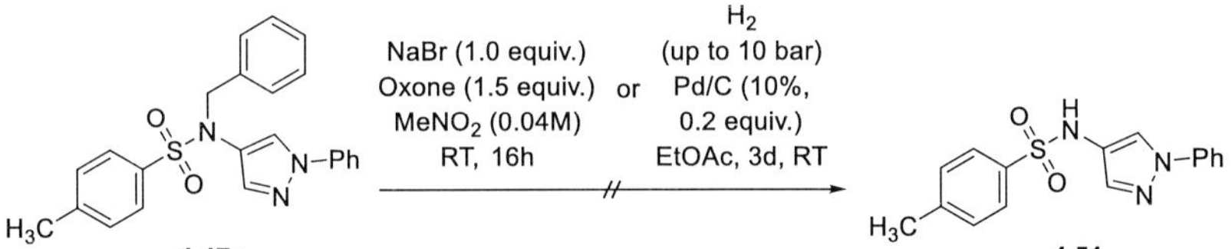

Scheme 4.14 Unsuccessful deprotection of the *N*-benzyl using oxidative debenzylation via *in situ* generated radicals or a standard hydrogenation approach.

Since the debenzylation of *N*-benzyl groups is a quite common procedure, several different protocols have been reported in literature.[57] These preliminary results just indicate that suitable conditions have not yet been found. Alternatively a more easily removable group such as a *para*-methoxy substituted benzyl group could be used instead.

4.4 Syntheses of aminopyrazoles via strained cyclic ynamides

In pursuit of reaction conditions that could tolerate 5-substituted sydnones and would thus allow access to a much wider scope of pyrazoles, the option to perform the reaction completely copper-free was investigated. The strain-promoted alkyne azide cycloaddition (SPAAC), – related to the more famous CuAAC[58] – is well known for its efficacy despite its lack of copper. The SPAAC reaction was initially discovered by WITTIG and KREBS in 1961[59] and then popularized in the early 2000s by the BERTOZZI group[60] when they showed that it would allow the reaction to proceed in a bio-orthogonal manner (see Scheme **4.15**).

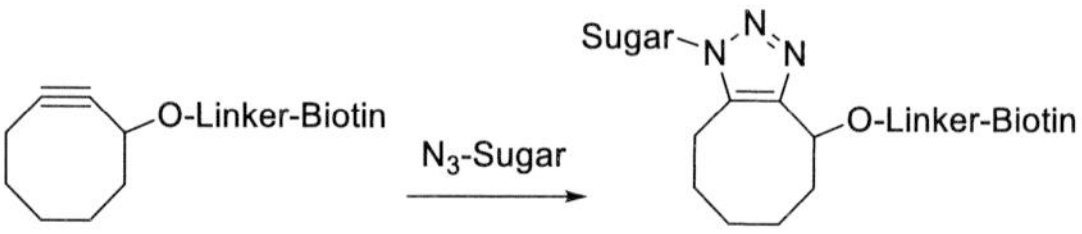

4.52   **4.53**
                                             Bertozzi, 2004

Scheme 4.15 Early reports of the use of SPAAC reactions to label biological samples.

Since the SPAAC reaction does not require use of cytotoxic copper it has gotten considerable attention in the last decade.[13,14,61-63] Although typically the SPAAC is performed with azides, these can also be replaced by sydnones to create pyrazoles in a bio-orthogonal manner.[64]

As discussed before, to date no reactions have been reported between ynamides and sydnones, nor has literature been published detailing cyclic ynamides reacting with sydnones. In fact, only two prior examples of cyclic strained ynamides have been reported in literature. The first synthesis encompasses the preparation of 3-azacyclohexyne **4.56** and was reported in 1988 by WENTRUP et al.[65] after using flash vacuum pyrolysis – brief and intense heating in absence of oxygen, followed by rapid cooling – on compound **4.54**.

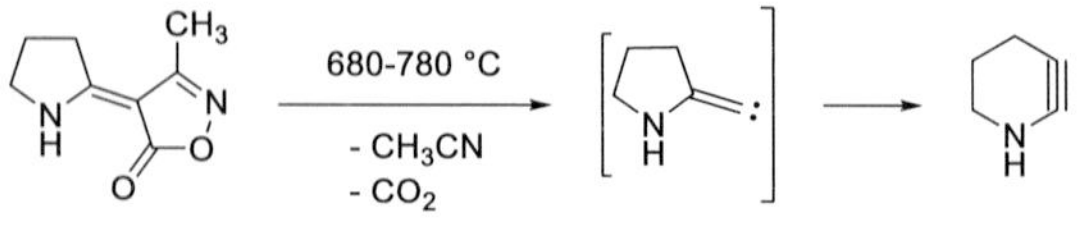

4.54   **4.55**   **4.56**
                                             Wentrup, 1988

Scheme 4.16 Synthesis of 3-azacyclohexyne via flash vacuum pyrolysis.

They found 3-azacyclohexyne **4.56** was unstable above −150 °C and therefore synthetically useless, but they were able to detect the presence of the triple bond using IR. However, recently

DANHEISER and coworkers managed to prepare a suitable *N*-tosyl-azacyclohexyne precursor **4.60** and trap the *in situ* formed azacyclohexyne.[66] Their precursor is synthesized using methods typically employed for the preparation of aryne species[67-72] and they suggest that the *N*-tosyl-azacyclohexyne might actually exhibit significant zwitterionic character. Although several heterocyclic arynes have been reported,[73-75] these are the only two examples where the nitrogen atom is located directly next to the strained alkyne bond.

Using DANHEISER's procedure the synthesis of strained cyclic ynamides was investigated and *N*-tosyl-azacyclohexyne precursor **4.60** was prepared in similar yield.

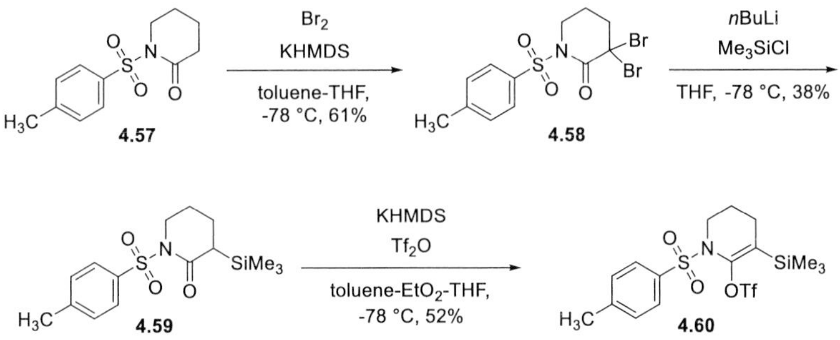

Scheme 4.17 Synthesis of the *N*-tosyl-azacyclohexyne precursor **4.60** following DANHEISER's procedure.

From precursor **4.60** the strained cyclic ynamide can be obtained *in situ* by addition of caesium fluoride and directly used in the desired cycloaddition. The strain-promoted sydnone ynamide cycloaddition tolerated a wide range of substitutions on the C-4 position of the sydnone (see Scheme **4.18**). After a small optimization study it was found that the reaction proceeds smoothest with the sydnone in slight excess. In reactions where the strain-activated ynamide was in excess, more complex reaction mixtures were observed, probably due to side-reactions. Although the reaction proceeds smoothly at room temperature, consuming all ynamide in a few hours, no kinetic studies to determine the reaction constants has been undertaken yet. Initial observations regarding the regioselective outcome of this cycloaddition suggest that C-4 unsubstituted sydnones (resulting in pyrazoles **4.62** and **4.63**) favor the 4,3-product **4.61a**. The reactions with aromatically C-4 substituted sydnones gave no clear preference in product ratio. Interestingly, from these preliminary results it seems that C-4 amide substituted sydnones tend to produce the 3,4-regio-isomer **4.61b** preferentially. However, due to the rather small number of experiments performed it is hard to truly conclude this. Additional experiments, molecular modelling and DFT calculations could help understand this behavior and are currently under investigation.

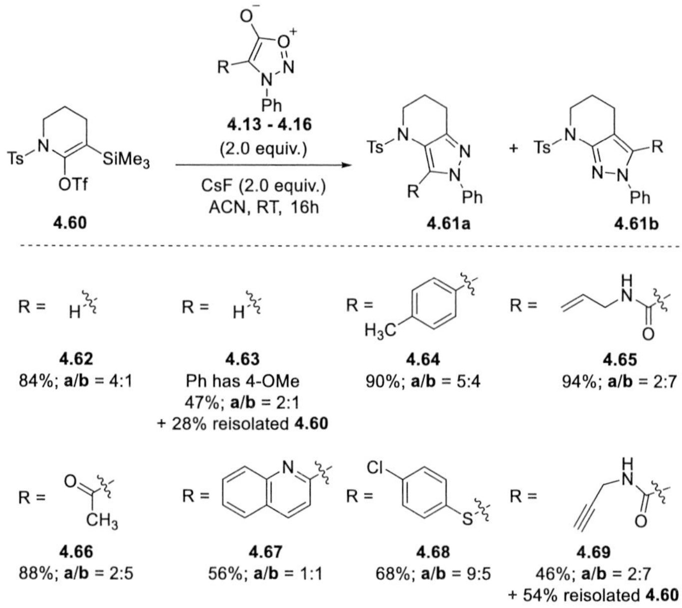

Scheme 4.18 *In situ* formation of *N*-tosyl-azacyclohexyne and subsequent pyrazole formation.

Azacyclooctynes

Most SPAAC procedures typically employ 8-membered rings, since smaller rings are typically unstable. These 8-membered rings are often further activated by addition of adjacent fluorine atoms and benzene rings in order to speed up the cycloaddition. Alternatively heteroatoms have been incorporated into the ring to increase the ring strain or electronic issues, especially sulfur[76-78] seems quite popular, although nitrogen[79,80] and oxygen have been investigated as well (Figure **4.3**). However, no reports of 8-membered strained ynamides – where the nitrogen atom is directly adjacent to the alkyne – have been published. As mentioned before: only two prior examples of cyclic ynamides have been reported at all, both 6-membered.

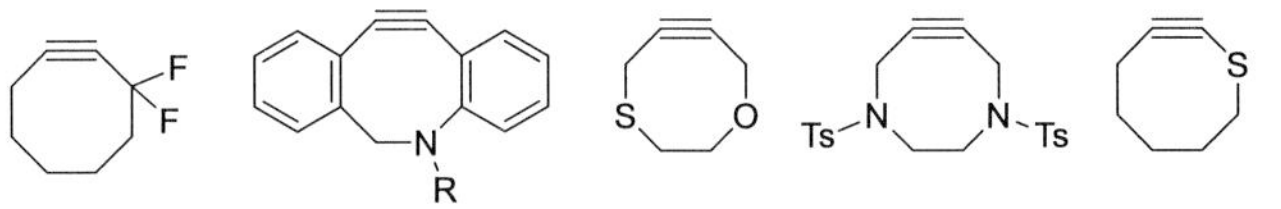

4.70 DIFO   4.71 DIBAC   4.72   4.73   4.74

Figure 4.3 Two well-studied and several heterocyclic cyclooctynes.

Cyclooctynes are accessible either by bromination of a double bond **4.75** and subsequent elimination, or by formation of a vinyl triflate **4.79** from ketones and subsequent elimination (Scheme **4.19**). Especially the latter conditions are often used to prepare cyclooctynes.[81]

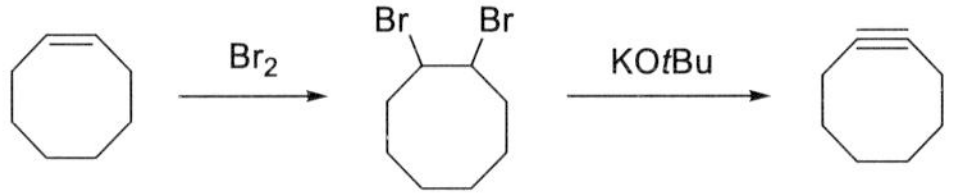

4.75   4.76   4.77

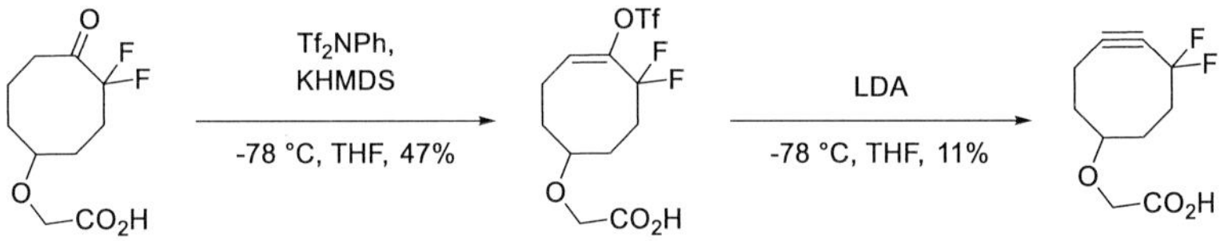

4.78   4.79   4.80 Bertozzi, 2006

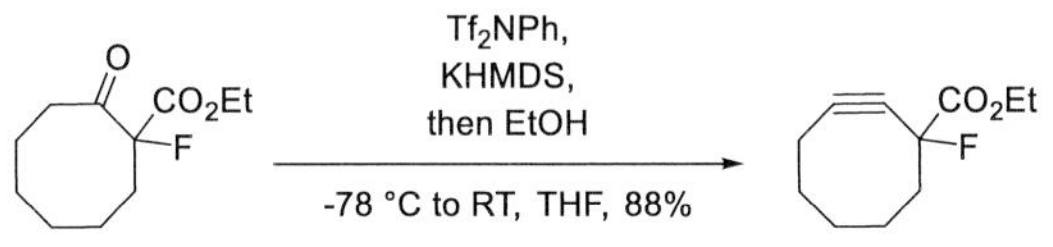

4.81   4.82 Schultz, 2010

Scheme 4.19 Different routes to cyclooctynes.

In literature the vinyl triflate is sometimes quenched with LDA[81,82] or *in situ* eliminated by stirring in methanol or ethanol while warming to room temperature.[83,84] Initial experiments on tosylated azocan-2-one **4.84** that involved quenching with LDA, did not result in the isolation of any product. When, however, a reaction control of the cooled reaction mixture was done with ESI-MS, where a small sample is first diluted in methanol, the cyclic ynamide product peaks could be identified – albeit only as $[2M+K]^+$. To date, isolation of the cyclic ynamide proved a challenge. Since its reactivity may limit its lifetime considerably it was decided to attempt to trap the 3-azacyclooctyne *in situ* with a sydnone in order to prove its existence. By addition of 2.5 equivalents of phenylsydnone **4.3** to the reaction mixture the expected pyrazole **4.87** product could be isolated in 40% yield as a mixture of regio-isomers. Investigations into the isolation of the 3-azacyclooctyne are currently ongoing.

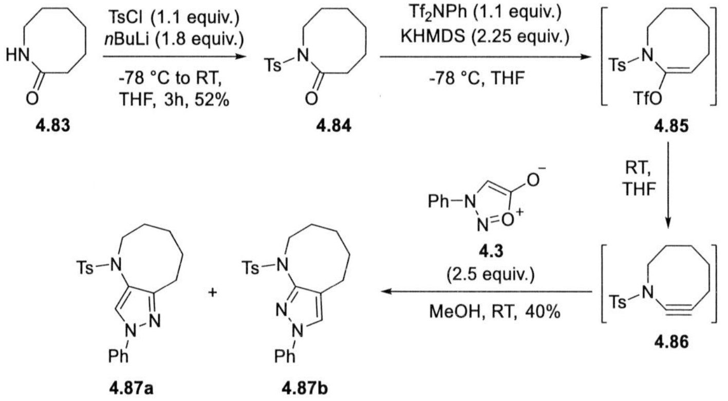

Scheme 4.20 Trapping of 3-azacyclooctyne as pyrazoles.

4.5 Conclusion and outlook

For the first time the scope of sydnone-alkyne cycloaddition has been expanded beyond the typical electron-deficient alkynes to the more electron-rich sulfonyl ynamides. In doing so a novel access to 3- and 4-aminopyrazoles has been described. Using terminal sulfonyl ynamides, C-4 unsubstituted sydnones, copper sulfate, 1,10-phenanthroline and sodium ascorbate, 4-aminopyrazoles are accessible in reasonable yields. Employing other copper additives – especially Cu(OTf)$_2$ – resulted in the facile conversion of the ynamide into an amide. Internal ynamides were reluctant to react at all. X-ray crystallography confirmed the regiochemistry of the obtained pyrazoles. Using two equivalents of potassium diphenylphosphide the tosyl could be removed in moderate yield. Although initial attempts to remove the benzyl group failed, a large array of possible deprotection procedures remain to be tested.

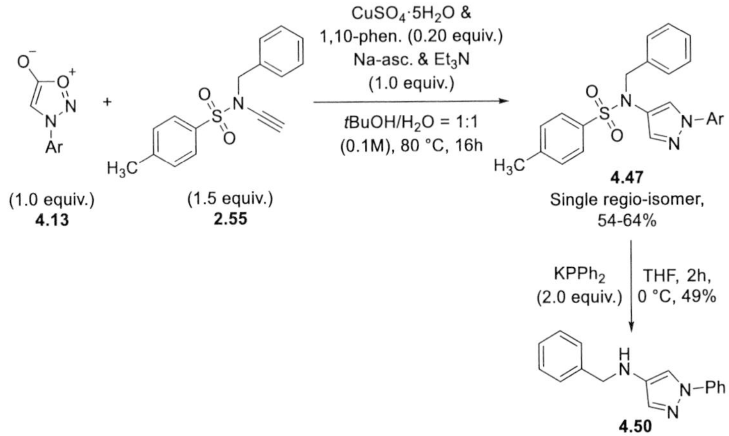

Scheme 4.21 Copper-catalyzed cycloaddition reaction between terminal sulfonyl ynamides and C-4 unsubstituted sydnones yield 1,4-pyrazoles, that could be detosylated in moderate yield.

A copper-free alternative using *in situ* prepared strained cyclic ynamides was found to tolerate a much wider array of sydnones. Several diversely C-4 substituted sydnones underwent the cycloaddition smoothly. The pyrazole products were always obtained as a mixture of regio-isomers and preliminary observations suggest that C-4 unsubstituted sydnones favor formation of the 4,3-products **4.61b**, aromatically C-4 substituted sydnones had no clear preference and C-4 amide substituted sydnones tend to produce large quantities of the 3,4-regio-isomer **4.61a**.

However, the rather small number of experiments performed can hardly confirm or deny any of these observations to be generally applicable rules. Additional experiments, molecular modelling and DFT calculations need to be undertaken in order to understand this behavior.

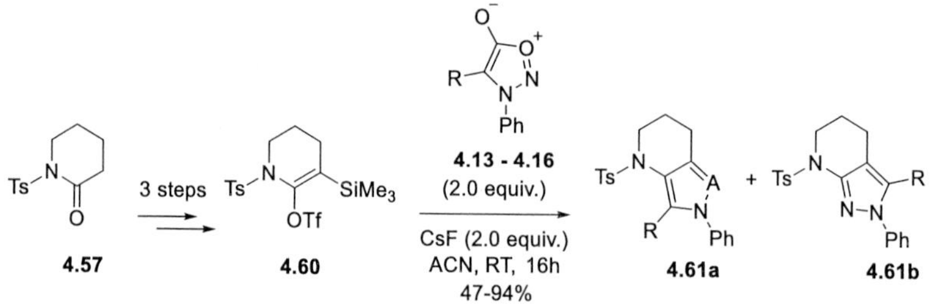

Scheme 4.22 *In situ* prepared strained cyclic ynamides produce a mixture of regio-isomers, but tolerate C-4 substituted sydnones.

Since most strain-promoted cycloadditions employ 8-membered instead of 6-membered rings, initial attempts to synthesize an 8-memberd ynamide were undertaken. Although cyclooctynes that contain heteroatoms in the ring are not unheard of, no reports of 3-azacyclooctynes were found. Likely because of issues with stability, as the nitrogen atom is believed to activate the triple bond considerably. And indeed, the isolation of the 3-azacyclooctyne has proven a challenge so far, but its existence has been confirmed by direct trapping with a sydnone. Since these experiments are merely a proof of concept, several additional experiments can be planned to explore the scope and reactivity of the 3-azacyclooctynes – cycloadditions with azides, oxime chlorides or ethyl diazoacetate and even Diels-Alder reactions. Attempts to isolate and characterize the 3-azacyclooctynes are currently ongoing.

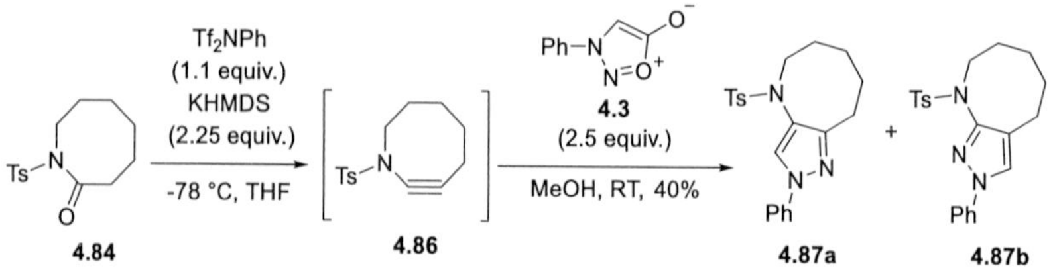

Scheme 4.23 Trapping of 3-azacyclooctyne as pyrazoles.

4.6 Experimental

General remarks

Nuclear magnetic resonance spectroscopy (NMR): ^{1}H-NMR spectra were recorded using the following devices: *Bruker* Avance 300 (300 MHz), *Bruker* Avance 400 (400 MHz) or a *Bruker* Avance DRX 500 (500 MHz); ^{13}C-NMR spectra were recorded using the following devices: *Bruker* Avance 300 (75 MHz), *Bruker* Avance 400 (100 MHz) or a *Bruker* Avance DRX 500 (125 MHz). All measurements were carried out at room temperature. The following solvents from *Eurisotop* were used: chloroform-d_1, methanol-d_4, acetone-d_6, DMSO-d_6. Chemical shifts δ were expressed in parts per million (ppm) and referenced to chloroform (^{1}H: δ = 7.26 ppm, ^{13}C: δ = 77.00 ppm), methanol (^{1}H: δ = 3.31 ppm, ^{13}C: δ = 49.00 ppm), acetone (^{1}H: δ = 2.05 ppm, ^{13}C: δ = 30.83 ppm) or DMSO (^{1}H: δ = 2.50 ppm, ^{13}C: δ = 39.43 ppm).[85] The signal structure is described as follows: s = singlet, d = doublet, t = triplet, q = quartet, quin = quintet, bs = broad singlet, m = multiplet, dt = doublet of triplets, dd = doublet of doublets, td = triplet of doublets, ddd = doublet of doublet of doublets, tt = triplet of triplets, dq = doublet of quintets, ddt = doublet of doublet of triplets. The spectra were analyzed according to first order and all coupling constants are absolute values and expressed in Hertz (Hz). The multiplicities of the signals of ^{13}C-NMR spectra were determined using characteristic chemical shifts and DEPT (Distortionless Enhancement by Polarization Transfer) and are described as follows: "+" = primary or tertiary (positive DEPT-135 signal), "–" = secondary (negative DEPT-135 signal), "C_q" = quarternary carbon atoms (no DEPT signal).

Infrared spectroscopy (IR): IR spectra were recorded on a *Bruker* IFS 88 using ATR (Attenuated Total Reflection). The intensities of the peaks are given as follows: vs = very strong (0 - 10% transmission), s = strong (11 - 40% transmission), m = medium (41 - 70% transmission), w = weak (71 - 90% transmission), vw = very weak (91 - 100% transmission).

Mass spectrometry (EI-MS, ESI-MS, FAB-MS): Mass spectra were recorded on a *Finnigan* MAT 95 using EI-MS (Electron ionization mass spectrometry) at 70 eV or FAB-MS (Fast Atom Bombardment Mass Spectroscopy) with 3-NBA as matrix. In special cases an *Agilent Technologies* 6230 TOF-LC/MS was used to record ESI-TOF-MS spectra in the positive mode. The molecular fragments are stated as the ratio of mass over charge m/z. The intensities of the

signals are given in percent relative to the intensity of the base signal (100%). The molecular ion is abbreviated $[M]^+$ for EI-MS and ESI-MS, the protonated molecular ion is abbreviated $[M+H]^+$ for FAB-MS.

Thin layer chromatography (TLC): Analytical thin layer chromatography was carried out using silica coated aluminium plates (silica 60, F_{254}, layer thickness: 0.25 mm) with fluorescence indicator by *Merck*. Detection proceeded under UV light at $\lambda = 254$ nm. For development phosphomolybdic acid solution (5% phosphomolybdic acid in ethanol, dip solution); potassium permanganate solution (1.00 g potassium permanganate, 2.00 g acetic acid, 5.00 g sodium bicarbonate in 100 mL water, dip solution) was used followed by heating in a hot air stream. Preparative thin layer chromatography was carried out using either the silica coated aluminium plates (silica 60, F_{254}, layer thickness: 0.25 mm) or the silica coated PSC glass plates (silica 60, F_{254}, layer thickness: 2.0 mm) by *Merck*.

Analytical balance: Used device: *Sartorius* Basic, model LA310S, range from 0.1 mg to 310.0 g.

Microwave: Reactions heated using microwave irradiation were carried out in a single mode *CEM* Discover LabMate microwave operated with *CEM*'s Synergy software. This instrument works with a constantly focused power source (0 - 300W). Irradiation can be adjusted via power- or temperature control. The temperature was monitored with an infrared sensor.

Solvents and reagents: Solvents of technical quality were distilled prior to use. Solvents of p.a. (*per analysi*) quality were commercially purchased (*Acros, Fisher Scientific, Sigma Aldrich*) and used without further purification. Absolute solvents were either commercially purchased as absolute solvents stored over molecular sieves under argon atmosphere or freshly prepared by distillation of p.a. quality over a drying agent (dichloromethane from calcium hydride, THF from sodium using benzophenone as indicator, and diethyl ether from sodium) and stored under argon atmosphere. Reagents were commercially purchased (*ABCR, Acros, Alfa Aesar, Fluka, J&K, Sigma Aldrich, TCI, VWR*) and used without further purification if not stated otherwise.

Reactions: For reactions with air- or moisture-sensitive reagents, the glassware with a PTFE coated magnetic stir bar was heated under high vacuum with a heat gun, filled with argon and closed with a rubber septum. All solution phase reactions were performed using the typical

Schlenk procedures with argon as inert gas. Liquids were transferred with V2A-steel cannulas. Solids were used as powders if not indicated otherwise. Reactions at low temperatures were cooled in flat Dewar flasks from *Isotherm*. The following cooling mixtures were used: 0 °C (Ice/Water), 0 to −10 °C (Ice/Water/NaCl), −10 to −78 °C (acetone/dry ice *or* isopropanol/dry ice). In general, solvents were removed at preferably low temperatures (40 °C) under reduced pressure. If not stated otherwise, solutions of ammonium chloride, sodium chloride and sodium hydrogen carbonate are aqueous, saturated solutions. Reaction progress of liquid phase reaction was checked with thin-layer chromatography (TLC). Crude products were purified according to literature procedures by preparative TLC or flash column chromatography using Silica gel 60 (0.063×0.200 mm, 70-230 mesh ASTM) (*Merck*), Geduran® Silica gel 60 (0.040×0.063 mm, 230-400 mesh ASTM) (*Merck*) or Celite® (*Fluka*) and sea sand (calcined, purified with hydrochloric acid, *Riedel-de Haën*) as stationary phase. Eluents (mobile phase) were p.a. and volumetrically measured.

General remark: Synthesis and characterization of the sydnones used in this chapter have been performed by Júlia Comas-Barceló, from the University of Sheffield in the UK.

General procedure GP4.1: 4-aminopyrazole synthesis using copper sulfate

A 10 mL closed vial was charged with the respective sydnone (1.00 equiv., 0.200 mmol), *N*-benzyl-*N*-ethynyl-4-methylbenzenesulfonamide (86.0 mg, 1.50 equiv. 0.300 mmol), triethylamine (28 μl, 1.00 equiv., 0.200 mmol), sodium ascorbate (79.0 mg, 2.00 equiv., 0.400 mmol), 1,10-phenanthroline (7.2 mg, 0.20 equiv., 0.040 mmol) and copper(II) sulfate pentahydrate (10.0 mg, 0.200 equiv., 0.040 mmol) in a mixture of water (1.0 mL) and *t*BuOH (1.0 mL) to give an orange solution that was stirred at 80 °C for 16h. After TLC (*c*Hex/EtOAc mixtures) showed full conversion of the sydnone the reaction was quenched with a 0.05M aqueous solution of EDTA and extracted with CH_2Cl_2 (3 × 10 mL). The combined organic layers were washed with water (1 × 10 mL), brine (1 × 10 mL), dried over anhydrous magnesium sulfate and concentrated under reduced pressure. The crude products were purified by recrystallization from MeOH or by flash column chromatography.

General procedure GP4.2: 4-aminopyrazole synthesis using strained ynamides

A 10 mL vial was charged with the respective sydnone (0.100 mmol, 2.00 equiv.) and 1-tosyl-3-(trimethylsilyl)-1,4,5,6-tetrahydropyridin-2-yl trifluoromethanesulfonate (22.9 mg, 0.050 mmol, 1.00 equiv.) in MeCN (0.05M) to give a colorless solution. After stirring for 5 minutes Caesium fluoride (15.2 mg, 0.100 mmol, 2.00 equiv.) was added rapidly in one portion. The reaction mixture was stirred at RT for 16h and the conversion was checked with ESI-MS. The crude reaction mixture was purified employing preparative TLC or flash column chromatography (cHex/EtOAc mixtures).

N-benzyl-4-methyl-*N*-(1-phenyl-1H-pyrazol-4-yl)benzenesulfonamide (4.47a)

This compound was prepared according to general procedure **GP4.1** using 3-phenyl-3H-1,2,3-oxadiazol-1-ium-5-olate (32.4 mg, 0.20 mmol) and was isolated after recrystallization from MeOH as colorless needles (52.0 mg, 0.129 mmol, 64%). **^{1}H-NMR** (400 MHz, CDCl$_3$): δ = 7.76 (s, 1H), 7.58 - 7.63 (m, 2H), 7.51 - 7.55 (m, 2H), 7.35 - 7.41 (m, 2H), 7.21 - 7.32 (m, 9H), 4.66 (s, 2H), 2.42 (s, 3H) ppm. − **^{13}C-NMR** (101 MHz, CDCl$_3$): δ = 143.8 (C_q), 139.6 (C_q), 137.2 (s), 135.7 (C_q), 135.0 (C_q), 129.6 (+, 2 × CH_{Ar}), 129.3 (+, 2 × CH_{Ar}), 128.5 (+, 2 × CH_{Ar}), 128.2 (+, 2 × CH_{Ar}), 127.8 (+, CH_{Ar}), 127.5 (+, 2 × CH_{Ar}), 126.7 (+, CH_{Ar}), 124.5 (+, NCH), 123.6 (C_q), 118.7 (+, NCH), 54.7 (−, CH_2), 21.5 (+, CH_3) ppm. − **IR** (ATR): ṽ = 1597 (w), 1495 (m), 1453 (w), 1391 (w), 1364 (w), 1343 (m), 1200 (w), 1089 (m), 1050 (m), 978 (w) cm^{-1}. − **MS** (EI), *m/z* (%): 403 (44) [M]$^+$, 316 (11), 248 (100). − **HRMS** (EI, C$_{23}$H$_{21}$O$_2$N$_3$^{32}S): calc. 403.1350; found 403.1349.

N-benzyl-4-methyl-*N*-(1-(4-(trifluoromethyl)phenyl)-1H-pyrazol-4-yl)benzene-sulfonamide (4.47b)

This compound was prepared according to general procedure GP4.1 using 3-(4-(trifluoromethyl)phenyl)-3H-1,2,3-oxadiazol-1-ium-5-olate (46.0 mg, 0.200 mmol) and was isolated after flash column chromatography (cHex/EtOAc = 9:1) as colorless crystals (54.0 mg, 0.115 mmol, 57%). **^{1}H-NMR** (300 MHz, CDCl$_3$): δ = 7.84 (s, 1H), 7.67 (s, 4H), 7.62 (d, *J* = 8.1 Hz, 3H), 7.27 - 7.38 (m, 7H), 4.67 (s, 2H), 2.44 (s, 3H) ppm. −

^{13}**C-NMR** (101 MHz, CDC$_3$): δ = 144.1 (C_q), 142.0 (C_q), 137.9 (+, CH_{Ar}), 135.5 (C_q), 135.0 (C_q), 129.7 (+, 2 × CH_{Ar}), 128.6 (+, CH_{Ar}), 128.6 (+, CH_{Ar}), 128.2 (+, 2 × CH_{Ar}), 128.0 (+, 2 × CH_{Ar}), 127.9 (+, CH_{Ar}), 127.7 (+, CH_{Ar}), 127.5 (+, 2 × CH_{Ar}), 126.7 (+, CH_{Ar}), 126.7 (+, q, CH_{Ar}, J = 4.2 Hz), 124.5 (C_q), 124.3 (+, CH_{Ar}), 123.8 (C_q, q, CF_3, J = 272.2 Hz), 118.4 (+, NCH), 54.5 (−, CH_2), 21.6 (+, CH_3) ppm. − ^{19}**F-NMR** (377 MHz, CDCl$_3$): δ −66.64 (s, 3F) ppm. − **IR** (ATR): ṽ = 3333 (vw), 1702 (vw), 1614 (w), 1523 (w), 1494 (w), 1440 (w), 1353 (w), 1321 (m), 1163 (m), 1110 (m), 1069 (w), 1020 (w), 943 (w) cm^{-1}. − **MS** (EI), m/z (%): 471 (11) [M]$^+$, 315 (55), 238 (100). − **HRMS** (EI, C$_{24}$H$_{20}$N$_3$O$_2$F$_3$^{32}S): calc. 471.1223; found 471.1222.

N-benzyl-*N*-(1-(4-fluorophenyl)-1H-pyrazol-4-yl)-4-methylbenzenesulfonamide (4.47c)

This compound was prepared according to general procedure GP4.1 using 3-(4-fluorophenyl)-3H-1,2,3-oxadiazol-1-ium-5-olate (36.0 mg, 0.200 mmol) and was isolated after recrystallization from MeOH as colorless crystals (45.0 mg, 0.107 mmol, 53%). ^{1}**H-NMR** (400 MHz, CDCl$_3$): δ = 7.68 (s, 1H), 7.59 (d, J = 8.3 Hz, 2H), 7.44 - 7.51 (m, 2H), 7.28 (s, 1H), 7.19 - 7.25 (m, 7H), 7.03 - 7.10 (m, 2H), 4.63 (s, 2H), 2.41 (s, 3H) ppm. − ^{13}**C-NMR** (101 MHz, CDCl$_3$): δ = 161.2 (d, C_{Ar}-F, J = 246 Hz), 143.9 (C_q), 137.1 (+, CH_{Ar}), 135.7 (C_q), 135.0 (C_q), 129.7 (+, CH_{Ar}), 128.5 (+, CH_{Ar}), 128.3 (+, CH_{Ar}), 127.9 (+, CH_{Ar}), 127.5 (+, CH_{Ar}), 124.8 (+, CH_{Ar}), 123.7 (C_q), 120.6 (+, CH_{Ar}), 120.5 (+, CH_{Ar}), 116.3 (+, CH_{Ar}), 116.1 (+, CH_{Ar}), 54.7 (−, CH_2), 21.6 (+, CH_3) ppm. − **IR** (ATR): ṽ = 3110 (vw), 1514 (m), 1454 (w), 1406 (w), 1364 (w), 1233 (w), 1162 (w), 1090 (m), 1027 (m) cm^{-1}. − **MS** (EI), m/z (%): 422 (32) [M+H]$^+$, 307 (15), 266 (22), 154 (100). − **HRMS** (EI, C$_{23}$H$_{20}$N$_3$O$_2$F^{32}S): calc. 421.1255; found 421.1256.

N-benzyl-*N*-(1-(4-methoxyphenyl)-1H-pyrazol-4-yl)-4-methylbenzenesulfonamide (4.47d)

This compound was prepared according to general procedure GP4.1 using 3-(4-methoxybenzyl)-3H-1,2,3-oxadiazol-1-ium-5-olate (38.0 mg, 0.200 mmol) and was isolated after recrystallization from MeOH as colorless crystals (49.0 mg, 0.113 mmol, 57%). **¹H-NMR** (400 MHz, CDCl$_3$): δ = 7.62 (d, J = 8.3 Hz, 2H), 7.45 (d, J = 8.8 Hz, 2H), 7.21 - 7.32 (m, 9H), 6.92 (m, 2H), 4.65 (s, 2H), 3.82 (s, 3H), 2.44 (s, 3H) ppm. − **¹³C-NMR** (101 MHz, CDCl$_3$): δ = 158.4 (C_q), 143.8 (C_q), 135.8 (C_q), 135.1 (C_q), 129.7 (+, CH$_{Ar}$), 128.5 (+, CH$_{Ar}$), 128.3 (+, CH$_{Ar}$), 127.8 (+, CH$_{Ar}$), 127.6 (+, CH$_{Ar}$), 120.4 (+, CH$_{Ar}$), 114.5 (+, CH$_{Ar}$), 55.6 (+, CH$_3$), 54.8 (−, CH$_2$), 21.6 (+, CH$_3$) ppm. − **IR** (ATR): ṽ = 2922 (vw), 1748 (w), 1700 (w), 1595 (w), 1514 (w), 1494 (w), 1452 (w) cm^{-1}. − **MS** (FAB), *m/z* (%): 434 (100) [M+H]$^+$, 278 (60), 153 (97). − **HRMS** (FAB, C$_{24}$H$_{24}$N$_3$O$_3$^{32}S): calc. 434.1533; found 434.1535.

N-benzyl-1-phenyl-1H-pyrazol-4-amine (4.50)

A 10 mL closed vial was charged with *N*-benzyl-4-methyl-*N*-(1-phenyl-1H-pyrazol-4-yl)benzenesulfonamide (30.0 mg, 0.074 mmol, 1.00 equiv.) in dry THF (1.0 mL) and stirred at 0 °C for 5 min before potassium diphenylphosphide (0.30 mL, 0.149 mmol, 2.00 equiv.) was added slowly. The reaction mixture was stirred at 0 °C for 2h and subsequently diluted with 1M HCl and stirred for 5 min. The suspension was quenched with sat. aq. NaHCO$_3$ and the aqueous layer extracted with CH$_2$Cl$_2$ (3 × 20 mL). The combined organic layers were dried over MgSO$_4$, filtered and concentrated under reduced pressure. Preparative TLC (*c*Hex/EtOAc/Et$_3$N = 90:5:1) yielded the pure product as white solid (9.0 mg, 0.036 mmol, 49%). **¹H-NMR** (300 MHz, CDCl$_3$): δ = 7.58 (d, J = 7.7 Hz, 2H), 7.27 - 7.45 (m, 9H), 7.22 (m, 1H), 4.25 (s, 2H) ppm. − **¹³C-NMR** (101 MHz, CDCl$_3$): δ = 140.3 (C_q), 131.9 (+, CH$_{Ar}$), 129.3 (+, CH$_{Ar}$), 128.7 (+, CH$_{Ar}$), 128.0 (+, 2 × CH$_{Ar}$), 127.6 (+, CH$_{Ar}$), 125.7 (+, CH$_{Ar}$), 118.2 (+, 2 × CH$_{Ar}$), 52.2 (−, CH$_2$) ppm. Two quaternary signals are missing due to low sensitivity. − **IR** (ATR): ṽ = 3307 (w), 3057 (vw), 2920 (w), 2848 (w), 1662 (w), 1594 (m), 1493 (m),

1393 (m), 1259 (w), 1170 (w), 1101 (w), 1073 (w), 1045 (w) cm^{-1}. − **MS** (EI), *m/z* (%): 249.2 (100) [M]$^+$, 158.1 (45), 104.1 (71). − **HRMS** (EI, C$_{16}$H$_{15}$N$_3$): calc. 249.1260; found 249.1259.

1-Tosylpiperidin-2-one (4.57)

A 250 mL round-bottomed flask was charged with piperidin-2-one (6.00 g, 60.5 mmol, 1.00 equiv.) in dry THF (100 mL) to give a yellow solution. The reaction mixture was cooled to −78 °C and stirred for 10 min before *n*-butyllithium (29.1 mL, 2.5M in hexanes, 72.6 mmol, 1.20 equiv.) was added dropwise over 15 min. The reaction mixture was stirred at −78 °C for 1h before a solution of 4-methylbenzenesulfonyl chloride (12.69 g, 66.6 mmol, 1.10 equiv.) in dry THF (50 mL) was added via dropping funnel over 10 min. The reaction mixture was allowed to slowly reach room temperature over 1.5h and subsequently transferred to a separatory funnel and diluted with 100 mL of sat. aq. NH$_4$Cl solution and 50 mL of water. The aqueous layer was separated and extracted with EtOAc (2 × 200 mL). The combined organic layers were washed with 200 mL of sat. aq. NaHCO$_3$ solution, dried over MgSO$_4$, filtered, and concentrated to give a white solid. Trituration with Et$_2$O (2 × 50 mL) afforded a white solid which was collected by removal of liquid (9.30 g, 36.7 mmol, 61%). **^{1}H-NMR** (400 MHz, CDCl$_3$): δ = 7.85 - 7.92 (m, 2H), 7.29 (m, 2H), 3.89 (t, *J* = 6.1 Hz, 2H), 2.40 (s, 3H), 2.37 - 2.40 (m, 2H), 1.84 - 1.93 (m, 2H), 1.71 - 1.80 (m, 2H) ppm. − **^{13}C-NMR** (101 MHz, CDCl$_3$): δ = 170.1 (C$_q$), 144.6 (C$_q$), 136.0 (C$_q$), 129.2 (+, 2 × CH$_{Ar}$), 128.5 (+, 2 × CH$_{Ar}$), 46.8 (−, CH$_2$), 34.0 (−, CH$_2$), 23.2 (−, CH$_2$), 21.5 (+, CH$_3$), 20.3 (−, CH$_2$) ppm. − **IR** (ATR): ṽ = 2951 (vw), 1681 (m), 1594 (w), 1445 (w), 1382 (w), 1346 (m), 1280 (w), 1260 (w), 1151 (m), 1117 (m), 1088 (m), 1040 (m) cm^{-1}. − **MS** (FAB), *m/z* (%): 254 (100) [M]$^+$, 154.9 (14). − **HRMS** (FAB, C$_{12}$H$_{16}$NO$_3$F$_3$^{32}S): calc. 254.0845; found 254. 0847.

3,3-Dibromo-1-tosylpiperidin-2-one (4.58)

A 250 mL three-neck round-bottomed flask was charged with 1-tosylpiperidin-2-one (3.00 g, 11.84 mmol, 1.00 equiv.) in dry THF (100 mL) and cooled to −78 °C. A solution of KHMDS (59.2 mL, 0.5M in toluene, 29.6 mmol, 2.50 equiv.) was added dropwise with the aid of a dropping funnel. Bromine (2.2 mL, 42.6 mmol, 3.60 equiv.) was added dropwise and the reaction was stirred for 25 min at $-78\,°C$. The yellow reaction mixture was transferred to a separatory funnel and washed with sat. aq. sodium thiosulfate solution. The aqueous layer was extracted with Et_2O (2×150 mL) and the combined organic layers were washed with sat. aq. NaCl (1×50 mL) and subsequently dried over $MgSO_4$, filtered and concentrated under reduced pressure. Trituration with Et_2O (6×50 mL) afforded a white solid (2.99 g, 7.27 mmol, 61%). ^{1}H-NMR (400 MHz, $CDCl_3$): $\delta = 7.88$ - 7.95 (m, 2H), 7.35 – 7.33 (m, 2H), 4.04 (t, $J = 6.2$ Hz, 2H), 2.87 - 2.95 (m, 2H), 2.44 (s, 3H), 2.11 - 2.19 (m, 2H) ppm. – ^{13}C-NMR (101 MHz, $CDCl_3$): $\delta = 162.9$ (C_q), 145.4 (C_q), 134.3 (C_q), 129.5 (+, $2 \times CH_{Ar}$), 129.0 (+, $2 \times CH_{Ar}$), 58.9 (C_q), 46.7 (−, CH_2), 45.6 (−, CH_2), 21.9 (−, CH_2), 21.7 (+, CH_3) ppm. – IR (ATR): $\tilde{v} = 2914$ (vw), 1699 (m), 1594 (w), 1483 (w), 1433 (w), 1360 (w), 1274 (w), 1163 (m), 1110 (m), 1085 (w) cm^{-1}. – MS (FAB), m/z (%): 410.0/411.9/413.9 (40/85/42) [M+H]$^+$, 154 (100). – HRMS (FAB, $C_{12}H_{14}NO_3{}^{79}Br^{81}Br^{32}S$): calc. 411.9035; found 411.9037.

1-Tosyl-3-(trimethylsilyl)piperidin-2-one (4.59)

A 250 mL 3-neck round-bottomed flask was charged with 3,3-dibromo-1-tosylpiperidin-2-one (2.95 g, 7.18 mmol, 1.00 equiv.) and chlorotrimethylsilane (1.82 mL, 14.4 mmol, 2.00 equiv.) in dry THF (120 mL) and the reaction mixture was cooled to $-78\,°C$. A solution of n-butyllithium (3.01 mL, 2.5M in hexanes, 7.53 mmol, 1.05 equiv.) was added dropwise over 5 min and the reaction mixture was stirred at $-78\,°C$ for 1h. A second equivalent of n-butyllithium (3.01 mL, 2.5M in hexanes, 7.53 mmol, 1.05 equiv.) was added dropwise over 5 min and the reaction mixture was stirred at $-78\,°C$ for another 2h. The reaction mixture was transferred to a separatory funnel and diluted with a solution of sat. aq. NH_4Cl. The aqueous layer was extracted with EtOAc (3×50 mL) and the combined organic layers were washed with sat. aq. NaCl, dried over $MgSO_4$, filtered and concentrated under reduced pressure. The crude oil was purified using flash chromatography using acetone-deactivated silica gel (e.g. silica gel is mixed with acetone (ca. 10 mL/g) and this slurry is used to pack the column. Before the crude is applied the column is flushed with two column volumes of cHex) to yield a white crystalline solid (886 mg, 2.72 mmol, 38%). ^{1}H-NMR (400 MHz, $CDCl_3$): $\delta = 7.88$ - 7.93 (m, 2H), 7.29 (d, $J = 7.8$ Hz, 2H), 3.77 - 3.97 (m, 2H), 2.41 (s, 3H), 2.08 - 2.14 (m, 1H), 1.83 - 2.00

(m, 2H), 1.62 - 1.78 (m, 2H), 0.04 (s, 9H). − ^{13}C-NMR (101 MHz, CDCl$_3$): δ = 172.7 (C_q), 144.3 (C_q), 136.7 (C_q), 129.1 (+, 2 × CH_{Ar}), 128.5 (+, 2 × CH_{Ar}), 47.6 (−, CH_2), 36.3 (+, CH), 23.7 (−, CH_2), 23.0 (−, CH_2), 21.6 (+, CH_3), -2.3 (+, 3 × CH_3) ppm. − IR (ATR): ṽ = 2952 (w), 1664 (w), 1596 (w), 1343 (w), 1276 (w), 1241 (w), cm^{-1}. − MS (FAB), m/z (%): 326.1 (100) [M]$^+$, 454.1 (26), 256.0 (35). − HRMS (FAB, C$_{15}$H$_{24}$NO$_3$^{32}S^{28}Si): calc. 326.1241; found 326.1242.

1-Tosyl-3-(trimethylsilyl)-1,4,5,6-tetrahydropyridin-2-yl trifluoromethanesulfonate (4.60)

A 250 mL three-neck round-bottomed flask was charged with Et$_2$O (40 mL) and cooled to −78 °C. A solution of KHMDS (5.44 mL, 0.5M in toluene, 2.72 mmol, 1.00 equiv.) was added rapidly. A solution of 1-tosyl-3-(trimethylsilyl)piperidin-2-one (886 mg, 2.72 mmol, 1.00 equiv.) in THF (20 mL) was added dropwise over 5 min and the reaction mixture was stirred for 1h at −78 °C. Trifluoromethanesulfonic anhydride (0.506 mL, 2.99 mmol, 1.10 equiv.) was added slowly over 1 min and after 5 min the reaction mixture turned to a colorless solution. After a total of 10 min the reaction mixture was transferred to a separatory funnel and diluted with a sat. aq. NaHCO$_3$ solution. The aqueous layer was extracted with Et$_2$O (2 × 150 mL). The combined organic layers were washed with sat. aq. NaCl (1 × 25 mL). The organic layer was dried over MgSO$_4$, filtered and concentrated under reduced pressure. The crude was purified using flash chromatography using acetone-deactivated silica gel (e.g. silica gel is mixed with acetone (ca. 10 mL/g) and this slurry is used to pack the column. Before the crude is applied the column is flushed with two column volumes of cHex, column ran isocratic cHex/EtOAc = 20:1) to yield a clear oil (648 mg, 1.42 mmol, 52%). ^{1}H-NMR (300 MHz, CDCl$_3$): δ = 7.71 - 7.73 (d, J = 8.7 Hz, 2H), 7.31 - 7.33 (d, J = 7.9 Hz, 2H), 3.58 (m, 2H), 2.44 (s, 3H), 1.71 - 1.78 (m, 2H), 1.17 - 1.33 (m, 2H), 0.21 (s, 9H) ppm. − ^{13}C-NMR (101 MHz, CDCl$_3$): δ = 144.9 (C_q), 141.2 (C_q), 135.2 (C_q), 129.8 (+, 2 × CH_{Ar}), 128.0 (+, 2 × CH_{Ar}), 126.1 (C_q), 120.0 (C_q), 116.8 (C_q), 47.8 (−, CH_2), 26.9 (−, CH_2), 24.8 (−, CH_2), 21.7 (+, CH_3), 20.1 (−, CH_2), -1.4 (+, 3 × CH_3) ppm. − IR (ATR): ṽ = 2956 (vw), 1627 (w), 1410 (m), 1367 (m), 1200 (m), 1172 (s), 1127 (s), 1049 (m) cm^{-1}. − MS (EI), m/z (%): 457 (100) [M]$^+$, 286 (72). − HRMS (EI, C$_{22}$H$_{16}$NO$_5$F$_3$^{32}S^{28}Si): calc. 457.0655; found 457.0655.

2-phenyl-4-tosyl-4,5,6,7-tetrahydro-2H-pyrazolo[4,3-b]pyridine (4.62a)

2-phenyl-7-tosyl-4,5,6,7-tetrahydro-2H-pyrazolo[3,4-b]pyridine (4.62b)

These compounds were prepared according to general procedure GP4.2 using 3-phenyl-3H-1,2,3-oxadiazol-1-ium-5-olate (16.0 mg, 0.10 mmol) and were isolated after preparative TLC (cHex/EtOAc = 4:1) as colorless crystals (15.0 mg, 0.042 mmol, 84%) Ratio by ^{1}H-NMR **A/B** = 4:1. **Isomer (B):** ^{1}**H-NMR** (400 MHz, CDCl$_3$): δ = 8.20 (s, 1H), 7.63 - 7.69 (m, 3H), 7.43 - 7.48 (m, 2H), 7.17 - 7.23 (m, 4H), 3.61 - 3.65 (m, 2H), 3.39 (t, J = 6.4 Hz, 1H), 2.86 (s, 1H), 2.67 (t, J = 6.4 Hz, 2H), 2.39 (s, 3H) ppm. – **Isomer (A):** ^{1}**H-NMR** (400 MHz, CDCl$_3$): δ = 7.95 - 7.99 (m, 2H), 7.58 - 7.63 (m, 2H), 7.54 (s, 1H), 7.41 (m, 2H), 7.27 - 7.25 (m, 3H), 3.82 - 3.87 (m, 2H), 2.53 - 2.59 (m, 2H), 2.38 (s, 3H), 1.94 (m, 2H) ppm. – ^{13}**C-NMR** (101 MHz, CDCl$_3$): δ = 147.5 (C_q), 143.6 (C_q), 140.1 (C_q), 135.9 (C_q), 129.8 (+, CH_{Ar}), 129.5 (+, CH_{Ar}), 129.3 (+, CH_{Ar}), 129.2 (+, CH_{Ar}), 128.2 (+, CH_{Ar}), 127.2 (+, CH_{Ar}), 125.4 (+, CH_{Ar}), 123.3 (+, CH_{Ar}), 118.5 (+, CH_{Ar}), 117.9 (+, CH_{Ar}), 108.0 (C_q), 46.9 (−, CH_2), 46.2 (−, CH_2), 29.7 (−, CH_2), 22.8 (−, CH_2), 21.6 (+, CH_3), 21.4 (−, CH_2), 18.9 (−, CH_2) ppm. – **IR** (ATR): ṽ = 2922 (vw), 2852 (vw), 1685 (w), 1596 (w), 1490 (m), 1459 (w), 1391 (w), 1349 (m), 1284 (m), 1222 (w), 1161 (m), 1088 (m) cm^{-1}. – **MS** (EI), m/z (%): 353.2 (100) [M]$^+$, 198.1 (89). – **HRMS** (EI, C$_{19}$H$_{19}$N$_3$O$_2$^{32}S): calc.353.1193; found 353.1193.

2-(4-methoxyphenyl)-4-tosyl-4,5,6,7-tetrahydro-2H-pyrazolo[4,3-b]pyridine (4.63a)

2-(4-methoxyphenyl)-7-tosyl-4,5,6,7-tetrahydro-2H-pyrazolo[3,4-b]pyridine (4.63b)

These compounds were prepared according to general procedure GP4.2 using 3-(4-methoxyphenyl)-3H-1,2,3-oxadiazol-1-ium-5-olate (36.0 mg, 0.188 mmol) and were isolated as a mixture after flash column chromatography (cHex/EtOAc = 9:1 → 1:1) as colorless solids (17.0 mg, 0.044 mmol, 47%) Ratio by ^{1}H-NMR **A/B** = 2:1. Another fraction yielded unreacted 1-tosyl-3-(trimethylsilyl)-1,4,5,6-tetrahydropyridin-2-yl trifluoromethanesulfonate (12.0 mg, 0.026 mmol, 28%) **Isomer (A):** ^{1}H-NMR (400 MHz, CDCl$_3$): δ = 7.95 (d, J = 8.3 Hz, 2H), 7.48 - 7.53 (m, 2H), 7.43 (s, 1H), 7.23 - 7.26 (m, 2H), 6.89 - 6.95 (m, 2H), 3.83 (s, 3H), 3.59 - 3.66 (m, 2H), 2.54 (t, J = 6.4 Hz, 2H), 2.38 (s, 3H), 1.92 (dd, J = 6.2, 4.7 Hz, 2H) ppm. − **Isomer (B):** ^{1}H-NMR (400 MHz, CDCl$_3$): δ = 8.10 (s, 1H), 7.64 (d, J = 8.3 Hz, 2H), 7.54 - 7.59 (m, 2H), 7.22 - 7.25 (m, 2H), 6.95 - 6.99 (m, 2H), 3.85 (s, 3H), 3.79 - 3.82 (m, 2H), 2.65 (t, J = 6.4 Hz, 2H), 2.39 (s, 3H), 1.63 - 1.72 (m, 2H) ppm. − ^{13}C-NMR (101 MHz, CDCl$_3$): δ = 158.1 (C_q), 157.5 (C_q), 147.1 (C_q), 144.0 (C_q), 143.6 (C_q), 141.1 (C_q), 136.0 (C_q), 134.6 (C_q), 134.0 (C_q), 129.8 (+, CH_{Ar}), 129.7 (+, CH_{Ar}), 129.2 (+, CH_{Ar}), 128.2 (+, CH_{Ar}), 127.2 (+, CH_{Ar}), 123.4 (+, CH_{Ar}), 121.6 (+, CH_{Ar}), 120.2 (+, CH_{Ar}), 119.6 (+, CH_{Ar}), 117.9 (+, CH_{Ar}), 114.5 (+, CH_{Ar}), 114.4 (+, CH_{Ar}), 107.5 (C_q), 55.6 (+, CH_3), 55.5 (+, CH_3), 46.9 (−, CH_2), 46.2 (−, CH_2), 29.7 (−, CH_2), 26.9 (−, CH_2), 22.8 (−, CH_2), 21.6 (−, CH_2), 21.3 (+, CH_3), 20.9 (−, CH_2), 18.9 (−, CH_2) ppm. − **IR** (ATR): ṽ = 2919 (m), 2849 (w), 1595 (w), 1513 (s), 1461 (m), 1336 (w), 1300 (w), 1242 (m) cm^{-1}. − **MS** (EI), m/z (%): 383.3 (57) [M]$^+$, 228.2 (100). − **HRMS** (EI, C$_{20}$H$_{21}$N$_3$O$_3$^{32}S): calc. 383.1298; found 383.1299.

2-phenyl-3-(p-tolyl)-4-tosyl-4,5,6,7-tetrahydro-2H-pyrazolo[4,3-b]pyridine (4.64a)

2-phenyl-3-(p-tolyl)-7-tosyl-4,5,6,7-tetrahydro-2H-pyrazolo[3,4-b]pyridine (4.64b)

These compounds were prepared according to general procedure GP4.2 using 3-phenyl-4-(p-tolyl)-3H-1,2,3-oxadiazol-1-ium-5-olate (23.0 mg, 0.092 mmol) and were isolated after preparative TLC (cHex/EtOAc = 4:1) as colorless solids (10.0 mg (A) and 8.0 mg (B), 0.041 mmol, 80%). **Isomer (A):** ^{1}**H-NMR** (400 MHz, CDCl$_3$): δ = 7.41 (d, J = 8.3 Hz, 2H), 7.26 - 7.31 (m, 3H), 7.16 - 7.21 (m, 4H), 7.05 - 7.16 (m, 4H), 3.69 - 3.76 (m, 2H), 2.68 (t, J = 7.1 Hz, 2H), 2.40 (s, 3H), 2.34 (s, 3H), 1.59 - 1.68 (m, 2H) ppm. − ^{13}**C-NMR** (101 MHz, CDCl$_3$): δ = 143.5 (C_q), 143.4 (C_q), 140.2 (C_q), 138.0 (C_q), 136.5 (C_q), 136.4 (C_q), 129.9 (+, CH_{Ar}), 129.4 (+, CH_{Ar}), 128.9 (+, CH_{Ar}), 128.6 (+, CH_{Ar}), 127.6 (+, CH_{Ar}), 127.3 (+, CH_{Ar}), 125.6 (+, CH_{Ar}), 119.0 (C_q), 47.5 (−, CH_2), 29.7 (−, CH_2), 21.6 (+, CH_3), 21.5 (+, CH_3), 20.4 (−, CH_2), 19.5 (−, CH_2) ppm. − **IR** (ATR): ṽ = 2918 (w), 2849 (w), 1692 (w), 1594 (w), 1519 (w), 1498 (m), 1448 (w), 1344 (m), 1286 (m), 1161 (w), 1161 (m), 1088 (m) cm^{-1}. − **MS** (EI), m/z (%): 443.3 (7) [M]$^+$, 288.2 (50), 194.1 (100). − **HRMS** (EI, $C_{26}H_{25}N_3O_2{}^{32}S$): calc.443.1662; found 443.1660. − **Isomer (B):** ^{1}**H-NMR** (400 MHz, CDCl$_3$): δ = 8.00 (d, J = 8.3 Hz, 2H), 7.28 (d, J = 8.3 Hz, 3H), 7.16 - 7.25 (m, 4H), 7.10 (m, J = 7.8 Hz, 2H), 6.99 (m, J = 8.1 Hz, 2H), 3.83 - 3.91 (m, 2H), 2.50 (t, J = 6.3 Hz, 2H), 2.40 (s, 3H), 2.33 (s, 3H), 1.92 (dt, J = 11.2, 5.9 Hz, 2H) ppm. − ^{13}**C-NMR** (101 MHz, CDCl$_3$): δ = 146.8 (C_q), 143.5 (C_q), 140.1 (C_q), 138.4 (C_q), 138.2 (C_q), 136.2 (C_q), 129.2 (+, CH_{Ar}), 129.1 (+, CH_{Ar}), 129.0 (+, CH_{Ar}), 128.5 (+, CH_{Ar}), 128.4 (+, CH_{Ar}), 127.2 (+, CH_{Ar}), 126.2 (+, CH_{Ar}), 124.4 (+, CH_{Ar}), 106.3 (C_q), 46.8 (−, CH_2), 29.7 (−, CH_2), 22.9 (−, CH_2), 21.6 (+, CH_3), 21.3 (+, CH_3), 19.7 (−, CH_2) ppm. − **IR** (ATR): ṽ = 2918 (w), 2849 (w), 1693 (w), 1594 (w), 1519 (w), 1498 (m), 1448 (w), 1344 (m), 1286 (m), 1161 (w), 1161 (m), 1088 (m) cm^{-1}. − **MS** (EI), m/z (%): 443.3 (77) [M]$^+$, 288.2 (100). − **HRMS** (EI, $C_{26}H_{25}N_3O_2{}^{32}S$): calc.443.1662; found 443.1664.

N-allyl-2-phenyl-4-tosyl-4,5,6,7-tetrahydro-2H-pyrazolo[4,3-b]pyridine-3-carboxamide (4.65a)

N-allyl-2-phenyl-7-tosyl-4,5,6,7-tetrahydro-2H-pyrazolo[3,4-b]pyridine-3-carboxamide (4.65b)

These compounds were prepared according to general procedure GP4.2 using 4-(allylcarbamoyl)-3-phenyl-3H-1,2,3-oxadiazol-1-ium-5-olate (20.0 mg, 0.083 mmol) and were isolated after preparative TLC (*c*Hex/EtOAc = 4:1) as colorless solids (14.0 mg (B) and 4.0 mg (A), 0.040 mmol, 94%). **Isomer (A):** ^{1}**H-NMR** (400 MHz, CDCl$_3$): δ = 7.64 (d, J = 8.3 Hz, 2H), 7.33 - 7.54 (m, 5H), 7.30 (d, J = 8.1 Hz, 2H), 5.93 (ddt, J = 17.1, 10.3, 5.7, 5.7 Hz, 1H), 5.27 (dq, J = 17.1, 1.5 Hz, 1H), 5.17 (dq, J = 10.2, 1.4 Hz, 1H), 4.04 (tt, J = 5.8, 1.5 Hz, 2H), 3.63 - 3.69 (m, 3H), 2.51 (t, J = 7.1 Hz, 2H), 2.45 (s, 3H), 1.45 - 1.55 (m, 2H) ppm. − ^{13}**C-NMR** (101 MHz, CDCl$_3$): δ = 159.4 (C_q), 144.7 (C_q), 143.3 (C_q), 140.3 (C_q), 135.0 (C_q), 133.6 (+, CH_{Ar}), 131.4 (+, CH_{Ar}), 129.9 (+, CH_{Ar}), 128.9 (+, CH_{Ar}), 128.1 (+, CH_{Ar}), 128.1 (+, CH_{Ar}), 124.5 (+, CH_{Ar}), 119.9 (C_q), 116.9 (−, CH_2), 47.3 (−, CH_2), 42.4 (−, CH_2), 29.7 (−, CH_2), 21.7 (+, CH_3), 19.8 (−, CH_2), 18.6 (−, CH_2) ppm. **Isomer (B):** ^{1}**H-NMR** (400 MHz, CDCl$_3$): δ = 7.87 - 7.93 (m, 2H), 7.34 - 7.47 (m, 5H), 7.27 (s, 1H), 7.25 (s, 1H), 5.66 - 5.73 (m, 1H), 5.51 (app. t., 1H), 4.97 - 5.10 (m, 2H), 3.87 (tt, J = 5.7, 1.5 Hz, 2H), 3.78 - 3.85 (m, 2H), 2.68 (t, J = 6.4 Hz, 2H), 2.40 (s, 3H), 1.87 - 1.96 (m, 2H) ppm. − ^{13}**C-NMR** (101 MHz, CDCl$_3$): δ = 159.7 (C_q), 146.9 (C_q), 143.9 (C_q), 139.5 (C_q), 135.6 (C_q), 133.0 (C_q), 133.0 (+, CH), 129.3 (+, 2 × CH_{Ar}), 129.1 (+, 2 × CH_{Ar}), 128.2 (+, 2 × CH_{Ar}), 128.0 (+, CH_{Ar}), 124.5 (+, 2 × CH_{Ar}), 117.0 (−, CH_2), 110.1 (C_q), 46.7 (−, CH_2), 41.9 (−, CH_2), 22.3 (−, CH_2), 21.6 (+, CH_3), 19.7 (−, CH_2) ppm. − **IR** (ATR): $\tilde{v}$ = 3394 (vw), 2923 (vw), 1662 (w), 1494 (w), 1354 (m), 1256 (m) cm^{-1}. − **MS** (FAB), *m/z* (%): 437.2 (8) [M+H]$^+$, 132.9 (100). − **HRMS** (FAB, C$_{23}$H$_{25}$N$_4$O$_3$^{32}S): calc. 437.1642; found 437.1640.

1-(2-phenyl-4-tosyl-4,5,6,7-tetrahydro-2H-pyrazolo[4,3-b]pyridin-3-yl)ethan-1-one (4.66a)

1-(2-phenyl-7-tosyl-4,5,6,7-tetrahydro-2H-pyrazolo[3,4-b]pyridin-3-yl)ethan-1-one (4.66b)

This inseparable mixture of compounds was prepared according to general procedure GP4.2 using 4-acetyl-3-phenyl-3H-1,2,3-oxadiazol-1-ium-5-olate (19.0 mg, 0.092 mmol) and was isolated after preparative TLC (cHex/EtOAc = 5:1) as colorless solids (16.0 mg, 0.040 mmol, 88%) Ratio by ^{1}H-NMR: **A/B** = 2:5. **^{1}H-NMR** (400 MHz, CDCl$_3$): δ = 7.90 (d, J = 8.3 Hz, 2H, major product), 7.52 (d, J = 8.3 Hz, 2H, minor product), 7.37 - 7.49 (m, 5H), 7.28 - 7.35 (m, 3H), 7.25 (m, 3H), 3.78 - 3.87 (m, 2H, major product), 3.70 - 3.78 (m, 2H, minor product), 2.73 (t, J = 6.4 Hz, 2H, major product), 2.57 - 2.63 (m, 2H, minor product), 2.56 (s, 3H, minor product), 2.43 (s, 3H, minor product), 2.40 (s, 3H, major product), 2.14 (s, 3H, major product), 1.90 - 2.01 (m, 2H, major product), 1.36 (m, 2H, minor product) ppm. $-$ ^{13}C-NMR (101 MHz, CDCl$_3$): δ = 189.8 (C_q), 146.9 (C_q), 143.9 (C_q), 140.5 (C_q), 137.6 (C_q), 135.6 (C_q), 129.9 (+, CH$_{Ar}$), 129.2 (+, CH$_{Ar}$), 129.0 (+, CH$_{Ar}$), 128.9 (+, CH$_{Ar}$), 128.5 (+, CH$_{Ar}$), 128.4 (+, CH$_{Ar}$), 128.3 (+, CH$_{Ar}$), 127.9 (+, CH$_{Ar}$), 125.7 (s+, CH$_{Ar}$ 125.6 (+, CH$_{Ar}$), 111.8 (C_q), 46.7 ($-$, CH$_2$), 46.5 ($-$, CH$_2$), 30.6 (+, CH$_3$), 30.2 (+, CH$_3$), 22.3 ($-$, CH$_2$), 21.6 (+, CH$_3$), 21.6 (+, CH$_3$), 20.7 ($-$, CH$_2$), 20.4 ($-$, CH$_2$), 18.3 ($-$, CH$_2$) ppm. $-$ **IR** (ATR): $\tilde{v}$ = 2925 (w), 1674 (m), 1595 (w), 1552 (w), 1497 (m), 1459 (w), 1436 (w), 1355 (w), 1341 (m), 1301 (w) cm^{-1}. $-$ **MS** (EI), m/z (%): 395.3 (77) [M]$^+$, 240.2 (100). $-$ **HRMS** (EI, C$_{21}$H$_{21}$N$_3$O$_3^{32}$S): calc. 395.1298; found 395.1300.

2-(2-phenyl-4-tosyl-4,5,6,7-tetrahydro-2H-pyrazolo[4,3-b]pyridin-3-yl)quinoline (4.67a)

2-(2-phenyl-7-tosyl-4,5,6,7-tetrahydro-2H-pyrazolo[3,4-b]pyridin-3-yl)quinoline (4.67b)

A

B

These compounds were prepared according to general procedure GP4.2 using 3-phenyl-4-(quinolin-2-yl)-3H-1,2,3-oxadiazol-1-ium-5-olate (28.0 mg, 0.096 mmol) and were isolated after preparative TLC (cHex/EtOAc = 4:1) as colorless solids (6.5 mg and 6.5 mg, 0.027 mmol, 56%). Due to low signal intensities in the NOESY and HSQC-NMR experiments it could not be determined which fraction was which isomer. **Isomer 1:** **^{1}H-NMR** (400 MHz, CDCl$_3$) δ = 8.20 (br. s., 1H), 7.84 (m, 1H), 7.76 (m, 1H), 7.61 (m, 3H), 7.43 - 7.52 (m, 1H), 7.26 (s, 5H), 7.12 - 7.22 (m, 2H), 3.74 - 3.83 (m, 2H), 2.79 (t, J = 6.8 Hz, 2H), 2.37 (s, 3H), 1.70 (br. s., 2H) ppm. –

^{13}C-NMR (101 MHz, CDCl$_3$): δ = 129.6 (+, CH_{Ar}), 128.8 (+, CH_{Ar}), 127.8 (+, CH_{Ar}), 125.6 (+, CH_{Ar}), 123.7 (+, CH_{Ar}), 47.0 (–, CH_2), 29.7 (–, CH_2), 21.5 (+, CH_3), 20.7 (–, CH_2) ppm. Due to low signal intensities several quaternary carbon were not detected. – **IR** (ATR): ṽ = 2920 (m), 2851 (w), 1730 (vw), 1597 (w), 1499 (w), 1458 (w), 1359 (w), 1344 (m), 1289 (w), 1259 (w), 1158 (m), 1088 (m), 1064 (m), 1002 (m) cm^{-1}. – **MS** (FAB), m/z (%): 481.2 (4) [M+H]$^+$, 325.0 (16), 132.8 (100). – **HRMS** (EI, $C_{28}H_{25}N_4O_2{}^{32}S$): calc. 481.1693; found 481.1692. **Isomer 2:** **^{1}H-NMR** (400 MHz, CDCl$_3$): δ = 8.08 (d, J = 7.8 Hz, 1H), 7.95 - 8.02 (m, 2H), 7.75 - 7.86 (m, 2H), 7.63 (t, J = 7.3 Hz, 1H), 7.27 - 7.31 (m, 7H), 7.26 (br. s., 2H), 7.06 (d, J = 8.5 Hz, 1H), 3.84 - 3.94 (m, 2H), 2.81 (br. s., 2H), 2.41 (s, 3H), 1.95 - 2.04 (m, 3H) ppm. – **^{13}C-NMR** (101 MHz, CDCl$_3$): δ = 147.5 (C_q), 143.8 (C_q), 135.7 (C_q, 129.3 (C_q), 128.9 (C_q), 128.3 (+, CH_{Ar}), 127.6 (+, CH_{Ar}), 127.1 (+, CH_{Ar}), 126.9 (+, CH_{Ar}), 124.7 (+, CH_{Ar}), 122.3 (+, CH_{Ar}), 46.8 (–, CH_2), 29.7 (–, CH_2), 22.7 (–, CH_2), 21.6 (+, CH_3) ppm. – **IR** (ATR): ṽ = 2919 (w), 2850 (w), 1730 (vw), 1595 (w), 1493 (w), 1458 (w), 1359 (w), 1289 (w), 1259 (w), 1162 (m), 1088 (w), 1007 (m) cm^{-1}. – **MS** (FAB), m/z (%): 481.2 (23) [M+H]$^+$, 325.1 (18), 132.8 (100). – **HRMS** (EI, $C_{28}H_{25}N_4O_2{}^{32}S$): calc. 481.1693; found 481.1692.

3-((4-chlorophenyl)thio)-2-phenyl-4-tosyl-4,5,6,7-tetrahydro-2H-pyrazolo[4,3-b]pyridine (4.68a)

3-((4-chlorophenyl)thio)-2-phenyl-7-tosyl-4,5,6,7-tetrahydro-2H-pyrazolo[3,4-b]pyridine (4.68b)

These compounds were prepared according to general procedure GP4.2 using 4-((4-chlorophenyl)thio)-3-phenyl-3H-1,2,3-oxadiazol-1-ium-5-olate (25.0 mg, 0.083 mmol) and were isolated after preparative TLC (cHex/EtOAc = 4:1) as colorless solids (9.0 mg (A) and 5.0 mg (B), 0.029 mmol, 68%). **Isomer (A):** ^{1}**H-NMR** (400 MHz, CDCl$_3$): δ = 7.68 (d, J = 8.3 Hz, 2H), 7.29 - 7.43 (m, 7H), 6.97 - 7.03 (m, 2H), 6.72 - 6.77 (m, 2H), 3.62 - 3.67 (m, 2H), 2.70 (t, J = 6.9 Hz, 2H), 2.42 (s, 3H), 1.61 - 1.66 (m, 2H) ppm. − ^{13}**C-NMR** (101 MHz, CDCl$_3$): δ = 144.1 (C_q), 143.0 (C_q), 139.4 (C_q), 136.4 (C_q), 133.5 (C_q), 130.2 (C_q), 129.7 (+, CH_{Ar}), 129.4 (+, CH_{Ar}), 128.7 (+, CH_{Ar}), 128.6 (+, CH_{Ar}), 128.0 (+, CH_{Ar}), 127.9 (+, CH_{Ar}), 125.4 (+, CH_{Ar}), 121.3 (+, CH_{Ar}), 113.7 (C_q), 47.2 (−, CH_2), 29.7 (−, CH_2), 21.6 (+, CH_3), 20.6 (−, CH_2), 19.5 (−, CH_2) ppm. − **IR** (ATR): ṽ = 2922 (w), 2852 (w), 1729 (w), 1596 (w), 1542 (m), 1497 (w), 1474 (w), 1357 (w) cm^{-1}. − **MS** (EI), m/z (%): 495.3/497.3 (10/4) [M]$^+$, 340.2 (100), 246.1 (72). − **HRMS** (EI, C$_{25}$H$_{22}$N$_3$O$_2$^{32}S$_2$^{35}Cl): calc. 495.0837; found 495.0839. − **Isomer (B):** ^{1}**H-NMR** (400 MHz, CDCl$_3$): δ = 7.95 (m, J = 8.3 Hz, 2H), 7.27 - 7.45 (m, 7H), 7.12 - 7.18 (m, 2H), 6.81 - 6.88 (m, 2H), 3.81 - 3.88 (m, 2H), 2.43 (s, 3H), 2.37 - 2.42 (m, 2H), 1.86 - 1.95 (m, 2H) ppm. − ^{13}**C-NMR** (CDCl$_3$): δ = 147.3 (C_q), 143.9 (C_q), 139.1 (C_q), 135.9 (C_q), 133.5 (C_q), 132.3 (C_q), 129.4 (+, CH_{Ar}), 129.2 (+, CH_{Ar}), 128.6 (+, CH_{Ar}), 128.3 (+, CH_{Ar}), 128.0 (+, CH_{Ar}), 127.6 (+, CH_{Ar}), 126.9 (+, CH_{Ar}), 124.9 (+, CH_{Ar}), 114.1 (C_q), 47.0 (−, CH_2), 22.2 (−, CH_2), 21.6 (+, CH_3), 19.3 (−, CH_2) ppm. − **IR** (ATR): ṽ = 2920 (w), 2850 (w), 1729 (w), 1595 (w), 1542 (m), 1497 (w), 1474 (w), 1357 (w) cm^{-1}. − **MS** (EI), m/z (%): 495.3/497.3 (77/34) [M]$^+$, 431.3 (21), 340.2 (100) 196.1 (49). − **HRMS** (EI, C$_{25}$H$_{22}$N$_3$O$_2$^{32}S$_2$^{35}Cl): calc. 495.0837; found 495.0837.

2-p1henyl-*N*-(prop-2-yn-1-yl)-4-tosyl-4,5,6,7-tetrahydro-2H-pyrazolo[4,3-b]pyridine-3-carboxamide (4.69a)

2-phenyl-*N*-(prop-2-yn-1-yl)-7-tosyl-4,5,6,7-tetrahydro-2H-pyrazolo[3,4-b]pyridine-3-carboxamide (4.69b)

These compounds were prepared according to general procedure GP4.2 using 3-phenyl-4-(prop-2-yn-1-ylcarbamoyl)-3H-1,2,3-oxadiazol-1-ium-5-olate (44.0 mg, 0.179 mmol) and were isolated as a mixture after preparative TLC (*c*Hex/EtOAc = 4:1) as colorless solids (18.0 mg, 0.041 mmol, 46%) Ratio by ^{1}H-NMR **A/B** = 2:7. Another fraction yielded unreacted 1-tosyl-3-(trimethylsilyl)-1,4,5,6-tetrahydropyridin-2-yl trifluoromethanesulfonate (22.0 mg, 0.048 mmol, 54%). **^{1}H-NMR** (400 MHz, CDCl$_3$): δ = 7.90 (d, J = 8.3 Hz, 2H, major product), 7.62 (d, J = 8.3 Hz, 2H, minor product), 7.55 (t, J = 5.1 Hz, 1H, minor product), 7.34 - 7.52 (m, 7H), 7.29 (d, J = 8.1 Hz, 1H), 7.25 (m, 2H), 5.65 (t, J = 5.1 Hz, 1H, major product), 4.19 (dd, J = 5.3, 2.5 Hz, 2H, minor product), 4.04 (dd, J = 5.4, 2.7 Hz, 2H, major product), 3.79 - 3.85 (m, 2H, major product), 3.62 - 3.68 (m, 2H, minor product), 2.68 (t, J = 6.4 Hz, 2H, major product), 2.51 (t, J = 7.2 Hz, 2H, minor product), 2.43 (s, 3H, minor product), 2.40 (s, 3H, major product), 2.26 (t, J = 2.7 Hz, 1H), 2.18 (t, J = 2.5 Hz, 1H, major product), 1.88 - 1.96 (m, 2H, major product), 1.46 - 1.54 (m, 1H, minor product) ppm. − ^{13}**C-NMR** (101 MHz, CDCl$_3$): δ = 159.3 (C_q), 159.2 (C_q), 146.9 (C_q), 144.8 (C_q), 144.0 (C_q), 143.3 (C_q), 139.4 (C_q), 135.6 (C_q), 134.9 (C_q), 132.1 (C_q), 130.0 (+, CH_{Ar}), 129.3 (+, CH_{Ar}), 129.2 (+, CH_{Ar}), 128.9 (+, CH_{Ar}), 128.2 (+, CH_{Ar}), 128.1 (+, CH_{Ar}), 128.1 (+, CH_{Ar}), 124.5 (+, CH_{Ar}), 124.4 (+, CH_{Ar}), 120.1 (C_q), 110.6 (C_q), 79.0 (+, CH), 78.4 (+, CH), 72.0 (+, CH), 71.8 (+, CH), 47.4 (−, CH_2), 46.7 (−, CH_2), 29.7 (s), 29.2 (−, CH_2), 22.2 (−, CH_2), 21.6 (+, CH_3), 19.8 (−, CH_2), 19.7 (−, CH_2), 18.6 (−, CH_2) ppm. − **IR** (ATR): $\tilde{v}$ = 3295 (w), 2923 (w), 1665 (w), 1595 (w), 1565 (m), 1494 (w), 1460 (w), 1339 (m), 1302 (w), 1165 (w) cm^{-1}. − **MS** (EI), *m/z* (%): 434.3 (100) [M]$^+$, 280.2 (77), 279.2 (91). − **HRMS** (EI, C$_{23}$H$_{22}$N$_4$O$_3$^{32}S): calc. 434.1407; found 434.1406.

1-Tosylazocan-2-one (4.84)

 A 250 mL round-bottomed flask was charged with azocan-2-one (5.00 g, 39.3 mmol, 1.00 equiv.) in dry THF (100 mL) to give a yellow solution. The reaction mixture was cooled to −78 °C and stirred for 10 min before *n*-butyllithium (29.1 mL, 2.5M in hexanes, 72.6 mmol, 1.85 equiv.) was added dropwise over 15 min. The reaction mixture was stirred at −78 °C for 1h before a solution of 4-methylbenzenesulfonyl chloride (8.24 g, 43.2 mmol, 1.10 equiv.) in dry THF (30 mL) was added via dropping funnel over 10 min. The reaction mixture was allowed to slowly reach room temperature over 1.5h and subsequently transferred to a separatory funnel and diluted with 100 mL of sat. aq. NH_4Cl solution and 50 mL of H_2O. The aqueous layer was separated and extracted with EtOAc (2 × 200 mL). The combined organic layers were washed with 200 mL of sat. aq. $NaHCO_3$ solution, dried over $MgSO_4$, filtered, and concentrated to give a white solid. Trituration with Et_2O (2 × 50 mL) and recrystallization from CH_2Cl_2/MeOH mixtures the pure product crystallized as large colorless cubical crystals (5.79 g, 20.6 mmol, 52.4%). **^{1}H-NMR** (400 MHz, CDCl$_3$): δ = 7.91 (d, *J* = 8.3 Hz, 2H), 7.30 (d, *J* = 8.3 Hz, 2H), 4.04 - 4.12 (m, 2H), 2.46 - 2.52 (m, 2H), 2.42 (s, 3H), 1.88 (t, *J* = 5.7 Hz, 2H), 1.72 - 1.80 (m, 2H), 1.43 - 1.59 (m, 4H) ppm. − **^{13}C-NMR** (101 MHz, CDCl$_3$): δ = 174.8 (*C*$_q$), 144.5 (*C*$_q$), 136.4 (*C*$_q$), 129.1 (+, 2 × *C*H$_{Ar}$), 129.0 (+, 2 × *C*H$_{Ar}$), 46.0 (−, *C*H$_2$), 36.3 (−, *C*H$_2$), 31.0 (−, *C*H$_2$), 28.4 (−, *C*H$_2$), 26.1 (−, *C*H$_2$), 23.7 (−, *C*H$_2$), 21.6 (+, *C*H$_3$) ppm. − **IR** (ATR): ṽ = 2916 (vw), 2866 (vw), 1685 (w), 1594 (w), 1483 (w), 1448 (w), 1349 (w), 1307 (w), 1243 (w), 1168 (w) cm^{-1}. − **MS** (FAB), *m/z* (%): 282 (100) [M+H]$^+$. − **HRMS** (FAB, C$_{14}$H$_{20}$NO$_3$^{32}S): calc. 282.1158; found 282.1157.

2-phenyl-4-tosyl-4,5,6,7,8,9-hexahydro-2H-pyrazolo[4,3-b]azocine (4.87a)

2-phenyl-9-tosyl-4,5,6,7,8,9-hexahydro-2H-pyrazolo[3,4-b]azocine (4.87b)

A 25 mL Schleck flask was charged with 1-tosylazocan-2-one (28.0 mg, 0.100 mmol, 1.00 equiv.) in THF (1.0 mL) to give a colorless solution. The reaction mixture was cooled to -78 °C and solution of KHMDS (0.45 mL, 0.225 mmol, 0.5M in THF, 2.25 equiv.) was added dropwise over 5 min and the reaction mixture was stirred at −78 °C for 30 min. N-Phenyl-bis(trifluoromethanesulfonimide) (39.1 mg, 0.110 mmol, 1.10 equiv.) was added slowly and dropwise as solution in THF (1.0 mL). The reaction mixture was stirred at −78 °C for 60 min. Next 3-phenyl-3H-1,2,3-oxadiazol-1-ium-5-olate (40.1 mg, 0.250 mmol, 2.50 equiv.) was added as solution MeOH (1.0 mL). The reaction mixture was stirred at RT for 16h and the crude concentrated and purified using preparative TLC (cHex/EtOAc = 7:1) to yield a white powder (15.0 mg, 0.040 mmol,

40%). Ratio by ^{1}H-NMR **A/B** = 2:3. **^{1}H-NMR** (400 MHz, CDCl$_3$): δ = 7.95 - 8.00 (m, 2H), 7.68 - 7.72 (m, 3H), 7.68 - 7.68 (m, 1H), 7.65 (s, 1H), 7.51 - 7.59 (m, 5H), 7.38 - 7.45 (m, 5H), 7.33 - 7.38 (m, 4H), 7.27 - 7.33 (m, 6H), 3.62 (t, J = 5.8, 3H, major isomer), 3.49 (t, J = 5.8, 2H, minor isomer), 2.66 - 2.75 (m, 2H, minor isomer), 2.53 - 2.60 (m, 3H, major isomer), 2.46 (s, 3H, minor isomer), 2.44 (s, 4H, major isomer), 1.70 - 1.78 (m, 6H), 1.61 - 1.68 (m, 5H) ppm. – **^{13}C-NMR** (101 MHz, CDCl$_3$): δ = 154.3 (C_q), 147.6 (C_q), 143.5 (C_q), 143.1 (C_q), 139.9 (C_q), 139.8 (C_q), 137.5 (C_q), 137.1 (C_q), 133.9 (C_q), 129.7 (+, CH_{Ar}), 129.6 (+, CH_{Ar}), 129.6 (+, CH_{Ar}), 129.5 (+, CH_{Ar}), 129.4 (+, CH_{Ar}), 129.3 (+, CH_{Ar}), 128.2 (+, CH_{Ar}), 127.4 (+, CH_{Ar}), 127.3 (+, CH_{Ar}), 127.1 (+, CH_{Ar}), 126.4 (+, CH_{Ar}), 126.2 (+, CH_{Ar}), 125.6 (+, CH_{Ar}), 125.1 (+, CH_{Ar}), 123.5 (+, CH_{Ar}), 122.8 (C_q), 120.1 (C_q), 118.8 (+, CH_{Ar}), 118.3 (+, CH_{Ar}), 52.5 (−, CH_2), 51.8 (−, CH_2), 30.6 (−, CH_2), 29.7 (−, CH_2), 29.2 (−, CH_2), 28.0 (−, CH_2), 26.7 (−, CH_2), 26.3 (−, CH_2), 26.1 (−, CH_2), 26.0 (−, CH_2), 26.0 (−, CH_2), 25.9 (−, CH_2), 25.9 (−, CH_2), 22.8 (−, CH_2), 21.6 (+, CH_3), 21.6 (+, CH_3) ppm. – **IR** (ATR): ṽ = 2925 (w), 2853 (w), 1732 (w), 1597 (m), 1365 (m), 1335 (m), 1154 (m) cm^{-1}. – **MS** (FAB), m/z (%): 382.3 (85) [M+H]$^+$, 226.1 (45), 154.0 (81), 136.0 (100). – **HRMS** (FAB, $C_{21}H_{24}N_3O_2{}^{32}S$): calc. 382.1584; found 382.1582.

X-ray Crystallographic data

All X-ray crystallographic work in this chapter has been performed by Dr. Martin NIEGER, the Laboratory of Inorganic Chemistry, University of Helsinki (Finland). All X-ray images shown in this chapter are depicted with displacement parameters drawn at 50 % probability level.

N-benzyl-4-methyl-N-(1-phenyl-1H-pyrazol-4-yl)benzene-sulfonamide – SB872_HY – (4.47a)

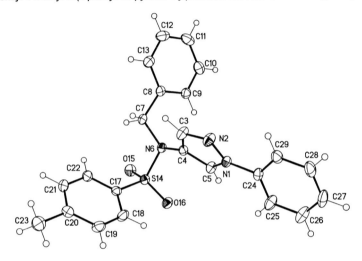

Crystal data

$C_{23}H_{21}N_3O_2S$	$F(000) = 848$
$M_r = 403.49$	$D_x = 1.364$ Mg m^{-3}
Monoclinic, $P2_1/n$ (no.14)	Cu $K\alpha$ radiation, $\lambda = 1.54178$ Å
$a = 16.8632$ (5) Å	Cell parameters from 9905 reflections
$b = 5.6827$ (2) Å	$\theta = 3.1–72.0°$
$c = 20.8486$ (6) Å	$\mu = 1.67$ mm^{-1}
$\beta = 100.486$ (1)°	$T = 123$ K
$V = 1964.52$ (11) Å^3	Needles, colourless
$Z = 4$	$0.20 \times 0.04 \times 0.02$ mm

Data collection

Bruker D8 VENTURE diffractometer with Photon100 detector	3835 independent reflections
Radiation source: INCOATEC microfocus sealed tube	3333 reflections with $I > 2\sigma(I)$
Detector resolution: 10.4167 pixels mm^{-1}	$R_{int} = 0.039$
rotation in ϕ and ω, 1°, shutterless scans	$\theta_{max} = 72.0°$, $\theta_{min} = 3.1°$
Absorption correction: multi-scan SADABS (Sheldrick, 2014)	$h = -20 \rightarrow 20$
$T_{min} = 0.834$, $T_{max} = 0.971$	$k = -7 \rightarrow 7$
20213 measured reflections	$l = -25 \rightarrow 25$

Refinement

Refinement on F^2	Secondary atom site location: difference Fourier map
Least-squares matrix: full	Hydrogen site location: difference Fourier map
$R[F^2 > 2\sigma(F^2)] = 0.035$	H-atom parameters constrained
$wR(F^2) = 0.087$	$w = 1/[\sigma^2(F_o^2) + (0.0332P)^2 + 1.3646P]$ where $P = (F_o^2 + 2F_c^2)/3$
$S = 1.04$	$(\Delta/\sigma)_{max} = 0.001$
3835 reflections	$\Delta)_{max} = 0.24$ e Å^{-3}
264 parameters	$\Delta)_{min} = -0.37$ e Å^{-3}
0 restraints	Extinction correction: SHELXL2014/7 (Sheldrick 2014, Fc*=kFc[1+0.001xFc$^2\lambda^3$/sin(2θ)]$^{-1/4}$
Primary atom site location: structure-invariant direct methods	Extinction coefficient: 0.00065 (11)

174

Pyrazoles

N-benzyl-4-methyl-N-(1-(4-(trifluoromethyl)phenyl)-1H-pyrazol-4-yl)benzene-sulfonamide −
SB873_HY − (4.47b)

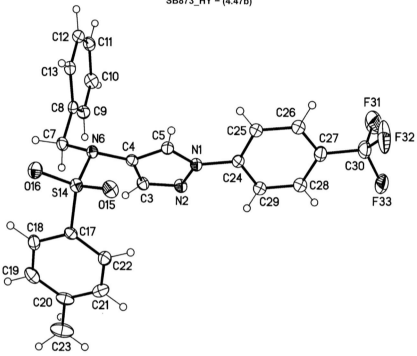

Crystal data

$C_{24}H_{20}F_3N_3O_2S$	$Z = 2$
$M_r = 471.49$	$F(000) = 488$
Triclinic, $P\text{-}1$ (no.2)	$D_x = 1.470$ Mg m^{-3}
$a = 5.9736$ (3) Å	Cu $K\alpha$ radiation, $\lambda = 1.54178$ Å
$b = 9.6554$ (5) Å	Cell parameters from 8072 reflections
$c = 19.0149$ (9) Å	$\theta = 4.7–72.1°$
$\alpha = 83.760$ (2)°	$\mu = 1.83$ mm^{-1}
$\beta = 86.840$ (2)°	$T = 123$ K
$\gamma = 77.757$ (2)°	Needles, colourless
$V = 1064.86$ (9) Å^3	$0.20 \times 0.09 \times 0.03$ mm

Data collection

Bruker D8 VENTURE diffractometer with Photon100 detector	4112 independent reflections
Radiation source: INCOATEC microfocus sealed tube	3650 reflections with $I > 2\sigma(I)$
Detector resolution: 10.4167 pixels mm^{-1}	$R_{int} = 0.029$
rotation in ϕ and ω, 1°, shutterless scans	$\theta_{max} = 72.1°$, $\theta_{min} = 2.3°$
Absorption correction: multi-scan *SADABS* (Sheldrick, 2014)	$h = -7\rightarrow7$
$T_{min} = 0.782$, $T_{max} = 0.942$	$k = -11\rightarrow11$
11969 measured reflections	$l = -22\rightarrow22$

Refinement

Refinement on F^2	Primary atom site location: structure-invariant direct methods
Least-squares matrix: full	Secondary atom site location: difference Fourier map
$R[F^2 > 2\sigma(F^2)] = 0.038$	Hydrogen site location: difference Fourier map
$wR(F^2) = 0.094$	H-atom parameters constrained
$S = 1.03$	$w = 1/[\sigma^2(F_o^2) + (0.038P)^2 + 0.786P]$ where $P = (F_o^2 + 2F_c^2)/3$
4112 reflections	$(\Delta/\sigma)_{max} = 0.001$
292 parameters	$\Delta\rangle_{max} = 0.36$ e Å^{-3}
40 restraints	$\Delta\rangle_{min} = -0.40$ e Å^{-3}

4.7 References

[1]   J. C. Earl, A. W. Mackney, *J. Chem. Soc.*, **1935**, 899-900, *204, The action of acetic anhydride on N-nitrosophenylglycine and some of its derivatives.*

[2]   W. Baker, W. D. Ollis, V. D. Poole, *J. Chem. Soc.*, **1950**, (1), 1542-1551, *320, Cyclic meso-ionic compounds. Part III. Further properties of the sydnones and the mechanism of their formation.*

[3]   J. Applegate, K. Turnbull, *Synthesis*, **1988**, *1988* (12), 1011-1012, *The Efficient Synthesis of 3-Arylsydnones Under Neutral Conditions.*

[4]   H.-J. Tien, J.-C. Yeh, S.-C. Wu, *J. Chin. Chem. Soc.*, **1992**, *39* (5), 443-447, *Acetylation and Debromination of Sydnones Accelerated by Ultrasound.*

[5]   H. Kato, M. Ohta, *Bull. Chem. Soc. Jpn.*, **1962**, *35* (8), 1418-1419, *Decomposition of the Sydnone Ring.*

[6]   F. Dumitraşcu, C. Drâghici, D. Dumitrescu, L. Tarko, D. Râileanu, *Liebigs Ann.*, **1997**, *1997* (12), 2613-2616, *Direct Iodination of Sydnones and Their Cycloadditions to Form 5-Iodopyrazoles.*

[7]   K. Turnbull, D. M. Krein, S. A. Tullis, *Synth. Commun.*, **2003**, *33* (13), 2209-2214, *Alkynylation of 4-Bromo-3-phenylsydnone with Aryl Acetylenes.*

[8]   D. L. Browne, J. B. Taylor, A. Plant, J. P. A. Harrity, *J. Org. Chem.*, **2009**, *74* (1), 396-400, *Cross Coupling of Bromo Sydnones: Development of a Flexible Route toward Functionalized Pyrazoles.*

[9]   H.-J. Tien, G.-M. Fang, S.-T. Lin, L.-L. Tien, *J. Chin. Chem. Soc.*, **1992**, *39* (1), 107-110, *A Facile One-Pot Synthesis of 4-Acyl and 4-(1-Hydroxyethyl)Sydnones.*

[10]   D. L. Browne, J. P. A. Harrity, *Tetrahedron*, **2010**, *66* (3), 553-568, *Recent developments in the chemistry of sydnones.*

[11]   A. W. Brown, J. P. A. Harrity, *J. Org. Chem.*, **2015**, *80* (4), 2467-2472, *Direct Arylation of Sydnones with Aryl Chlorides toward Highly Substituted Pyrazoles.*

[12]   S. Kolodych, E. Rasolofonjatovo, M. Chaumontet, M.-C. Nevers, C. Créminon, F. Taran, *Angew. Chem. Int. Ed.*, **2013**, *52* (46), 12056-12060, *Discovery of Chemoselective and Biocompatible Reactions Using a High-Throughput Immunoassay Screening.*

[13]   Craig S. McKay, M. G. Finn, *Chemistry & Biology*, **2014**, *21* (9), 1075-1101, *Click Chemistry in Complex Mixtures: Bioorthogonal Bioconjugation.*

[14]   O. Boutureira, G. J. L. Bernardes, *Chem. Rev.*, **2015**, *115* (5), 2174-2195, *Advances in Chemical Protein Modification.*

[15]   M. Kawase, H. Sakagami, N. Motohashi, *Top. Heterocycl. Chem.*, **2009**, *16* (Bioactive Heterocycles VII), 135-152, *The chemistry of bioactive mesoionic heterocycles.*

[16]   S. K. Bhosale, S. R. Deshpande, R. D. Wagh, A. S. Dhake, *J. Chem. Pharm. Res.*, **2015**, *7* (5), 1247-1263, *Biological activities of 1, 2, 3-oxadiazolium-5-olate derivatives.*

[17]   B. V. Badami, *Resonance*, **2006**, *11* (10), 40-48, *Mesoionic compounds. An unconventional class of aromatic heterocycles.*

[18]   S. K. Bhosale, S. R. Deshpande, R. D. Wagh, *J. Chem. Pharm. Res.*, **2012**, *4* (2), 1185-1199, *Mesoionic sydnone derivatives: an overview.*

[19]   R. Chandrasekhar, M. J. Nanjan, *Mini-Rev. Med. Chem.*, **2012**, *12* (13), 1359-1365, *Sydnones. A brief review.*

[20]   R. Huisgen, R. Grashey, H. Gotthardt, R. Schmidt, *Angew. Chem. Int. Ed.*, **1962**, *1* (1), 48-49, *1,3-Dipolar Additions of Sydnones to Alkynes. A New Route into the Pyrazole Series.*

[21]   R. Huisgen, *Angew. Chem. Int. Ed.*, **1963**, *2* (10), 565-598, *1,3-Dipolar Cycloadditions. Past and Future.*

[22]   S. Specklin, E. Decuypere, L. Plougastel, S. Aliani, F. Taran, *J. Org. Chem.*, **2014**, *79* (16), 7772-7777, *One-Pot Synthesis of 1,4-Disubstituted Pyrazoles from Arylglycines via Copper-Catalyzed Sydnone–Alkyne Cycloaddition Reaction.*

[23]   J. Comas-Barceló, R. S. Foster, B. Fiser, E. Gomez-Bengoa, J. P. A. Harrity, *Chem. Eur. J.*, **2015**, *21* (8), 3257-3263, *Cu-Promoted Sydnone Cycloadditions of Alkynes: Scope and Mechanism Studies.*

[24]   L. Knorr, *Ber. Chem.*, **1883**, *16* (2), 2597-2599, *Einwirkung von Acetessigester auf Phenylhydrazin.*

[25]   J. Kempson, *Name Reactions in Heterocyclic Chemistry II; Knorr pyrazole synthesis*; John Wiley & Sons, Inc., 2011.

[26]   H. v. Pechmann, *Ber. Chem.*, **1898**, *31* (3), 2950-2951, *Pyrazol aus Acetylen und Diazomethan.*

[27]   R. J. Mullins, *Name Reactions in Heterocyclic Chemistry II; Pechmann pyrazole synthesis*; John Wiley & Sons, Inc., 2011.

[28]   E. Arbačiauskienė, G. Vilkauskaitė, G. A. Eller, W. Holzer, A. Šačkus, *Tetrahedron,* **2009**, *65* (37), 7817-7824, *Pd-catalyzed cross-coupling reactions of halogenated 1-phenylpyrazol-3-ols and related triflates.*

[29]   S. D. Lindell, B. A. Moloney, B. D. Hewitt, C. G. Earnshaw, P. J. Dudfield, J. E. Dancer, *Bioorg. Med. Chem. Lett.* , **1999**, *9* (14), 1985-1990, *The design and synthesis of inhibitors of adenosine 5'-monophosphate deaminase.*

[30]   S. Ishibuchi, H. Morimoto, T. Oe, T. Ikebe, H. Inoue, A. Fukunari, M. Kamezawa, I. Yamada, Y. Naka, *Bioorg. Med. Chem. Lett.* , **2001**, *11* (7), 879-882, *Synthesis and structure–activity relationships of 1-Phenylpyrazoles as xanthine oxidase inhibitors.*

[31]   C. Lamberth, *Heterocycles,* **2007**, *71*, 1467-1502, *Pyrazole Chemistry in Crop Protection.*

[32]   C. B. Vicentini, C. Romagnoli, E. Andreotti, D. Mares, *J. Agric. Food Chem.,* **2007**, *55* (25), 10331-10338, *Synthetic Pyrazole Derivatives as Growth Inhibitors of Some Phytopathogenic Fungi.*

[33]   S. Fustero, R. Román, J. F. Sanz-Cervera, A. Simón-Fuentes, J. Bueno, S. Villanova, *J. Org. Chem.,* **2008**, *73* (21), 8545-8552, *Synthesis of New Fluorinated Tebufenpyrad Analogs with Acaricidal Activity Through Regioselective Pyrazole Formation.*

[34]   S. Kumari, S. Paliwal, R. Chauhan, *Synth. Commun.,* **2014**, *44* (11), 1521-1578, *Synthesis of Pyrazole Derivatives Possessing Anticancer Activity: Current Status.*

[35]   B. F. Abdel-Wahab, H. A. Mohamed, *Turk. J. Chem.,* **2012**, *36* (6), 805-826, *Synthetic access to benzazolyl (pyrazoles, thiazoles, or triazoles).*

[36]   H. K. Arora, S. Jain, *Pharm. Lett.,* **2013**, *5* (1), 340-354, *The synthetic development of pyrazole nucleus: from reflux to microwave.*

[37]   S. Dadiboyena, A. Nefzi, *Eur. J. Med. Chem.,* **2011**, *46* (11), 5258-5275, *Synthesis of functionalized tetrasubstituted pyrazolyl heterocycles – A review.*

[38]   S. Fustero, M. Sánchez-Roselló, P. Barrio, A. Simón-Fuentes, *Chem. Rev.,* **2011**, *111* (11), 6984-7034, *From 2000 to Mid-2010: A Fruitful Decade for the Synthesis of Pyrazoles.*

[39]   S. Rajappa, *Heterocycles,* **1977**, *7* (1), 507-527, *Recent developments in the synthesis of heterocycles from enamines and isothiocyanates.*

[40]   S. S. Rajput, S. N. Patel, S. B. Chaudhari, *World J. Pharm. Res.,* **2014**, *3* (10), 1151-1172, *Pyrazole: synthesis, reaction and biological activity.*

[41]   L. Yet, *Prog. Heterocycl. Chem.,* **2012**, *24*, 243-279, *Five-membered ring systems: with more than one N atom.*

[42]   G. Molteni, *ARKIVOC,* **2007**, (2), 224-246, *Silver(I) salts as useful reagents in pyrazole synthesis.*

[43]   M. Kopp, J.-C. Lancelot, P. Dallemagne, S. Rault, *J. Het. Chem.,* **2001**, *38* (5), 1045-1050, *Synthesis of novel pyrazolopyrrolopyrazines, potential analogs of sildenafil.*

[44]   T. M. Abu Elmaati, F. M. El-Taweel, *J. Het. Chem.,* **2004**, *41* (2), 109-134, *New trends in the chemistry of 5-aminopyrazoles.*

[45]   X. Liang, Y. Huang, J. Zang, Q. Gao, B. Wang, W. Xu, Y. Zhang, *Bioorg. Med. Chem., Design, synthesis and preliminary biological evaluation of 4-aminopyrazole derivatives as novel and potent JAKs inhibitors.*

[46]   A. A. Zabierek, K. M. Konrad, A. M. Haidle, *Tetrahedron Lett.,* **2008**, *49* (18), 2996-2998, *A practical, two-step synthesis of 1-alkyl-4-aminopyrazoles.*

[47]   W. Huang, S. Liu, B. Chen, X. Guo, Y. Yu, *RSC Adv.,* **2015**, *5* (41), 32740-32743, *Synthesis of polysubstituted 4-aminopyrazoles and 4-hydroxypyrazoles from vinyl azides and hydrazines.*

[48]   E. Aiello, S. Aiello, F. Mingoia, A. Bacchi, G. Pelizzi, C. Musiu, M. Giovanna Setzu, A. Pani, P. La Colla, M. Elena Marongiu, *Bioorg. Med. Chem.,* **2000**, *8* (12), 2719-2728, *Synthesis and antimicrobial activity of new 3-(1-R-3(5)-methyl-4-nitroso-1H-5(3)-pyrazolyl)-5-methylisoxazoles.*

[49]   L. Cecchi, V. Colotta, F. Melani, G. Palazzino, G. Filacchioni, C. Martini, G. Giannaccini, A. Lucacchini, *J. Pharm. Sci.,* **1989**, *78* (6), 437-442, *Synthesis of 1,5-diaryl-3-methyl-1 H-pyrazolo[4,5-c]isoquinolines and studies of binding to specific peripheral benzodiazepine binding sites.*

[50]   D. L. Browne, M. D. Helm, A. Plant, J. P. A. Harrity, *Angew. Chem. Int. Ed.,* **2007**, *46* (45), 8656-8658, *A Sydnone Cycloaddition Route to Pyrazole Boronic Esters.*

[51]   E. Decuypere, S. Specklin, S. Gabillet, D. Audisio, H. Liu, L. Plougastel, S. Kolodych, F. Taran, *Org. Lett.*, **2015**, *17* (2), 362-365, *Copper(I)-Catalyzed Cycloaddition of 4-Bromosydnones and Alkynes for the Regioselective Synthesis of 1,4,5-Trisubstituted Pyrazoles.*

[52]   P. Šimůnek, M. Svobodová, V. Bertolasi, V. Macháček, *Synthesis*, **2008**, (11), 1761-1766, *Facile and Straightforward Method Leading to Substituted 4-Amino-1-arylpyrazoles.*

[53]   M. Takahashi, H. Kikuchi, *Tetrahedron Lett.*, **1987**, *28* (19), 2139-2142, *Ring transformation reaction of 1,2,4,5-tetrazines to 4-aminopyrazoles by cyanotrimethylsilane.*

[54]   H. F. Anwar, M. H. Elnagdi, *ARKIVOC*, **2009**, (1), 198-250, *Recent developments in aminopyrazole chemistry.*

[55]   S. Yoshida, K. Igawa, K. Tomooka, *J. Am. Chem. Soc.*, **2012**, *134* (47), 19358-19361, *Nucleophilic Substitution Reaction at the Nitrogen of Arylsulfonamides with Phosphide Anion.*

[56]   K. Moriyama, Y. Nakamura, H. Togo, *Org. Lett.*, **2014**, *16* (14), 3812-3815, *Oxidative Debenzylation of N-Benzyl Amides and O-Benzyl Ethers Using Alkali Metal Bromide.*

[57]   P. G. M. Wuts, *Greene's Protective Groups in Organic Synthesis*; Wiley, 2014.

[58]   M. Meldal, C. W. Tornøe, *Chem. Rev.*, **2008**, *108* (8), 2952-3015, *Cu-Catalyzed Azide–Alkyne Cycloaddition.*

[59]   G. Wittig, A. Krebs, *Chem. Ber.*, **1961**, *94* (12), 3260-3275, *Zur Existenz niedergliedriger Cycloalkine, I.*

[60]   N. J. Agard, J. A. Prescher, C. R. Bertozzi, *J. Am. Chem. Soc.*, **2004**, *126* (46), 15046-15047, *A Strain-Promoted [3 + 2] Azide–Alkyne Cycloaddition for Covalent Modification of Biomolecules in Living Systems.*

[61]   J. Tummatorn, P. Batsomboon, R. J. Clark, I. V. Alabugin, G. B. Dudley, *J. Org. Chem.*, **2012**, *77* (5), 2093-2097, *Strain-Promoted Azide–Alkyne Cycloadditions of Benzocyclononynes.*

[62]   J. Dommerholt, O. van Rooijen, A. Borrmann, C. F. Guerra, F. M. Bickelhaupt, F. L. van Delft, *Nat. Commun.*, **2014**, *5*, 5378, *Highly accelerated inverse electron-demand cycloaddition of electron-deficient azides with aliphatic cyclooctynes.*

[63]   J. Dommerholt, F. P. J. T. Rutjes, F. L. Delft, *Top. Curr. Chem.*, **2016**, *374* (2), 1-20, *Strain-Promoted 1,3-Dipolar Cycloaddition of Cycloalkynes and Organic Azides.*

[64]   L. Plougastel, O. Koniev, S. Specklin, E. Decuypere, C. Creminon, D.-A. Buisson, A. Wagner, S. Kolodych, F. Taran, *Chem. Commun.*, **2014**, *50* (66), 9376-9378, *4-Halogeno-sydnones for fast strain promoted cycloaddition with bicyclo-[6.1.0]-nonyne.*

[65]   C. Wentrup, R. Blanch, H. Briehl, G. Gross, *J. Am. Chem. Soc.*, **1988**, *110* (6), 1874-1880, *Benzyne, cyclohexyne, and 3-azacyclohexyne and the problem of cycloalkyne versus cycloalkylideneketene genesis.*

[66]   S. F. Tlais, R. L. Danheiser, *J. Am. Chem. Soc.*, **2014**, *136* (44), 15489-15492, *N-Tosyl-3-Azacyclohexyne. Synthesis and Chemistry of a Strained Cyclic Ynamide.*

[67]   A. V. Dubrovskiy, N. A. Markina, R. C. Larock, *Org. Biomol. Chem.*, **2013**, *11* (2), 191-218, *Use of benzynes for the synthesis of heterocycles.*

[68]   C. M. Gampe, E. M. Carreira, *Angew. Chem. Int. Ed.*, **2012**, *51* (16), 3766-3778, *Arynes and Cyclohexyne in Natural Product Synthesis.*

[69]   Y. Himeshima, T. Sonoda, H. Kobayashi, *Chem. Lett.*, **1983**, *12* (8), 1211-1214, *Fluoride-induced 1,2-elimination of o-trimethylsilylphenyl triflate to benzyne under mild conditions.*

[70]   R. Sanz, *Org. Prep. Proc. Int.*, **2008**, *40* (3), 215-291, *Recent applications of aryne chemsitry to organic synthesis. A review.*

[71]   F. Shi, J. P. Waldo, Y. Chen, R. C. Larock, *Org. Lett.*, **2008**, *10* (12), 2409-2412, *Benzyne click chemistry: synthesis of benzotriazoles from benzynes and azides.*

[72]   P. M. Tadross, B. M. Stoltz, *Chem. Rev.*, **2012**, *112* (6), 3550-3577, *A Comprehensive History of Arynes in Natural Product Total Synthesis.*

[73]   A. E. Goetz, S. M. Bronner, J. D. Cisneros, J. M. Melamed, R. S. Paton, K. N. Houk, N. K. Garg, *Angew. Chem. Int. Ed.*, **2012**, *51* (11), 2758-2762, *An Efficient Computational Model to Predict the Synthetic Utility of Heterocyclic Arynes.*

[74]   A. E. Goetz, N. K. Garg, *J. Org. Chem.*, **2014**, *79* (3), 846-851, *Enabling the Use of Heterocyclic Arynes in Chemical Synthesis.*

[75]   T. C. McMahon, J. M. Medina, Y.-F. Yang, B. J. Simmons, K. N. Houk, N. K. Garg, *J. Am. Chem. Soc.*, **2015**, *137* (12), 4082-4085, *Generation and Regioselective Trapping of a 3,4-Piperidyne for the Synthesis of Functionalized Heterocycles.*

[76]   H. Schuhmacher, B. Beile, H. Meier, *Helv. Chim. Acta*, **2013**, *96* (2), 228-238, *1-Thiacyclooct-4-yne (5,6-didehydro-3,4,7,8-tetrahydro-2H-thiocin), and Its Sulfoxide and Its Sulfone.*

[77]   T. Hagendorn, S. Bräse, *RSC Adv.*, **2014**, *4* (30), 15493-15495, *A new route to dithia- and thiaoxacyclooctynes via Nicholas reaction.*

[78]   H. Meier, E. Stavridou, C. Storek, *Angew. Chem.*, **1986**, *98* (9), 838-839, *1-Thia-2-cyclooctyne: a strained cycloalkyne with a polarized triple bond.*

[79]   B. Gold, G. B. Dudley, I. V. Alabugin, *J. Am. Chem. Soc.*, **2013**, *135* (4), 1558-1569, *Moderating Strain without Sacrificing Reactivity: Design of Fast and Tunable Noncatalyzed Alkyne-Azide Cycloadditions via Stereoelectronically Controlled Transition State Stabilization.*

[80]   R. Ni, N. Mitsuda, T. Kashiwagi, K. Igawa, K. Tomooka, *Angew. Chem., Int. Ed.*, **2015**, *54* (4), 1190-1194, *Heteroatom-embedded Medium-Sized Cycloalkynes: Concise Synthesis, Structural Analysis, and Reactions.*

[81]   J. M. Baskin, J. A. Prescher, S. T. Laughlin, N. J. Agard, P. V. Chang, I. A. Miller, A. Lo, J. A. Codelli, C. R. Bertozzi, *Proc. Natl. Acad. Sci.*, **2007**, *104* (43), 16793-16797, *Copper-free click chemistry for dynamic in vivo imaging.*

[82]   N. J. Agard, J. M. Baskin, J. A. Prescher, A. Lo, C. R. Bertozzi, *ACS Chemical Biology*, **2006**, *1* (10), 644-648, *A Comparative Study of Bioorthogonal Reactions with Azides.*

[83]   V. Kostova, E. Dransart, M. Azoulay, L. Brulle, S.-K. Bai, J.-C. Florent, L. Johannes, F. Schmidt, *Bioorg. Med. Chem.*, **2015**, *23* (22), 7150-7157, *Targeted Shiga toxin–drug conjugates prepared via Cu-free click chemistry.*

[84]   M. K. Schultz, S. G. Parameswarappa, F. C. Pigge, *Org. Lett.*, **2010**, *12* (10), 2398-2401, *Synthesis of a DOTA−Biotin Conjugate for Radionuclide Chelation via Cu-Free Click Chemistry.*

[85]   H. E. Gottlieb, V. Kotlyar, A. Nudelman, *J. Org. Chem.*, **1997**, *62* (21), 7512-7515, *NMR Chemical Shifts of Common Laboratory Solvents as Trace Impurities.*

Chapter 5. Axially chiral dibenzo-1,3-diazepines

The research presented in this chapter has been conducted at the Nanotechnology and Molecular Science Synthesis Laboratory at the Queensland University of Technology in Brisbane, Australia (with Dr. Kye-Simeon MASTERS) and has been published as T. Wezeman, Y. Hu, S. Bräse, J. McMurtrie and K.-S. Masters, Aust. J. Chem. *2015*, **68**, 1859 – 1865.

5.0 Abstract & Graphical abstract

Non-symmetrical and axially-chiral dibenzo-1,3-diazepines have been prepared in four steps from commercially-available anilines. The key transformation was a novel Pd/CPhos-promoted direct arylation of the corresponding halo-substituted bis-anilines. These were themselves prepared by connection of a 2-haloaniline with N-(chloromethyl)anilines that were prepared *in situ* from their methyl sulfide precursors. This innovative and novel route to these highly interesting atropisomeric dibenzo-1,3-diazepines allows access with unprecedented ease to all sorts of new asymmetric chiral 7-membered N-heterocyclic carbenes that can be used as ligands, for catalysis or for screenings of potential drug candidates.

5.1 The benzodiazepine family and their potential

A short perspective on the history of benzodiazepines

The discovery of the benzodiazepine compound class – starting with the accidental discovery of chlordiazepoxide (**5.1**) (LIBRIUM) in 1955 by Leo STERNBACH[1] – had a tremendous effect on the field of medicine.[2] In the medical profession benzodiazepines have been found to enhance the effect of neurotransmitter gamma-aminobutyric acid (GABA) on the GABA$_A$ receptor and they are being used as minor tranquilizers, calming and sleeping aids, muscular relaxants, anti-epileptics and anxiolytics (reducing feelings of fear and anxiety). In the 1960s HOFFMANN-LA ROCHE synthesized several other diazepines and was responsible for their widespread production, to the extent that in 1977 benzodiazepines were the globally most prescribed medication.[3]

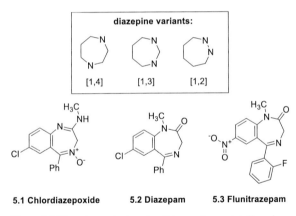

Figure 5.1 Diazepine variants and well-known benzo-1,4-diazepines.

The benzodiazepine family can easily be divided into several groups, depending on the position of the two nitrogen atoms in the 7-membered ring (see Figure **5.1**). Most medically relevant benzodiazepines share an aryl group at the C-5 position and are 1,4-diazepines. Some benzodiazepines are being produced and prescribed on an enormous scale (benzodiazepine prescriptions in the USA alone were 80.1 million in 2006[4]) and have to some degree even managed to become part of the "pop" culture, e.g. they are as well-known as aspirin, paracetamol and codeine. Perhaps the most famous 1,4-diazepine to date is diazepam (**5.2**), sold as 'VALIUM' and known for its anti-anxiety and sleep-inducing abilities – it was approved by the American Food and Drug Administration in 1963, only a few years after its predecessor LIBRIUM (**5.1**), and was found to be roughly 2.5 times as potent. Additionally diazepam is known for its euphoric effects and is abused as a mind-altering drug. Flunitrazepam (**5.3**), also known as ROHYPNOL or Roofies, is a 1,4-diazepine infamous for its abuse as a date rape drug.

A compound class that shares some of these very intriguing pharmaceutical and biological effects are the 1,3-diazepines – although those have only been explored in considerably less detail. It is safe to say that the benzodiazepines have a rich history in terms of biological activity. The development of novel compounds within this class can be considered highly interesting due their great potential.

The rise of the _N_-heterocyclic carbene

Besides the promise of interesting biological activity, the 1,3-benzodiazepine scaffold can be used to solve a very broad range of different issues. Amongst their many features, there is one rather unexplored field that is highly interesting to the community of synthetic organic chemists and shows great promise for producing new catalytic agents. The 2005 Nobel Prize in Chemistry, given to Yves CHAUVIN, Richard R. SCHROCK and Robert H. GRUBBS for their development of the metathesis reaction relies partly on _N_-heterocyclic carbenes. One of the most commonly used catalysts for metathesis reactions, the Grubbs type II catalyst (**5.4**) features an _N_-heterocyclic carbene (NHC) as a ligand. Considering the synthetic importance of NHC it is surprising to see that for the first 20 years after WANZLICK's early discoveries in the 1960s[5,6] the use of NHC ligands in metal complexes remained almost absent, due to issues with instability and reactivity. However, ever since 1991, when ARDUENGO reported the first isolation of a remarkably stable NHC **5.5**, the field has been steadily expanding.[7] The electron-rich, neutral σ-donating abilities of NHCs make them very attractive ligands for all types of reactions.[5,7-15]

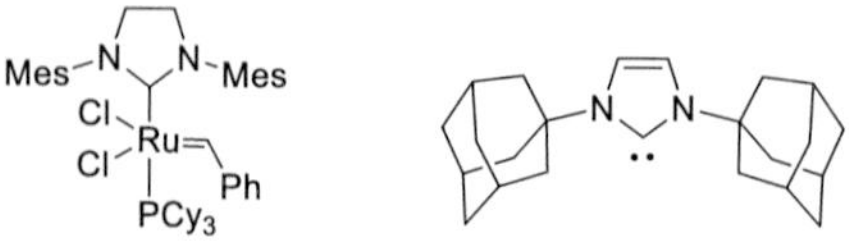

5.4 Grubbs II   **5.5 Arduengo**

Figure 5.2 Important _N_-heterocyclic carbenes.

In a search for different reactivities[16] the research community has explored a wide range of analogs, but the vast majority of NHCs are still monodentate and based upon imidazole or other 5-membered rings (*i.e.* '5NHCs') and only a few compound classes have been found to deviate from this pattern. For example: currently literature on 7-membered ring NHCs ('7NHCs) is close to non-existing, with the work of the STAHL group being the most notable exception. A few years ago they[17-21] investigated several 7-membered rings for their utility as NHCs, found that several dibenzo-1,3-diazepines were performing especially well as ligands, and filed a patent claim to support their discovery.[21] It must be noted that STAHL was not the first to grasp the concept of the 7NHC, as RIED and BRÄUTIGAM made some notable contributions to this field in the 1960s.[22]

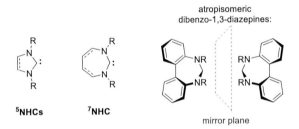

Figure 5.3 General structures of 5- and 7-membered ring NHCs and the axial chirality that can be found in dibenzo-1,3-diazepines.

One key feature that separates the dibenzo-1,3-diazepines from the crowd of NHCs is their special three-dimensional structure. Due to the axially twisted conformation of the dibenzo moiety the ligands are intrinsically chiral and offer enticing possibilities for enantioselective catalysis (see Figure **5.3**). However, aside from the few examples discussed below, this promising field has hardly been investigated – and the reason for that might be of synthetic origin.

A need for innovation – issues with existing diazepine syntheses

It can be argued that the main reason for the lack of examples of axially chiral diazepines is due to a distinctly slow rate of innovation in their synthesis; synthetic strategies have changed remarkably little since LE FEVRE initially reported his condensation reactions in 1929 (see Scheme **5.1**).[23] Via this type of chemistry simple symmetrical dimers from biaryls,[24] such as

2,2'-dinitro- or 2,2'-diamino-biphenyls, are accessible by linking of the two nitrogens.[2,24-28] These condensations of electrophiles to the diamino species often result in urea-type linked dibenzo-1,3-diazepines **5.9**, that can then be readily converted to their corresponding carbenes **5.10**.[12,29-31]

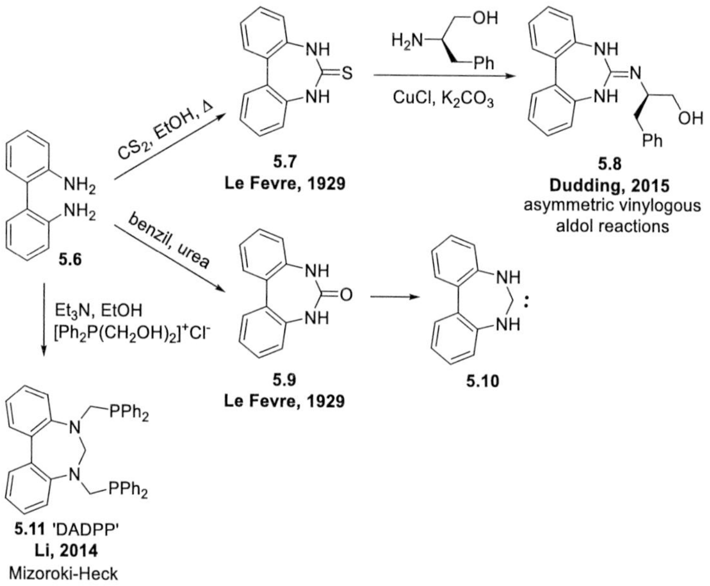

Scheme 5.1 Overview of 2,2'-diamino-biphenyl-based dibenzo-1,3-diazepine syntheses.

A guanidyl-based organocatalyst **5.8** based on LE FEVRE's work was computationally designed and synthesized by DUDDING and co-workers from thiourea **5.7** using copper chloride and potassium carbonate. They found it to be very useful in asymmetric vinylogous aldol reactions.[32] LI's diphospine dibenzodiazepine **5.11** "DADPP" ligand was found to be very valuable as ligand for the palladium-catalyzed Mizoroki-Heck reaction between electron-deficient aryl halides and *n*-butyl acrylate, with turn over numbers of up to 9.6 × 10⁷.[33]

Other routes based on 2,2'-dinitro- or 2,2'-diamino-biphenyls, such as the ones reported by STAHL in 2005 and 2009, rely on a different cyclization method: the formation of an amidinium tetrafluoroborate salt **5.13** and **5.16** with triethyl orthoformate and NH_4BF_4 to produce their final catalysts.[17,18,20] When STAHL first reported the preparation and use of his chiral

^{7}NHC–Pd^{II} salt **5.14**, (Scheme **5.2**), it categorically out-performed related ^{5}NHC catalysts in Wacker-type oxidative cyclization processes.[34]

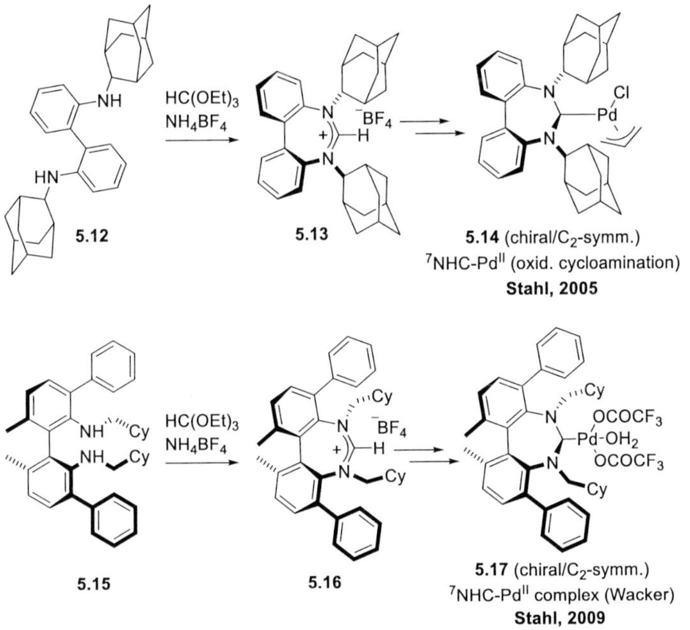
HC(OEt)$_3$
NH$_4$BF$_4$
5.12
5.13
5.14 (chiral/C$_2$-symm.)
^{7}NHC-Pd^{II} (oxid. cycloamination)
Stahl, 2005
HC(OEt)$_3$
NH$_4$BF$_4$
5.15
5.16
5.17 (chiral/C$_2$-symm.)
^{7}NHC-Pd^{II} complex (Wacker)
Stahl, 2009

Scheme 5.2 STAHL's ^{7}NHC–Pd^{II} salt syntheses from 2,2′-diamino-biphenyls with the formation of an amidinium tetrafluoroborate salt as key step.

Besides ligands for palladium-catalyzed reactions ^{7}NHCs have recently also been found to be useful for gold catalysis. A ^{7}NHC–Au^{I} complex **5.19** has been prepared by DUNSFORD *et al.* and found to catalyze the hydration of internal alkynes in a facile manner.[35]

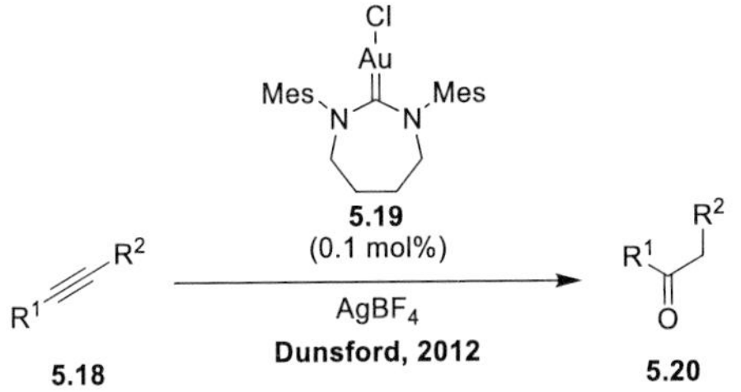
5.19
(0.1 mol%)
AgBF$_4$
Dunsford, 2012
5.18
5.20

Scheme 5.3 DUNSFORD's ^{7}NHC–Au^{I} complex catalyzing the hydration of internal alkynes.

Although the synthesis of diazepines based on 2,2'-dinitro- or 2,2'-diamino-biphenyls is a well-established and effective route to straightforward [7]NHCs, this route is unsuitable for the much more interesting non-symmetrical diazepines; their structures, reactivities and attributes can be tuned in a much more subtle manner than symmetrical 'homo-dimer' diazepines.

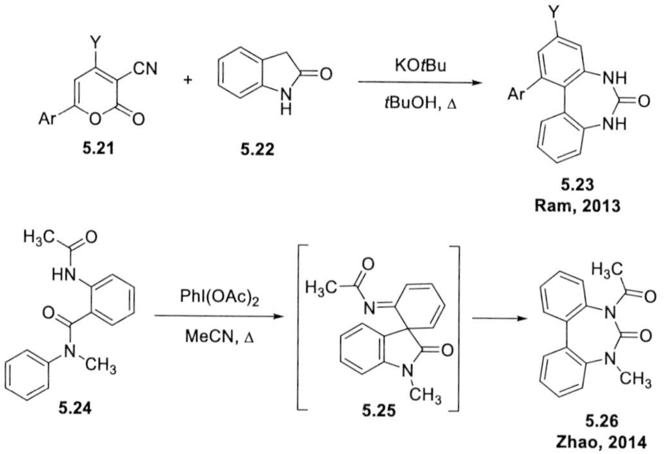

Scheme 5.4 Two rare examples of dibenzo-1,3-diazepine syntheses that are not based on 2,2'-dinitro- or 2,2'-diamino-biphenyls – although not the most practical for modular approaches.

Some of the few alternatives to the above mentioned routes reported in literature include a base-induced ring transformation as developed by RAM and coworkers (Scheme **5.4**). Although in principle this route allows access to unsymmetrical compounds **5.23**, it requires very specific precursors thus limiting the practical usefulness of the reaction.[31,36] Another recent development is a metal-free diazepin-2-one synthesis based on the cyclization of diamides **5.24** with the aid of hypervalent iodine complexes. Although this is a very interesting route it must be noted that they reported only one successful example where the internal *meta* position was substituted. The availability of this position for substituents is crucial in obtaining good axial chirality. A probable cause for the lack of examples of this type could be due to steric issues when forming the intermediate spiro compound **5.25**.

When designing axially chiral and unsymmetrical dibenzo-1,3-diazepines one must consider two main steps; the first step is the formation of the link between the two nitrogen atoms (the N−CH$_2$−N aminal) and the second the connection of the two arenes: the biaryl formation. In order to allow for a highly modular route it was chosen to explore the possibility of forming the aminal link first. Forming the biaryl bond first would pose issues with regioselectivity and require modified substrates, typically a sacrificial electrophile such as a stannane or boronate, to direct the reaction. Forming the aminal link first allows the use of the link to act as molecular tether, directing the subsequent CH activation to the *ortho* position. More on this technique and its alternatives is discussed in Chapter 5.3. However, the aminal formation is not trivial either, since a straightforward homodimerization of two anilines would not lead to an asymmetric product. In order to overcome this clear synthetic shortcoming a route that allows full synthetic flexibility and employs a modular approach needed to be developed.

5.2 Synthesis of axially chiral dibenzo-1,3-diazepines: aminal formation

The first step, involving the dimerization of two non-equivalent anilines was found to be a challenge of its own. At first glance it seems sensible to simply react aniline **5.30** with a dihalomethane, such as diiodomethane, forming the iodomethane-aniline [N−CH$_2$−I] **5.31**, which can then react in a second step with a different aniline. Giving this route a bit more thought one quickly realizes that there are some inherent issues one would like to avoid: The nitrogen atom in the iodomethane-aniline [N−CH$_2$−I] **5.31** activates the iodo group as a leaving group. In practical terms this means that the moment aniline **5.30** reacts it forms an even more reactive species, that will react with the first available nucleophile it encounters: a second aniline **5.30**. Therefore this route is not suitable for non-symmetrical couplings, since isolation of the intermediate iodomethyl is not favored. Attempts to form the phenol-based halomethyl **5.29** using ICH$_2$Cl instead of diiodomethane did not reduce the homo-dimerization to **5.27** significantly.[37]

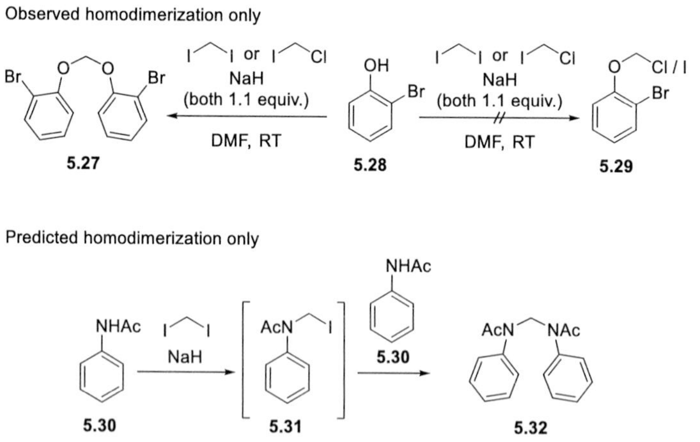

Observed homodimerization only

Predicted homodimerization only

Scheme 5.5 Issues when using dihalomethanes; only homodimers are accessible.[37]

Other classical methods to form homodimers of the desired type involve the condensation of the anilines with formaldehyde, but these procedures would lead to mixtures of aminals when performed with non-equivalent anilines.

Sometimes all you need is a strong vacuum

In the late 1980s BARLUENGA and his coworkers worked on the synthesis of related (1,3-bis-aryl) aminals. They reported three different methods for the unsymmetrical dimerization of anilines. Sadly, two of these methods involve the use of stoichiometric amounts of phenyllithium, something that would not be compatible with our desire to include haloarenes for the subsequent ringclosure. They also found that the phenyllithium routes mostly produced mixtures of A-A, A-B and B-B aminals, due to a side-reaction promoted by *in-situ* formed lithium methoxide. Their third route (shown in Scheme **5.6**) involves equimolar amounts of a *N*-(methoxymethyl)-aniline **5.32** and aniline, that react at room temperature but under strong vacuum to form the desired aminal product by elimination of methanol.[38] Sadly, attempts to apply this promising methodology to our substrates failed, likely due to the facile transformation of the *N*-(methoxymethyl)anilines to 1,3,5-triazines **5.34**, a side-reaction also reported by the BARLUENGA group.[39]

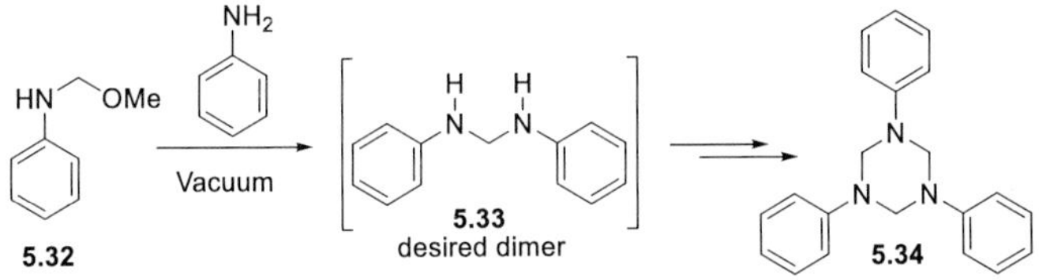

Scheme 5.6 BARLUENGA's aminal formation using vacuum, leading mostly to side-product **5.34**.

The acetal linker

Inspired by previous work from the MASTERS and BRÄSE groups,[37,40] where acetal links between two phenols are being prepared using a three-step procedure (see Scheme **5.7**), a similar approach, that would be applicable in the aminal setting, was designed. In their three-step procedure they first react phenol **5.35** with chloromethyl methyl sulfide under basic conditions at room temperature. Next this methyl sulfide **5.36** is converted into a methyl chloride **5.37** with use of sulfuryl chloride and finally reacted with another phenol **5.38**. This way asymmetrical products are easily accessible. If desired, subsequent ringclosure can be done in several ways, as will be discussed in Chapter 5.3. Ringclosure of dibenzo acetals **5.39** to dibenzo-1,3-dioxepines **5.40** was done with base-promoted homolytic aromatic substitution (BHAS) reactions.[37,41] The acetal-linked dibenzo-1,3-dioxepines **5.40** could be hydrolyzed to their *ortho,ortho*-diphenol counterparts **5.41** with hydrochloric acid in ethanol. This methodology has been used to selectively access these *ortho,ortho*-diphenols, since they are difficult to prepare using alternative approaches and feature in several interesting natural products.

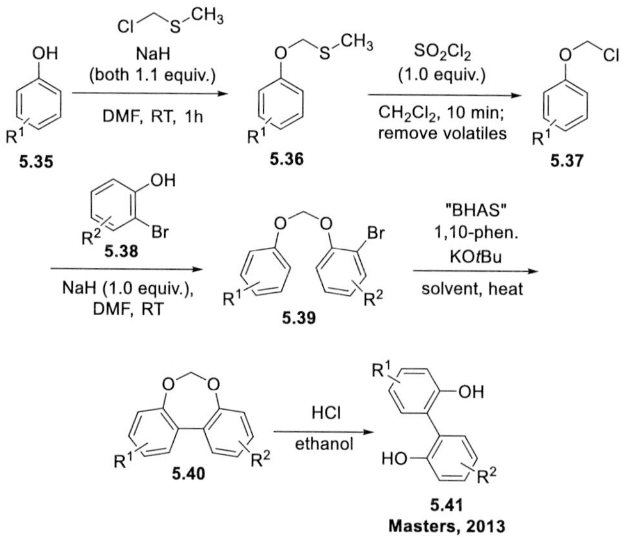

Scheme 5.7 Synthesis of unsymmetrical acetals **5.39** via methyl sulfide **5.36** and their subsequent cyclization and hydrolyzation.

Fortunately the acetal procedure described in Scheme **5.7** turned out to be adaptable to our situation, where an aminal link is desired. However, a few modifications to the established protocols were required. Initial experiments to form the methyl sulfide resulted in rather poor yields, so some reaction optimizations and temperature screens were performed. It was determined that the first step, the methyl sulfide formation, needs to be done at temperatures below zero and requires a strong base. Attempts to use unprotected anilines led to low conversions and would merely complicate subsequent steps. Therefore all anilines were acetylated before reacting them with chloromethyl methyl sulfide.

Table 5.1 Optimization of reaction conditions for the methyl sulfide formation.

Entry	Base (3.0 equiv.)	T [°C]	Yield [%][a]
1	K_2CO_3	90	-
2	Cs_2CO_3	90	-
3	Et_3N	90	-
4	DABCO	90	-
5	NaH	90	Traces
6	NaH	-5[b]	80[c]

a) Determined by ^{1}H-NMR; b) chloromethyl methyl sulfide was added dropwise over 10 minutes; c) isolated yield.

The resulting methyl sulfides **5.43** are then quantitatively transformed to their methyl chloride counterparts **5.44** by reaction with sulfuryl chloride at room temperature in 5-10 minutes. Gentle evaporation of volatiles (rotary evaporator, water bath at room temperature) delivered the pure products quantitatively (by ^{1}H-NMR). Typically the methyl chlorides were used directly and without delay for the subsequent reaction, due to their presumed sensitivity. Nonetheless, several appeared to be stable for at least 24h at room temperature in deuterated chloroform. The chlorides were added to a cold, vigorously stirred solution of potassium *tert*-butoxide and a N-acetyl-2-haloaniline **5.45**. Resulting aminal compounds **5.46** were purified on neutral aluminium oxide or via preparative HPLC (C_{18} column, water/THF) since observations and literature precedent[38] indicated they are not stable to acidic silica gel.

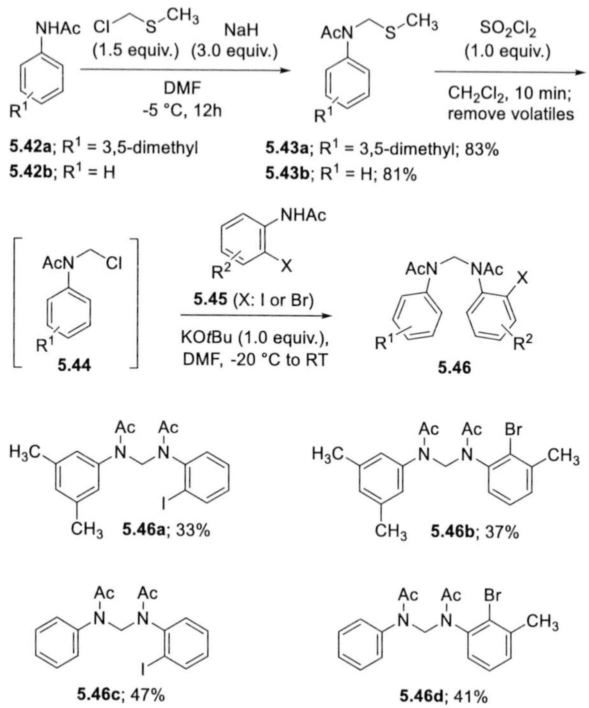

Scheme 5.8 Modular synthesis of methylene-linked bis-anilines (yields of isolated products).

When designing the methylene-linked bis-anilines **5.46** there are two clear routes that can be followed, depicted in Scheme **5.9**. One could either connect a methyl sulfide-bearing aniline **5.43** with a *ortho*-halo aniline **5.45**, e.g. Route A, or prepare a precursor that bears both the methyl sulfide linker and the halogen, like compound **5.47**, and connect with any aniline, e.g. route B. The latter route offers a clear and significant synthetic advantage since it would increase the substrate pool the chemist has to its disposal tremendously by allowing pairing with any aniline instead of being limited to only *ortho*-halo anilines. This same idea, route B, has been proven to work very efficiently with the phenol-based system.[40]

However, initial tests to form the methylene sulfide from *N*-Boc-protected 2-bromo anilines showed only very low conversion, as analyzed by GCMS (~10% product) and was therefore not investigated into greater extent. It remains however unclear if the reaction was unfavorable due to inherent issues with the *N*-Boc compound (e.g. steric or electronic effects) or if the observations only relate to a lack of optimization. Regardless; for the time being it was decided to continue with route A, since the methylene sulfide formation to *N*-acetyl **5.43** was found to proceed rather well.

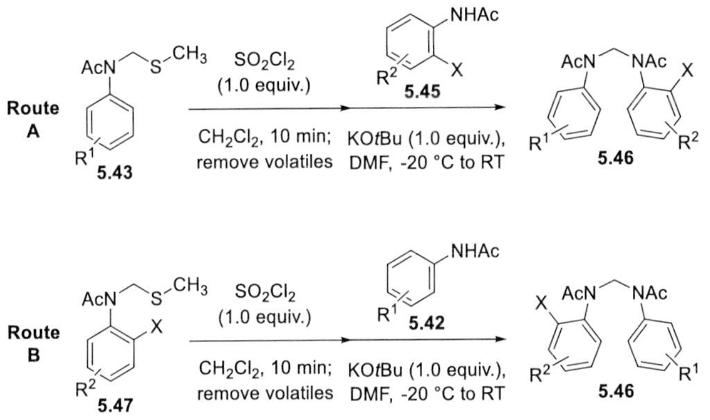

Scheme 5.9 Two possible routes to *ortho*-halogenated methylene-linked bis-anilines.

Later, in an effort to follow-up on my initial investigations, the MASTERS group found optimized conditions for the methylene sulfide formation on 2-bromo *N*-acetyl anilines **5.45** and was able to get the desired products **5.47** in decent yields. This indicates that most likely the Boc protecting group used at that stage hindered the reaction. Additionally, their unpublished work shows that the equivalents of chloromethyl methyl sulfide and sodium hydride can be reduced to 1.10 equivalents.

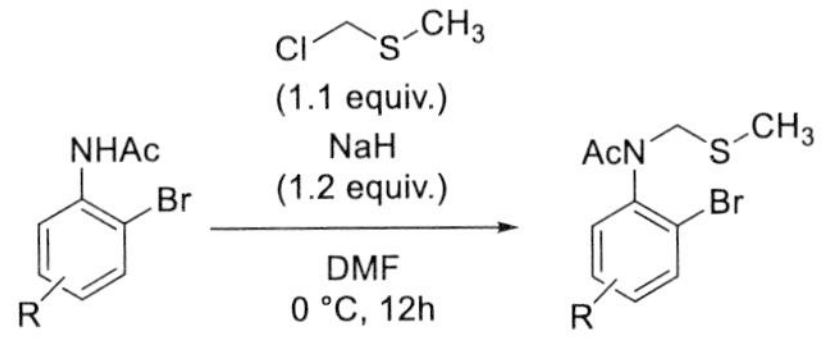

5.45a; R = 3-methyl   **5.47a**; R = 3-methyl; 73%
5.45b; R = 3,5-dimethoxy   **5.47b**; R = 3,5-dimethoxy; 70%

Scheme 5.10 Synthesis of methylene sulfide *ortho*-bromo anilines.

During the course of this work all anilines were *N*-protected before use to prevent unwanted side-reactions. For simplicity only the *N*-acetyl group was used as protecting group, but it must be noted that using different protecting groups may prove far more desirable at a later stage. This is because asymmetrically protected bis-anilines can also be orthogonally deprotectted, thus allowing for a modular approach to a wide range of asymmetrically N_1,N_2-substituted diazepines, that could all offer various reactivies as [7]NHCs or drug candidates.

5.3 Synthesis of axially chiral dibenzo-1,3-diazepines: biaryl bond formation

With a clear route toward (un)symmetrical aminals, all that remained was forming the ring. Fortunately biaryl bond formation has been a thoroughly studied field and there are several options available, with protocols ranging from more classical Ullmann reactions to modern transition metal catalysis and metal-free base-promoted homolytic aromatic substitution (BHAS) reactions.

Regardless of the route chosen to achieve the ring formation, one key requirement must be met. The reaction needs to be regiospecific in order to obtain the correct 7-membered ring and avoid intermolecular mixtures.

Biaryl bond formation

In the Ullmann reaction (Scheme **5.11**) – named after Fritz ULLMANN, who discovered it in 1901[42] – the regioselectivity comes from the positions of the two halogens on the separate aryl rings. Besides the need for two halogens, this approach suffers from two clear drawbacks: It usually requires quite harsh conditions and is infamous for its erratic yields. The Ullmann reaction is believed to involve the coupling of the two haloarenes via formation of an organocopper intermediate and subsequent nucleophilic aromatic substitution. Although in more recent years several improvements to the original protocol have been reported,[43-53] and even asymmetric dioxepines have been prepared based on Ullmann chemistry,[54] this procedure seemed less than ideal for the diazepine synthesis.

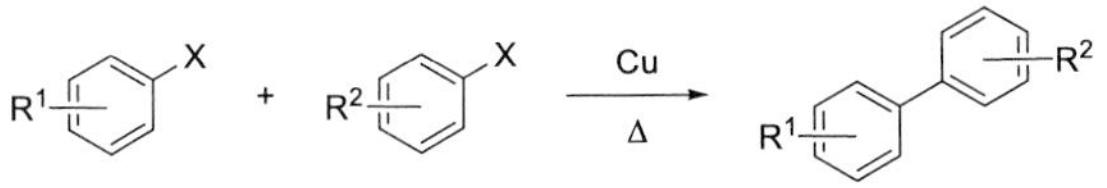

$$R^1 \text{—} X + R^2 \text{—} X \xrightarrow[\Delta]{Cu} R^1 \text{—} R^2$$

| 5.48 | 5.49 | 5.50 |

Scheme 5.11 The Ullmann reaction.

A very popular alternative – one could hardly imagine modern chemistry without it – is classic transition metal catalysis. These well-studied reactions, that are often palladium-catalyzed, typically employ a haloarene and stannanes (Stille), boronates (Suzuki), Grignard reagents (Kumada), organosilanes (Hiyama) or organozinc compounds (Negishi).[55-58]

Although these reactions are very frequently employed in many syntheses, the need for both a halogen and a sacrificial electrophile is a big disadvantage when considering a modular approach.

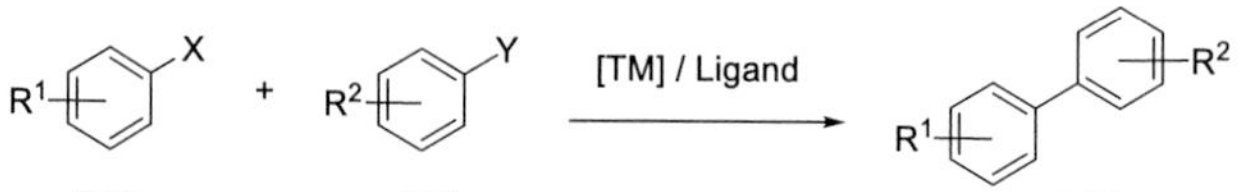

$$R^1\text{—}\underset{\textbf{5.48}}{\bigcirc}\text{—}X \;+\; R^2\text{—}\underset{\substack{\textbf{5.51}\\ Y = SnR_3, BR_2, MgBr, SiR_3, ZnR}}{\bigcirc}\text{—}Y \xrightarrow{\;[TM]\,/\,Ligand\;} R^1\text{—}\underset{\textbf{5.50}}{\bigcirc}\text{—}\bigcirc\text{—}R^2$$

Scheme 5.12 Different varieties of transition metal-catalyzed biaryl formations that rely on a haloarene and a specifically modified coupling partner. TM = transition metal.

One very elegant and steadily growing field of research is the use of the base-promoted homolytic aromatic substitution (BHAS) reaction. These are transition metal-free radical direct C–H arylation reactions usually involving two arenes, with only one of them bearing a halogen. Typical conditions include the use of potassium *tert*-butoxide in the presence of a catalytic amount of ligand, such as 1,10-phenanthroline. Especially the STUDER[59-63] and MASTERS[41,64] groups have worked on the development of BHAS[65,66] extensively in recent years.

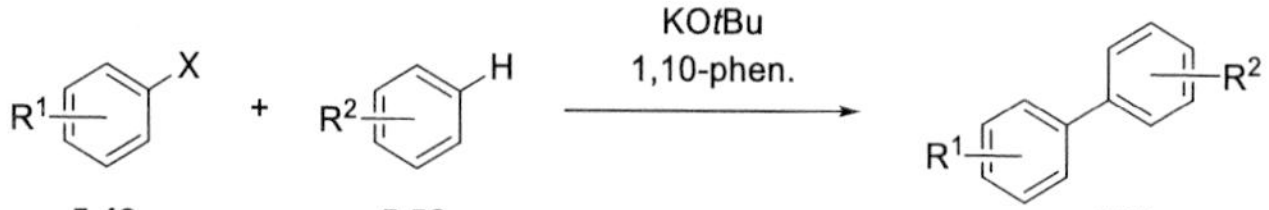

$$R^1\text{—}\underset{\textbf{5.48}}{\bigcirc}\text{—}X \;+\; R^2\text{—}\underset{\textbf{5.52}}{\bigcirc}\text{—}H \xrightarrow[\;1,10\text{-phen.}\;]{KO\textit{t}Bu} R^1\text{—}\underset{\textbf{5.50}}{\bigcirc}\text{—}\bigcirc\text{—}R^2$$

Scheme 5.13 Base-promoted homolytic aromatic substitution (BHAS).[67]

Although the metal-catalyzed methods discussed above offer regiocontrol through the positioning of the halogens and electrophiles on the arenes, the BHAS reaction does not have this advantage. Therefore, the design makes use of the concept of molecular tethering.

Molecular tethering relies on a covalently-bound tether – for example a -CH2- link – between the two heteroatoms on the aryl rings. When an intramolecular biaryl bond is formed this tether imparts regiocontrol. Previously this approach has been shown in the synthesis of biphenols/binaphthols and recently to establish an expeditious multi-bond forming BHAS reaction to form substituted dioxepines.[41]

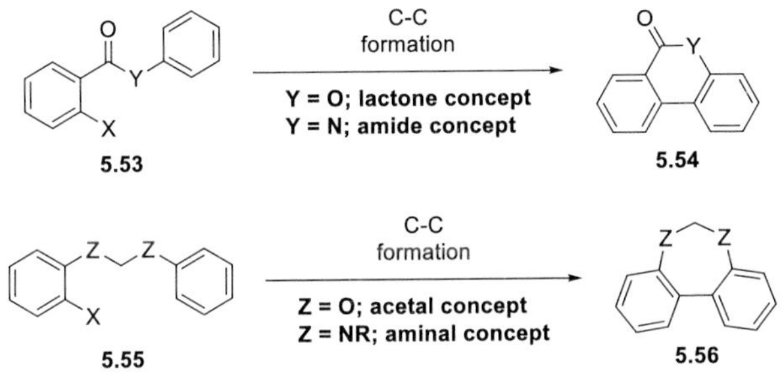

Scheme 5.14 Molecular tethering examples; the lactone,[68-70] the amide,[71] the acetal[37] and the aminal concept.

In the case of dibenzo-1,3-diazepine synthesis, the preparation of aminals **5.46** serve a dual purpose: It allows access to *N,N*-asymmetrical products in a modular way and acts as a tether for full regiocontrol.

Unfortunately several initial attempts to use BHAS reactions to close the diazepine ring were unsuccessful. However, it is believed that in the case of the BHAS reaction of aminal **5.46a** to diazepine **5.57a** a full investigation and detailed reaction conditions screening could certainly yield the desired product. However, to date this has not yet been attempted. Because the BHAS is a radical reaction an attempt was made to cyclize the product with the addition of a radical initiator, but this did not lead to the desired product either.

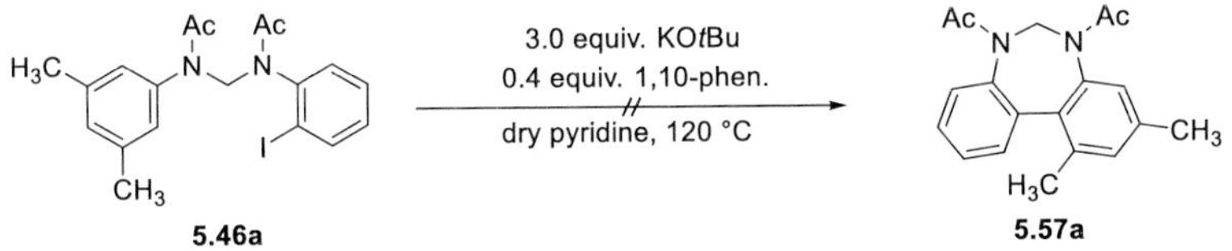

Scheme 5.15 Failed initial attempt to cyclize aminal **5.46a** using BHAS.

The main reason for the lack of further investigation into the BHAS reaction is that an equally direct and elegant alternative was found, using a palladium-catalyzed direct arylation approach.

Direct arylation

The palladium-catalyzed direct arylation approach has gotten a lot of attention by the research community over the last decades.[72,73] Its greatest advantage – it does not require a sacrificial electrophile – also has its challenges, especially in regioselectivity and predictability. Many groups have put great effort into finding suitable catalysts and ligands and to propose possible mechanistic reaction pathways. Especially the FAGNOU group has contributed tremendously to this field, with their results from the years 2004-2006 serving as a starting point for many recent scientific investigations.[74-77]

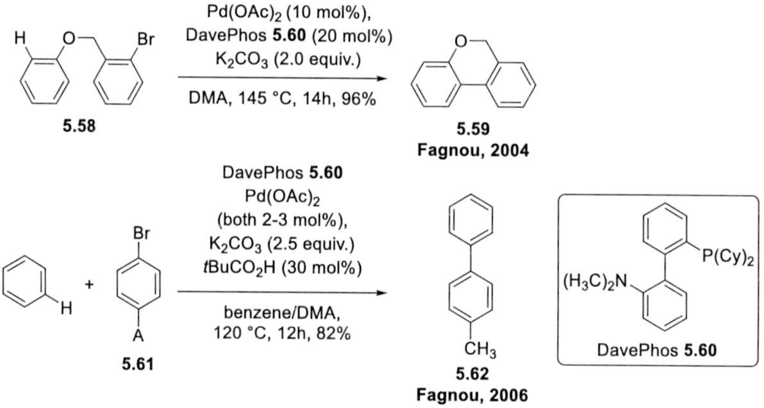

Scheme 5.16 FAGNOU's first results using a direct arylation reaction.[74,77]

Over the years, direct arylation[78-85] approaches have been employed to prepare polycyclic biaryls,[86] natural products,[87-89] and 6,7-dihydrodibenzo[b,d]oxepines.[74] For our work we started out by adapting the conditions from FAGNOU's colchicine work[90]: potassium carbonate in DMA at 145 °C, with a catalyst derived from Pd(OAc)₂ and CPhos (**5.67**) (both 10 mol%). Under these conditions, the reaction involves Pd^0 insertion to the $C_{sp2}-X$ bond, followed by concerted metalation-deprotonation assisted by the Pd^{II}-coordinated acetate. This leads to the *pseudo*-six-membered ring transition state of the concerted metalation-deprotonation event (CMD, see **5.65**, Scheme **5.17**) which underpins the mechanism of C–H direct arylation.[91] Over the last decade several possible mechanistic pathways for the direct arylation have been discussed,[77,86,91-96] and finally the CMD pathway was accepted as most likely. A pivalate, potassium carbonate or acetate is often believed to play an important role in the CMD *pseudo*-6-membered ring transition state. [91,97]

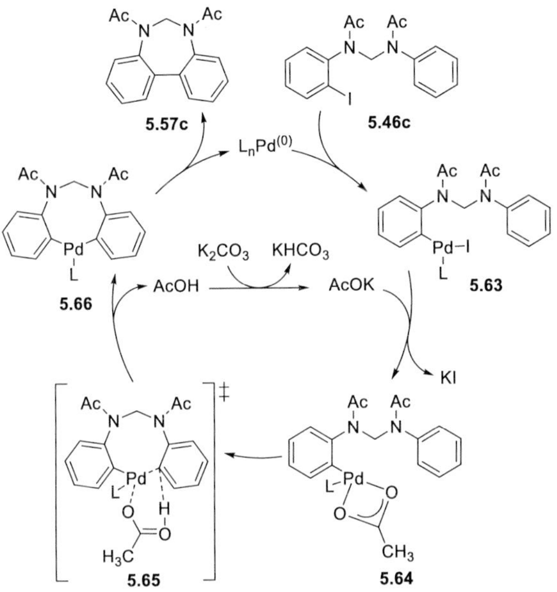

Scheme 5.17 Plausible concerted metalation-deprotonation mechanistic pathway of the direct arylation of the aminals to dibenzo-1,3-diazepines via an acetate-mediated *pseudo*-six-membered transition state.

Although the FAGNOU conditions typically included the use of the electron-rich and hindered biarylphosphine DavePhos (**5.60**),[98] we substituted this with another of BUCHWALD's ligands, the closely related CPhos (**5.67**),[99] which was readily at hand. We were pleased to see full conversion of the substrate in one hour, delivering diazepine **5.57a** in 78% yield (Scheme **5.18**). Cyclizations of the other prepared bis-anilines **5.46b-d** to diazepines **5.57b-d** were also successful under these conditions, although here the yields were relatively poor (29-39%). These initial and unoptimized reaction conditions show great promise and serve as a proof of concept, but it is highly advisable to conduct a short optimization study before continuing with future work.

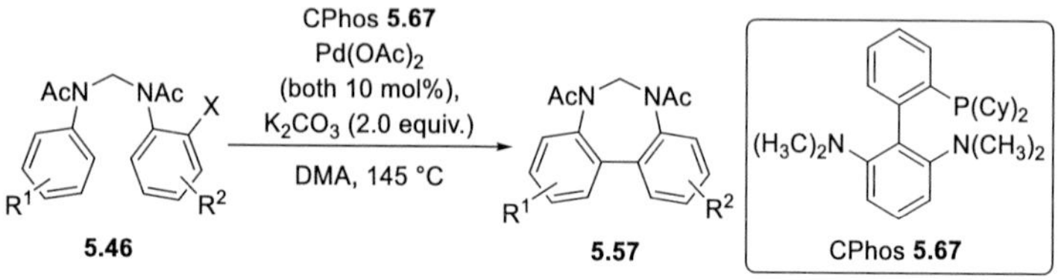

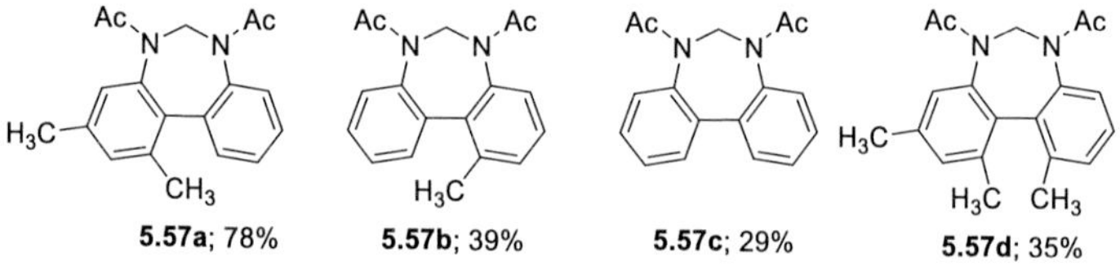

Scheme 5.18 Direct arylation of aminals to the desired dibenzo-1,3-diazepines.

Interestingly, structure-based searches suggest that dibenzo-1,3-diazepines such as **5.57a,b** and **d**, which are not C_2-symmetrical, had not yet been reported. After crystallizing both aminal **5.46a** and its cyclized product **5.57a**, X-ray crystallography by Prof. Dr. John MCMURTRIE from the Queensland University of Technology, using the Australian Synchrotron, confirmed the identity and solid-state configurations for aminal **5.46a** and diazepine **5.57a**. The axial torsion in the diazepine ring was clearly visible (Scheme **5.19** and Figure **5.5**).

Axial chirality

The most common case of chirality in organic chemistry is the chirality that arises when molecules possess stereogenic centers. However, a different type of chirality can also be found, even in case where compounds are symmetrical or possess no stereogenic centers whatsoever. If, due to sterical limitations, a part of the molecule is pushed out of its conventional spatial arrangement in such a way that it is no longer identical to its mirror image, the molecule is axially chiral. Due to steric reasons the otherwise free rotation is limited to such a degree that, on a practical timescale, there is practically no rotation. This is commonly encountered in biaryls, with the BINAP ligands being excellent examples (Figure **5.4**).

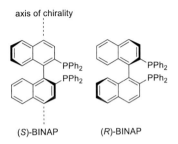

Figure 5.4 Axial chirality due to the twisted biaryl bond.

With the help of Prof. Dr. John McMURTRIE from the Queensland University of Technology crystal structures were obtained for the substrate **5.46a** and the chiral diazepine **5.57a**, Scheme **5.19**. In the solid-state structure of **5.46a** the iodine atom, which is required for the cyclization, points away from the site of C–C bond formation. In the structure of **5.57a** the aminal bridge constrains the rotation around the joined phenyl rings resulting in the axial chirality, which is typically observed for this class of compounds. The C1–N1···N2–C8 torsion angle is 79° resulting in a dihedral angle of 50° between the mean planes of the benzene rings.

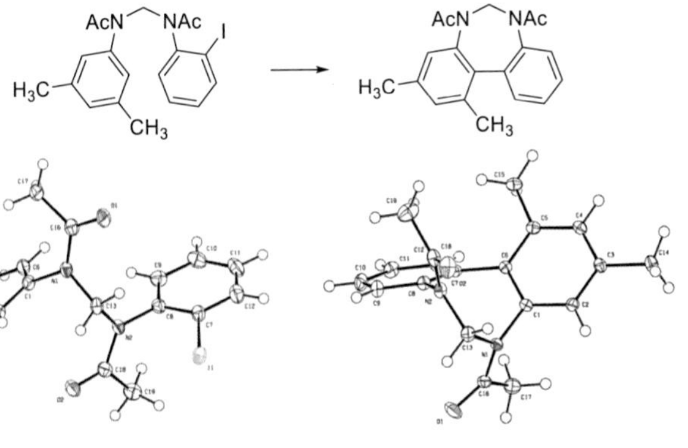

5.46a   **5.57a**

Scheme 5.19 Molecular structures of aminal **5.46a** and diazepine **5.57a**.

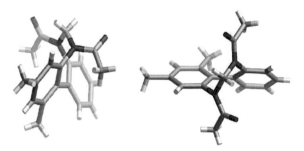

Figure 5.5 Molecular structure of **5.57a**, side view (left) and top view (right), clearly showing the twisted nature of the compound.

The axial torsion of dibenzo-1,3-diazepine **5.57a** is unambiguously visible in the molecular structures, as shown in Figure **5.5**. The current synthesis of the diazepines produces racemic mixtures of both atropisomers and the crystals are also achiral and contain both enantiomers of the molecule (only the *M*−enantiomer is shown). It is assumed that due to the lack of any other obvious molecular distortions in the solid-state structure, the solution-phase molecular structure will be similar. Although no definitive structure determination has been performed on the remaining diazepines **5.57c-d** it is reasonable to expect that they will have similar structural parameters.

5.4 Expanding the scope to mixed heteroacetals

Although it is always rewarding to see that the synthesis that one has designed works on the planned substrates, it even more interesting to try and test the limits of the methodology. Since it was clear that the aminal formation via the methyl sulfide precursor and the subsequent direct arylation work reasonable well on the *N,N*-species discussed above, it was decided to see what would happen if one were to mix things up. Since the MASTERS group had shown before that the methyl sulfide/acetal approach can easily facilitate the synthesis of *O,O*-species and their subsequent dioxepines via BHAS, as discussed in Chapter 5.2, working with *O,O*-species was not the most interesting thing to try.

Instead it was decided to prepare some mixed heteroacetals, with *N,O*- and *N,S*-heteroatoms as the goal, and attempt their subsequent cyclizations to the respective *N,O*- (oxazepine) and *N,S*- (thiazepine) analogues **5.69** and **5.71**.

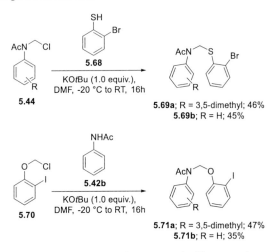

Scheme 5.20 Expanding the scope to mixed heteroacetals.

By combination of 2-bromothiophenol and the chloromethyl anilines **5.44**, prepared from their methyl sulfide derivatives, an unsubstituted and a 3,5-dimethyl *N,S*-acetal were accessible in moderate yields (Scheme **5.20**). For the *N,O*-linked compounds **5.71** it was decided to test whether the products would be equally accessible when the acetal tether originates from the phenol. An unsubstituted and a 3,5-dimethyl *N,O*-acetal were prepared by reacting chloromethyl ether **5.70**[37] with acetanilide in moderate yields (Scheme **5.20**). The clear advantage of an approach where the chloromethyl and the halogen are situated on the same building block is that it allows for a large variety of reaction partners.

The preparation of N,O-[100,101] and N,S-[102-104] 7-membered rings[105] would lead to the synthesis of N,O- and N,S-carbenes, which are useful as ligands in catalysis as well.[106]

The direct arylation of the N,X-acetals **5.69** and **5.71**, using the established Pd/CPhos protocol, successfully led to compounds **5.72** and **5.73** as evidenced by LC/GCMS and crude ^{1}H-NMR spectra. Sadly the isolation of the pure compounds from their complex reaction mixtures in sufficient quantities to do full characterizations has thus far proven a challenge – only the yields for the N,O-oxazepines **5.73** could be determined.

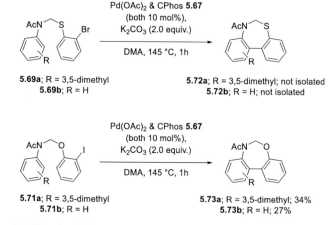

Scheme 5.21 Pd-catalyzed C–H arylation provides access to N,O-oxazepines and N,S-thiazepines.

From these preliminary results it is evident that, although the scope of the direct arylation reaction can be expanded to these mixed heteroacetals, a detailed reaction optimization is required to find conditions that allow adequate isolation and characterization of the products. Initial suggestions for improvement would be to decrease the temperature significantly. This should result is a slower reaction rate, but most likely also reduce the amount of side-reactions occurring.

5.5 Conclusion and outlook

The development of the aminal synthesis and its subsequent direct arylation provides for an unprecedented access to highly sophisticated, non-symmetrical and axially-chiral dibenzo-1,3-diazepines, -oxazepines and -thiazepines from simple, commercially available anilines. By synthesizing the methyl sulfide intermediates first and forming the chloromethyl *in situ*, directly before reacting it with its reaction partner, the issue of homodimerization can easily be avoided, allowing for a modular design. It was shown that the aminals and N,S-acetals are accessible via coupling of the N-(chloromethyl)anilines with 2-haloaniline or a 2-halothiophenol, but also that N,O-acetals can be prepared by nucleophilic substitution of 1-(chloromethoxy)-2-iodobenzene with an appropriate aniline.

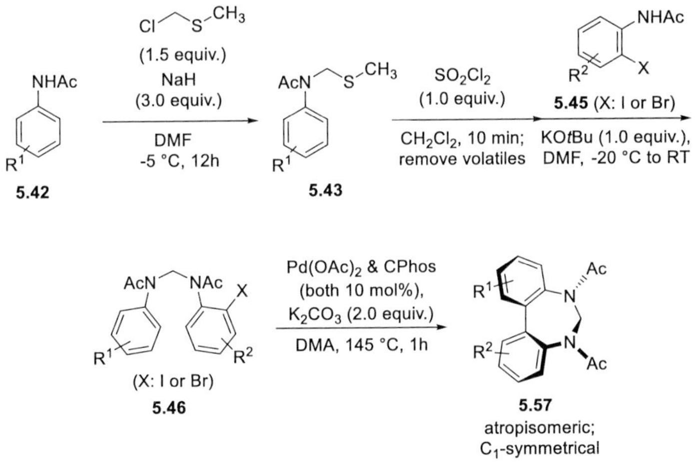

Scheme 5.22 Short summary of the methyl sulfide, aminal and the dibenzo-1,3-diazepine synthesis discussed in this chapter.

The transformation of the aminals **5.46** and the mixed acetals **5.69** and **5.71** to their corresponding cyclic dibenzo-1,3-diazepine, -oxazepines and -thiazepines is a Pd/CPhos-promoted direct arylation. As proven with X-ray crystallography, this innovative and novel route yielded atropisomeric dibenzo-1,3-diazepines **5.57**. The development of this methodology provides the basis for all sorts of new asymmetric chiral 7-membered N-heterocyclic carbenes that can be used as ligands, for catalysis or screened as potential drug candidates.

The first steps to explore the full potential of the modular synthesis of aminals and their subsequent direct arylation to diazepines have been taken. However, the results presented in this chapter are merely the initial results that show the proof of concept and confirm the presence of axial chirality in the final compounds. Not everything is fully optimized: Especially the direct arylation procedure of the aminals and the mixed heteroacetals could benefit from a detailed reaction optimization. Furthermore, the depth of the substrate scope should be increased in order to truly assess the versatility of the developed method. However, since this method is still in its infancy a lot of exciting advances can still be made.

Although in my work all aminals were formed by connecting a methyl chloride aniline with an *ortho*-halo aniline, the synthesis of a *N*-(chloromethyl)-2-haloaniline (such as **5.76**) would tremendously increase the range of reaction partners available. Luckily, recent results from the MASTERS group show that indeed the *N*-(methylsulfide)-2-bromoaniline **5.75** are accessible by the same chemistry. What remains now is to explore the aminal formation with this precursor and optimize the reaction conditions.

A suggestion that comes to mind is to employ anilines with different, orthogonal protecting groups. In the work presented here the main focus was primarily to explore different synthetic routes, reveal their potentials and lay the ground work. Using the small, simple and uncomplicated acetyl group was ideal to achieve this goal. However, a diazepine **5.79** with two different protecting groups on the nitrogen atoms can be substituted differently with ease, leading to a relatively simple and straightforward way to drastically increase the amount of ligands that can be tested for their activity. The different electronics and sterics of these side-groups could have a tremendous effect on the catalytic activity: changing the *N*-acetyl substituents to bulkier groups (such as the traditional mesityl group "Mes") would force these out of plane and likely change catalyst behavior (resting state, turn-over, selectivity *etc.*). But changing the side-groups to rather small methyl or isopropyl groups may allow a more facile [7]NHC formation and perhaps increase turn over numbers, since the active site is less sterically demanding.

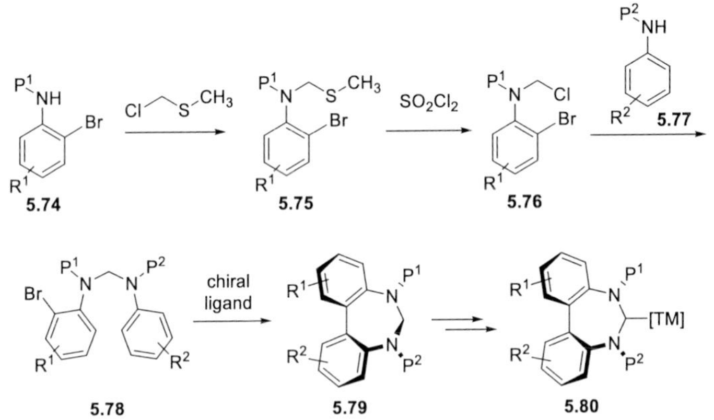

Scheme 5.23 Summary of suggestions for future work. Reacting an *ortho*-halo-*N*-protected methyl sulfide anilines with a second, orthogonally protected aniline, subsequent ringclosure using direct arylation with a chiral catalyst and finally formation of a ^{7}NHC-[TM] complex. TM = transition metal.

In order to provide the diazepines in an enantiopure manner, which is the next logical and desirable step, one would need to find or develop a suitable catalyst. All the direct arylation reactions presented in this chapter are performed using a rather straightforward cocktail of palladium and CPhos, one of BUCHWALD's phospine ligands. Therefore a logical step would be to try to obtain a chiral ligand that closely resembles CPhos. Scheme **5.24** depicts exactly this: A chiral, (*S*)-BINOL-based, phosphine ligand **5.81** and a short retrosynthetic pathway that the MASTERS group is planning to explore in the near future that is based on other ligand syntheses.[107-109]

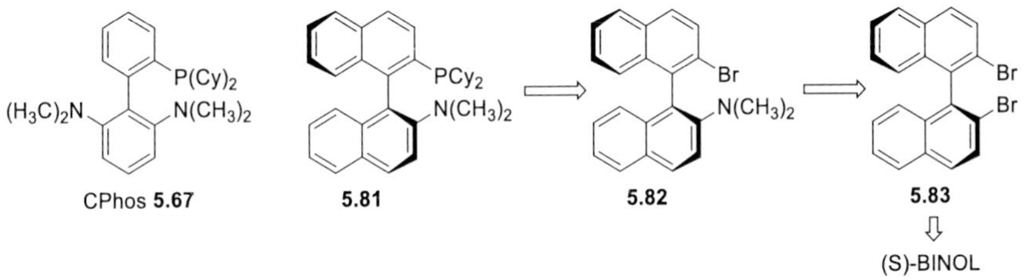

Scheme 5.24 Retrosynthesis of a possible chiral ligand closely resembling CPhos.

Lastly, all that remains is to actively screen the prepared compounds for their quality as ligands. In order for the dibenzo-1,3-diazepines **5.79** to be tested for their reactivity and selectivity as potential ligands they require transformation to their respective carbenes. This could be done analogous to the prior work of STAHL and co-workers with bromine (via the tribromide salts) to the amidinium monobromide salts[110] **5.84**. Treating these salts with a metal complex such as [Pd(allyl)Cl]₂ would likely deliver metal-NHC complexes **5.80**.[18]

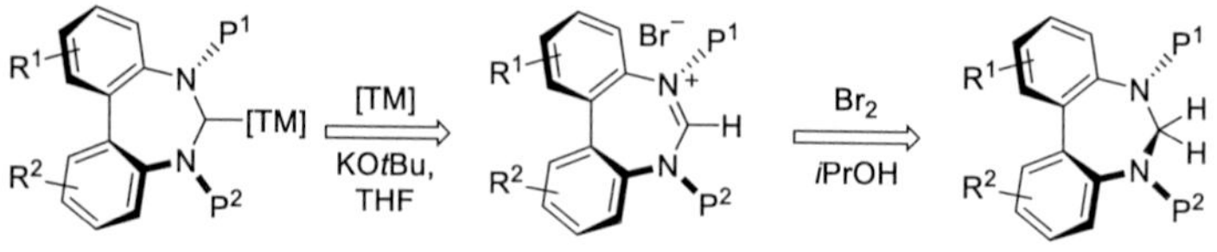

Scheme 5.25 Preparing ⁷NHCs (in analogy to STAHL's method) from diazepines.[5] TM = transition metal.

5.6 Experimental

General remarks:

Nuclear magnetic resonance spectroscopy (NMR): data were recorded on a *Varian* Infinity-Plus 400 spectrometer (^{1}H at 400 MHz; ^{13}C at 101 MHz) or a or a *Bruker* Avance DRX 600 (^{1}H at 600 MHz; ^{13}C at 150 MHz); All measurements were carried out at room temperature. The following solvents were used: chloroform-d_1, methanol-d_4, acetone-d_6, DMSO-d_6. Chemical shifts δ were expressed in parts per million (ppm) and referenced to chloroform (^{1}H: δ = 7.26 ppm, ^{13}C: δ = 77.00 ppm), methanol (^{1}H: δ = 3.31 ppm, ^{13}C: δ = 49.00 ppm), acetone (^{1}H: δ = 2.05 ppm, ^{13}C: δ = 30.83 ppm) or DMSO (^{1}H: δ = 2.50 ppm, ^{13}C: δ = 39.43 ppm).[111] The signal structure is described as follows: s = singlet, d = doublet, t = triplet, q = quartet, quin = quintet, bs = broad singlet, m = multiplet, dt = doublet of triplets, dd = doublet of doublets, td = triplet of doublets, ddd = doublet of doublet of doublets. The spectra were analyzed according to first order and all coupling constants are absolute values and expressed in Hertz (Hz).

Mass spectrometry (GC-MS, EI-MS): Mass spectra were recorded on the *Agilent* GCMS (HP6890 GC, HP 5973 Mass Spectrometer Detector) and HRMS on the *Agilent* QTOF-LC/MS 6520 using EI-MS (Electron ionization mass spectrometry).

Thin layer chromatography (TLC): Analytical and preparative thin layer chromatography was carried out using silica coated aluminium plates (silica 60, F254, layer thickness: 0.25 mm) with fluorescence indicator by *Grace*. Detection proceeded under UV light at λ = 254 nm and 366 nm. For development phosphomolybdic acid solution (5% phosphomolybdic acid in ethanol, dip solution); potassium permanganate solution (1.00 g potassium permanganate, 2.00 g acetic acid, 5.00 g sodium bicarbonate in 100 mL water, dip solution) was used followed by heating in a hot air stream.

Analytical balance: Used device: *Shimadzu* AUW120D.

Solvents and reagents: Solvents of technical quality were distilled prior to use. Solvents of p.a. (pro analysi) quality were commercially purchased (*Acros, Fisher Scientific, Sigma Aldrich*) and used without further purification. Absolute solvents were either commercially purchased as absolute solvents stored over molecular sieves under argon atmosphere or freshly prepared by the solvent purification system by *Innovative Technology*, model PS-micro, and stored under an argon atmosphere. Reagents were commercially purchased from industry companies (*ABCR, Acros, Alfa Aesar, Fluka, Sigma Aldrich, TCI, VWR*) and used without further purification unless stated otherwise.

Reactions: For reactions with air- or moisture-sensitive reagents, the glassware with a PTFE coated magnetic stir bar was heated under high vacuum with a heat gun, filled with argon and closed with a rubber septum. All reactions were performed using the typical Schlenk procedures with argon as inert gas. Liquids were transferred with V2A-steel cannulas. Solids were used as powders unless stated otherwise. Reactions at low temperatures were cooled in flat Dewar flasks. The following cooling mixtures were used: 0 °C (Ice/Water), 0 to −10 °C (Ice/Water/NaCl), −10 to −78 °C (acetone/liquid nitrogen or isopropanol/liquid nitrogen). In general, solvents were removed at preferably low temperatures (40 °C) under reduced pressure. Unless stated otherwise, solutions of ammonium chloride, sodium chloride and sodium hydrogen carbonate are aqueous, saturated solutions. Reaction progress of liquid phase reaction was checked with thin-layer chromatography (TLC). Crude products were purified according to literature procedures by preparative TLC or flash column chromatography using Silica gel Davisil (LC60A, 40-63 μm Grace) or Celite® (both *Sigma Aldrich*) and sea sand (calcined, purified with hydrochloric acid, *Merck*) as stationary phase. Eluents (mobile phase) were p.a. and volumetrically measured.

General procedure GP5.1: *N*-(methylthio)methylation of *N*-phenylacetamides

N-Phenylacetamide (1.00 equiv.) was dissolved in dry DMF (2.0 mL/mmol) at 0 °C under argon atmosphere. Sodium hydride (60% in mineral oil, 3.00 equiv.) was added in portions of 100 mg and the suspension was stirred at RT for 1h before the reaction mixture was cooled to –5 °C. Chloromethyl methyl sulfide (1.50 equiv. *Caution: very smelly compound*) was added dropwise over 30 minutes under vigorous stirring (Note: faster addition led to poorer results). After the addition was complete the reaction mixture was allowed to slowly reach RT while stirring overnight. The reaction was quenched with brine and extracted with EtOAc. The combined organic layers were washed with brine, dried over anhydrous sodium sulfate and concentrated under reduced pressure. Purification by flash chromatography (SiO₂, hexanes/EtOAc mixtures) afforded the *N*-((methylthio)methyl)-*N*-phenylacetamides.

General procedure GP5.2: formation of *N*,*N*-aminals and *N*,*S*-acetals

The corresponding *N*-((methylthio)methyl)-*N*-phenylacetamide (1.00 equiv.) was dissolved in dry CH₂Cl₂ (1.0 mL/mmol) at RT under argon atmosphere. Sulfuryl chloride (1M solution in CH₂Cl₂, 1.00 equiv.) was added and the solution was stirred for 1h at RT before the volatiles were gently removed under reduced pressure. Meanwhile a solution of the required *N*-acetylaniline or thiophenol (1.00 equiv.) and dry potassium *tert*-butoxide (2.00 equiv.) in dry DMF (2.0 mL/mmol) was prepared and stirred under argon atmosphere at −20 °C for 1h. The freshly prepared *N*-(chloromethyl)-*N*-phenylacetamide was dissolved in dry DMF and added dropwise to the reaction mixture. After the addition was complete the reaction was allowed to warm to RT over 10 minutes while stirring and quenched with brine as soon as completion was observed. The reaction mixture was extracted with CH$_2$Cl$_2$ and the combined organic layers were washed with brine and concentrated under reduced pressure. Purification of the aminals proceeded by preparative HPLC (C$_{18}$ column, gradient from 25-40% THF in water) and purification of the N,S-acetals via flash chromatography (SiO$_2$; hexanes/EtOAc mixtures).

General procedure GP5.3: formation of N,O-acetals

The corresponding N-(3,5-dimethylphenyl)acetamide (1.00 equiv.) and dry potassium tert-butoxide (1.00 equiv.) are charged in a round bottom flask with dry DMA (2.0 mL/mmol) at –20 °C under argon atmosphere. 1-(Chloromethoxy)-2-iodobenzene (1.00 equiv.) was then added and the reaction was stirred to RT overnight. After 16h, volatiles were removed under reduced pressure. Purification of the product by preparative thin layer chromatography (SiO$_2$, hexanes/EtOAc mixtures).

General procedure GP5.4: biaryl formation

Aryl halide (1.00 equiv.), potassium carbonate (2.00 equiv.), palladium acetate (0.10 equiv.) and CPhos (IUPAC: 2-(2-dicyclohexylphosphanylphenyl)-N1,N1,N3,N3-tetramethyl-benzene-1,3-diamine, 0.10 equiv.) were charged in a sealed vial. The setup was purged with argon three times before 2.0 mL anhydrous DMA was added. The reaction mixture was stirred at 145 °C for 1h or until GCMS confirmed full conversion of the starting material. The DMA was evaporated under reduced pressure or by gently blowing nitrogen into the flask and the crude was filtered over a short column (SiO$_2$) with EtOAc. Purification of the cyclized product was performed by flash column chromatography or preparative thin layer chromatography (SiO$_2$, hexanes/EtOAc mixtures).

N-(3,5-Dimethylphenyl)-N-((methylthio)methyl)acetamide (5.43a)

Prepared according to general procedure GP5.1 and isolated as yellow oil (1.79 g, 8.03 mmol, 83%). R_f (cHex/EtOAc = 2:1) = 0.44 − ^{1}H-NMR (400 MHz, CDCl$_3$): δ = 7.00 (s, 1H), 6.86 (s, 2H), 4.81 (s, 2H), 2.34 (s, 6H), 2.17 (s, 3H), 1.87 (s, 3H) ppm. − ^{13}C-NMR (101 MHz, CDCl$_3$): δ = 170.8, 141.8, 139.4, 129.9, 125.9, 52.4, 22.6, 21.2, 15.4 ppm. − **HRMS** (EI, C$_{12}$H$_{17}$NO^{32}SNa): calc. 246.0923; found: 246.0919.

N-((Methylthio)methyl)-N-phenylacetamide (5.43b)

Prepared according to general procedure GP5.1 and isolated as a yellow to orange oil that crystallized at –20 °C (2.37 g, 12.1 mmol, 81%). R_f (cHex/EtOAc = 2:1) = 0.37 − ^{1}H-NMR (400 MHz, CDCl$_3$): δ = 7.49 - 7.33 (m, 3H), 7.28 - 7.20 (m, 2H), 4.83 (s, 2H), 2.15 (s, 3H), 1.85 (s, 3H) ppm. − ^{13}C-NMR (101 MHz, CDCl$_3$): δ = 170.9, 142.0, 129.8, 128.6, 128.5, 52.7, 22.8, 15.5 ppm. − **HRMS** (EI, C$_{10}$H$_{13}$NO^{32}SNa): calc. 218.0610; found: 218.0602.

N-(3,5-Dimethylphenyl)-N-((N-(2-iodophenyl)acetamido)methyl)acetamide (5.46a)

N-(3,5-Dimethylphenyl)-N-((methylthio)methyl)acetamide (112 mg, 0.500 mmol, 1.00 equiv.) was dissolved in dry CH$_2$Cl$_2$ (2.0 mL) at RT under argon atmosphere. Sulfuryl chloride (0.500 mmol, 1.00 equiv., 0.500 mL, 1M solution in CH$_2$Cl$_2$) was added and the solution was stirred for 2h at RT before the volatiles were removed under reduced pressure. Meanwhile a solution of N-(2-iodophenyl)acetamide (124 mg, 0.480 mmol, 0.95 equiv.) and dry potassium $tert$-butoxide (56.0 mg, 0.500 mmol, 1.00 equiv.) in dry DMF was prepared and stirred under argon atmosphere at RT for 1h. The freshly prepared N-

(chloromethyl)-*N*-(3,5-dimethylphenyl)acetamide was dissolved in 1.0 mL dry DMF and added dropwise to the reaction mixture. After the addition was complete the reaction stirred at RT for 20 minutes. The reaction was quenched with brine (5 mL) and extracted with CH_2Cl_2 (3 × 20 mL). The combined organic layers were washed with brine (2 × 10 mL) and concentrated. Purification by HPLC (C_{18} column, 40% THF in water) afforded the *N*-(3,5-dimethylphenyl)-*N*-((*N*-(2-iodophenyl)acetamido)methyl)acetamide as a colorless oil (71.1 mg, 0.163 mmol, 33%). **Major rotamer**: ¹H-NMR (400 MHz, CDCl₃): δ = 7.91 (d, *J* = 7.3 Hz, 1H), 7.38 (t, *J* = 7.1 Hz, 1H), 7.09 (t, *J* = 7.0 Hz, 2H), 6.94 (s, 1H), 6.61 (s, 2H), 5.96 (d, *J* = 12.8 Hz, 1H), 5.13 (d, *J* = 12.9 Hz, 1H), 2.28 (s, 6H), 1.76 (s, 3H), 1.72 (s, 3H) ppm. **Minor rotamer**: ¹H-NMR (400 MHz, CDCl₃): δ = 7.86 (d, *J* = 7.6 Hz, 1H), 7.22 (t, *J* = 7.2 Hz, 1H), 6.99 (t, *J* = 7.4 Hz, 2H), 6.74 (d, *J* = 7.2 Hz, 1H), 6.49 (s, 2H), 5.67 (d, *J* = 14.1 Hz, 1H), 5.26 (d, *J* = 14.2 Hz, 1H), 2.22 (s, 6H), 1.84 (s, 3H), 1.73 (s, 3H) ppm. − ¹³C-NMR (101 MHz, CDCl₃): δ = 171.2, 170.3, 144.0, 142.1, 140.0, 139.8, 139.7, 139.1, 130.5, 130.2, 130.0, 129.6, 129.3, 128.9, 125.5, 125.4, 100.9, 59.7, 22.8, 22.6, 21.2, 21.1 ppm. − **HRMS** (EI, $C_{19}H_{21}IN_2O_2Na$): calc. 459.0540; found: 459.0539.

N-(2-Bromo-3-methylphenyl)-N-((N-(3,5-dimethylphenyl)acetamido)methyl)acetamide (5.46b)

N-(3,5-Dimethylphenyl)-*N*-((methylthio)methyl)acetamide (223 mg, 1.00 mmol, 1.00 equiv.) was dissolved in dry CH_2Cl_2 (2.0 mL) at RT under argon atmosphere. Sulfuryl chloride (1.00 mmol, 1.00 equiv., 1.00 mL, 1M solution in CH_2Cl_2) was added and the solution was stirred for 1h at RT before the volatiles were removed under reduced pressure. A mixture of *N*-(2-bromo-3-methylphenyl)acetamide (228 mg, 1.00 mmol, 1.00 equiv.) and dry potassium *tert*-butoxide (112 mg, 1.00 mmol, 1.00 equiv.) in dry DMF was prepared and stirred under argon atmosphere at −20 °C for 1h. The freshly prepared *N*-(chloromethyl)-*N*-(3,5-dimethylphenyl)acetamide was dissolved in 1.0 mL dry DMF and added dropwise to the reaction mixture. After the addition was complete the reaction was allowed to stir at RT for 20 minutes. The reaction was quenched with brine (5 mL) and extracted with CH_2Cl_2 (3 × 20 mL). The combined organic layers were washed with brine (2 × 10 mL) and concentrated. Purification by preparative TLC (neutral aluminium oxide) afforded the *N*-(2-bromo-3-methylphenyl)-*N*-((*N*-(3,5-dimethylphenyl)acetamido)methyl)-acetamide as clear oil (150 mg, 0.371 mmol, 37%). **Major rotamer:** ^{1}**H-NMR** (400 MHz, CDCl$_3$): δ = 7.29 - 7.19 (m, 2H), 6.94 - 6.93 (m, 2H), 6.59 (s, 2H), 5.97 (d, J = 13.0 Hz, 1H), 5.15 (d, J = 13.0 Hz, 1H), 2.44 (s, 3H), 2.28 (s, 6H), 1.77 (s, 3H), 1.73 (s, 3H) ppm. **Minor rotamer:** ^{1}**H-NMR** (400 MHz, CDCl$_3$): δ = 7.18 - 7.06 (m, 2H), 6.74 (s, 2H), 6.49 (s, 2H), 5.68 (d, J = 14.1 Hz, 1H), 5.26 (d, J = 14.1 Hz, 1H), 2.41 (s, 3H), 2.24 (s, 6H), 1.83 (s, 3H), 1.69 (s, 3H) ppm. – ^{13}**C-NMR** (101 MHz, CDCl$_3$): δ = 170.7, 139.2, 130.9, 129.8, 128.4, 127.7, 125.6, 117.7, 110.1, 59.5, 23.9, 22.8, 22.6, 21.3 ppm. – **HRMS** (EI, C$_{20}$H$_{23}$^{81}BrN$_2$O$_2$Na): calc. 427.0815; found: 427.0811.

N-(2-Iodophenyl)-N-((N-phenylacetamido)methyl)acetamide (5.46c)

N-((Methylthio)methyl)-*N*-phenylacetamide (97.6 mg, 0.500 mmol, 1.00 equiv.) was dissolved in dry CH$_2$Cl$_2$ (2.0 mL) at RT under argon atmosphere. Sulfuryl chloride (0.50 mmol, 0.50 mL, 1.00 equiv., 1M solution in CH$_2$Cl$_2$) was added and the solution was stirred for 1h at RT before the volatiles were removed under reduced pressure. Meanwhile a solution of *N*-(2-iodophenyl)acetamide (130 mg, 0.500 mmol, 1.00 equiv.) and dry potassium *tert*-butoxide (112 mg, 1.00 mmol, 2.00 equiv.) in dry DMF was prepared and stirred under argon atmosphere at RT for 1h. The freshly prepared *N*-(chloromethyl)-*N*-phenylacetamide was dissolved in dry DMF (1.0 mL) and added dropwise to the reaction mixture. After the addition was complete the reaction was allowed to stir at RT for 10 minutes. The reaction was quenched with brine (5 mL) and extracted with CH$_2$Cl$_2$ (3 × 20 mL). The combined organic layers were washed with brine (2 × 10 mL) and concentrated. Purification by HPLC (C$_{18}$ column, 40% THF in water) afforded the *N*-(2-iodophenyl)-*N*-((*N*-phenylacetamido)methyl)acetamide as a white foam (96.0 mg, 0.240 mmol, 47%). ^{1}**H-NMR** (400 MHz, CDCl$_3$): δ = 7.88 (d, J = 7.4 Hz, 1H), 7.45 - 7.23 (m, 4H), 7.12 - 6.99 (m, 4H), 5.92 (d, J = 12.9 Hz, 1H), 5.68 (d, J = 14.2 Hz, traces of the minor isomer, 1H), 5.28 (d, J = 14.2 Hz, traces of the minor isomer, 1H), 5.19 (d, J = 12.9 Hz, 1H), 1.75 (s, 3H), 1.68 (s, 3H) ppm. – ^{13}**C-NMR** (101 MHz, CDCl$_3$): δ = 171.0, 170.4, 169.2, 168.8, 168.4, 144.1, 143.67, 142.2, 141.6, 140.1, 139.7, 138.8, 130.2, 130.1, 130.0, 129.5, 129.4, 129.1, 128.9, 128.7, 128.6, 128.0, 127.9, 126.1, 123.6, 122.8, 119.7, 100.7, 99.4, 90.8, 63.4, 59.6, 53.5, 29.6,

24.6, 24.4, 22.7, 22.7, 21.7 ppm. − **HRMS** (EI, $C_{17}H_{17}IN_2O_2Na$): calc. 431.0227; found: 431.0227.

N-(2-Bromo-3-methylphenyl)-*N*-((*N*-phenylacetamido)methyl)acetamide (5.46d)

N-((Methylthio)methyl)-*N*-phenylacetamide (97.6 mg, 0.500 mmol, 1.00 equiv.) was dissolved in dry CH_2Cl_2 (2.0 mL) at RT under argon atmosphere. Sulfuryl chloride (0.50 mmol, 0.50 mL, 1.0 equiv., 1M solution in CH_2Cl_2) was added and the solution was stirred for 1h at RT before the volatiles were removed under reduced pressure. Meanwhile a solution of *N*-(2-bromo-3-methylphenyl)acetamide (114 mg, 0.500 mmol, 1.00 equiv.) and dry potassium *tert*-butoxide (112 mg, 1.00 mmol, 2.00 equiv.) in dry DMF was prepared and stirred under argon atmosphere at RT for 1h. The freshly prepared *N*-(chloromethyl)-*N*-phenylacetamide was dissolved in 1.0 mL dry DMF and added dropwise to the reaction mixture. After the addition was complete the reaction was allowed to stir at RT for 10 minutes. The reaction was quenched with brine (5 mL) and extracted with CH_2Cl_2 (3 × 20 mL). The combined organic layers were washed with brine (2 × 10 mL) and concentrated. Purification by HPLC (C_{18} column, 30% THF in water) afforded the *N*-(2-bromo-3-methylphenyl)-*N*-((*N*-phenylacetamido)-methyl)acetamide as white crystals (76.1 mg, 0.200 mmol, 41%). **¹H-NMR** (400 MHz, CDCl₃): δ = 8.19 - 8.06 (m, **minor rotamer**, 1H), 7.78 (s, **minor rotamer**, 1H), 7.50 - 7.43 (m, **minor isomer**, 1H), 7.45 - 7.29 (m, 3H), 7.29 - 7.16 (m, 2H), 7.16 - 7.02 (m, 2H), 6.94 (d, J = 5.8 Hz, 1H), 6.60 (d, J = 7.4 Hz, 1H), 5.94 (d, J = 12.9 Hz, **minor rotamer**, 1H), 5.74 (d, J = 14.2 Hz, 1H), 5.28 (t, J = 13.9 Hz, **minor rotamer**, 1H), 2.43 (s, 3H), 1.77 (s, 3H), 1.71 (s, 3H) ppm. − **¹³C-NMR** (101 MHz, CDCl₃): δ = 171.0, 170.8, 142.2, 141.1, 140.4, 130.9, 129.5, 128.2, 128.1, 128.0, 127.7, 126.7, 59.3, 23.8, 22.7, 22.4 ppm. − **HRMS** (EI, $C_{18}H_{19}{}^{79}BrN_2O_2Na$): calc. 397.0522; found: 397.0512.

1,1'-(1,3-Dimethyl-5*H*-dibenzo[d,f][1,3]diazepine-5,7(6*H*)-diyl)bis(ethan-1-one) (5.57a)

Reaction according to general procedure GP5.4. Stirred at 145 °C for 1h, purified by flash column chromatography (SiO$_2$, hexanes/EtOAc mixtures) followed by HPLC (C$_{18}$ column, 30% THF in water). Product isolated as colorless crystals (28.0 mg, 0.091 mmol, 78%). ^{1}H-NMR (400 MHz, CDCl$_3$): δ = 7.50 - 7.38 (m, 3H), 7.38 - 7.29 (m, 1H), 7.18 (s, 1H), 6.98 (s, 1H), 5.78 (dd, *J* = 45.1, 11.5 Hz, 2H), 2.36 (s, 6H), 1.63 (s, 3H), 1.60 (s, 3H) ppm. − ^{13}C-NMR (101 MHz, CDCl$_3$): δ = 171.0, 170.9, 139.7, 138.9, 138.9, 136.9, 136.4, 133.6, 132.7, 130.5, 129.6, 129.3, 128.9, 127.0, 63.1, 22.9, 22.7, 21.2, 20.4 ppm. − **HRMS** (EI, C$_{19}$H$_{20}$N$_2$O$_2$Na): calc. 331.1417; found: 331.1418.

1,1'-(1-Methyl-5*H*-dibenzo[d,f][1,3]diazepine-5,7(6*H*)-diyl)bis(ethan-1-one) (5.57b)

Reaction according to general procedure GP5.4. Stirred for at 145 °C 1h, product isolated as a colorless solid (7.0 mg, 0.024 mmol, 39%). ^{1}H-NMR (400 MHz, CDCl$_3$): δ = 7.50 - 7.41 (m, 3H), 7.40 - 7.31 (m, 3H), 7.20 - 7.15 (m, 1H), 5.80 (dd, *J* = 46.8, 11.6 Hz, 2H), 2.42 - 2.39 (m, 3H), 1.64 (s, 3H), 1.60 (s, 3H) ppm. − ^{13}C-NMR (101 MHz, CDCl$_3$): δ = 171.0, 170.8, 136.9, 132.0, 130.5, 129.9, 129.5, 129.3, 128.9, 126.4, 63.2, 23.0, 22.7, 20.5 ppm. − **HRMS** (EI, C$_{18}$H$_{18}$N$_2$O$_2$Na): calc. 317.1260; found: 317.1260.

1,1'-(5*H*-Dibenzo[d,f][1,3]diazepine-5,7(6H)-diyl)bis(ethan-1-one) (5.57c)

Reaction according to general procedure GP5.4. Stirred for 1h at 145 °C, product isolated as a slightly reddish solid (8.0 mg, 0.029 mmol, 29%). ^{1}H-NMR (400 MHz, CDCl$_3$): δ = 7.55 - 7.44 (m, 6H), 7.35 (t, *J* = 1.0 Hz, 1H), 7.34 - 7.32 (m, 1H), 5.94 (s, 2H), 1.58 (s, 6H) ppm. − ^{13}C-NMR (101 MHz, CDCl$_3$): δ = 171.1, 138.6, 138.2, 130.0, 129.8, 129.3, 128.7, 64.0, 22.9 ppm. − **HRMS** (EI, C$_{17}$H$_{16}$N$_2$O$_2$Na): calc. 303.1104; found: 303.1105.

1,1'-(1,3,11-Trimethyl-5*H*-dibenzo[d,f][1,3]diazepine-5,7(6H)-diyl)bis(ethan-1-one) (5.57d)

Reaction according to general procedure GP5.4 starting from **5.46b** (40.2 mg, 0.100 mmol). Reaction stirred at 145 °C for 2h, product isolated as a white solid (11.3 mg, 0.035 mmol, 35%). **¹H-NMR** (600 MHz, CDCl₃): δ = 7.38 - 7.37 (m, 2H), 7.20 - 7.19 (m, 2H), 7.01 (s, 1H), 5.69 (s, 2H), 2.41(s, 3H), 2.25 (s, 3H), 2.21 (s, 3H), 2.21 (s, 3H), 1.69 (s, 3H), 1.68 (s, 3H) ppm. – **¹³C-NMR** (150 MHz, CDCl₃): δ = 171.0, 164.6, 138.8, 138.6, 137.7, 137.3, 136.2, 133.0, 131.2, 129.3, 126.5, 125.9, 62.1, 22.6, 22.5, 21.1, 19.3, 19.2 ppm.

N-(3,5-Dimethylphenyl)-*N*-(((2-iodophenyl)thio)methyl)acetamide (5.69a)

N-(3,5-Dimethylphenyl)-*N*-((methylthio)methyl)acetamide (112 mg, 0.500 mmol, 1.00 equiv.) was dissolved in 2.0 mL dry CH₂Cl₂ at RT under argon atmosphere. Sulfuryl chloride (0.50 mmol, 0.50 mL, 1.0 equiv., 1M solution in CH₂Cl₂) was added and the solution was stirred for 10 min at RT before the volatiles were removed under reduced pressure. Meanwhile a mixture of 2-bromobenzenethiol (94.0 mg, 0.500 mmol, 1.00 equiv.) and dry potassium *tert*-butoxide (56.0 mg, 0.500 mmol, 1.00 equiv.) in dry dimethylacetamide at –20 °C was prepared and stirred under argon atmosphere for 10 min. The freshly prepared *N*-(chloromethyl)-*N*-(3,5-dimethylphenyl)acetamide was dissolved in 1.0 mL dry DMA and added dropwise to the reaction mixture. After the addition was complete the reaction was stirred to RT overnight before the volatiles were removed under reduced pressure. The reaction was quenched with brine (5 mL) and extracted with EtOAc (3 × 20 mL). The combined organic layers were washed with brine (1 × 10 mL) and concentrated. Purification by flash chromatography (hexanes/EtOAc = 9:1 → 4:1) afforded the product as a colorless oil (84.0 mg, 0.230 mmol, 46%). **¹H-NMR** (400 MHz, CDCl₃): δ = 7.51 (dd, *J* = 8.0, 1.0 Hz, 1H), 7.46 (dd, *J* = 7.9, 1.4 Hz, 1H), 7.19 (td, *J* = 7.8, 1.3 Hz, 1H), 7.02 (td, 1H), 6.94 (s, 1H), 6.85 (s, 1H), 6.73 (s, 2H), 5.21 (s, 2H), 2.28 (s, 6H), 1.83 (s, 3H) ppm. – **¹³C-NMR** (101 MHz, CDCl₃): δ = 170.8, 144.0, 141.4, 140.6, 139.5, 136.0, 133.1, 132.3, 131.4, 130.7, 130.1, 128.6, 127.8, 127.8, 126.80, 126.0, 122.1, 52.4, 22.8, 21.3 ppm.

N-(((2-Bromophenyl)thio)methyl)-*N*-phenylacetamide (5.69b)

N-(phenyl)-*N*-((methylthio)methyl)acetamide (97.5 mg, 0.500 mmol, 1.00 equiv.) was dissolved in 2.0 mL dry CH$_2$Cl$_2$ at RT under argon atmosphere. Sulfuryl chloride (0.50 mmol, 0.50 mL, 1.0 equiv., 1M solution in CH$_2$Cl$_2$) was added and the solution was stirred for 10 min at RT before the volatiles were removed under reduced pressure. Meanwhile a mixture of 2-bromobenzenethiol (94.0 mg, 0.500 mmol, 1.00 equiv.) and dry potassium *tert*-butoxide (56.0 mg, 0.500 mmol, 1.00 equiv.) in dry dimethylacetamide at –20 °C was prepared and stirred under argon atmosphere for 10 min. The freshly prepared *N*-(chloromethyl)-*N*-(phenyl)acetamide was dissolved in 1.0 mL dry DMA and added dropwise to the reaction mixture. After the addition was complete the reaction was stirred at RT overnight before the volatiles were removed under reduced pressure. The reaction was quenched with brine (5 mL) and extracted with EtOAc (3 × 20 mL). The combined organic layers were washed with brine (1 × 10 mL) and concentrated. Purification by flash chromatography (hexanes/EtOAc = 9:1 → 4:1) afforded the product as a colorless oil (75.0 mg, 0.220 mmol, 45%). **^{1}H-NMR** (400 MHz, CDCl$_3$): δ = 7.51 (dd, *J* = 8.0, 0.9 Hz, 1H), 7.45 (dd, *J* = 7.9, 1.4 Hz, 1H), 7.42 - 7.31 (m, 3H), 7.23 - 7.14 (m, 3H), 7.03 (td, *J* = 7.9, 1.2 Hz, 1H), 5.25 (s, 2H), 1.83 (s, 3H) ppm. – **^{13}C-NMR** (101 MHz, CDCl$_3$): δ = 170.8, 141.5, 133.2, 131.2, 129.8, 128.6, 128.5, 127.9, 52.3, 22.8 ppm.

N-(3,5-Dimethylphenyl)-*N*-((2-iodophenoxy)methyl)acetamide (5.71a)

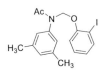

N-(3,5-dimethylphenyl)acetamide (100 mg, 0.610 mmol, 1.00 equiv.) and dry potassium *tert*-butoxide (72.0 mg, 0.610 mmol, 1.00 equiv.) are charged in a 10 mL round bottom flask with 2.0 mL dry DMA at –20 °C under argon atmosphere. 1-(Chloromethoxy)-2-iodobenzene (165 mg, 0.610 mmol, 1.00 equiv.) was added and the reaction was stirred at RT overnight before the volatiles were removed under reduced pressure. The product was purified by preparative TLC and isolated as a colorless oil (112 mg, 0.283 mmol, 47%). **^{1}H-NMR** (400 MHz, CDCl$_3$): δ = 7.77 (dd, *J* = 7.8, 1.5 Hz, 1H), 7.32 - 7.26 (m, 1H), 7.08 - 7.03 (m, 1H), 7.01 (s, 1H), 6.92 (s, 2H), 6.73 (td, *J* = 7.6, 1.3 Hz, 1H), 5.64 (s, 2H), 2.34 (s, 6H), 1.89 (d, *J* = 13.0 Hz, 3H) ppm. – **^{13}C-NMR** (101 MHz, CDCl$_3$): δ = 171.6, 155.4, 141.1, 139.7, 139.5, 130.2,

129.6, 126.3, 123.3, 113.8, 87.0, 75.2, 22.9, 21.3 ppm. – **HRMS** (EI, $C_{17}H_{18}INO_2Na$): calc. 418.0274; found: 418.0273.

N-((2-Iodophenoxy)methyl)-*N*-phenylacetamide (5.71b)

 N-(phenyl)acetamide (100 mg, 0.740 mmol, 1.00 equiv.) and dry potassium *tert*-butoxide (87.4 mg, 0.740 mmol, 1.00 equiv.) are charged in a 10 mL round bottom flask with 2.0 mL dry DMA at –20 °C under argon atmosphere. 1-(Chloromethoxy)-2-iodobenzene (199 mg, 0.740 mmol, 1.00 equiv.) was added and the reaction was stirred to RT overnight before the volatiles were removed under reduced pressure. The product was purified by preparative TLC and isolated as a colorless oil (95.8 mg, 0.26 mmol, 35%). **^{1}H-NMR** (400 MHz, CDCl$_3$): δ = 7.75 (dd, *J* = 7.8, 1.1 Hz, 1H), 7.47 - 7.34 (m, 3H), 7.31 (d, *J* = 7.1 Hz, 2H), 7.25 (t, *J* = 9.6 Hz, 1H), 7.03 (t, *J* = 12.1 Hz, 1H), 6.73 (td, *J* = 7.7, 1.0 Hz, 1H), 5.66 (s, 2H), 1.90 (s, 3H) ppm. – **^{13}C-NMR** (101 MHz, CDCl$_3$): δ = 171.4, 155.4, 141.3, 139.7, 129.8, 129.6, 128.7, 128.7, 123.4, 113.8, 87.0, 75.4, 23.0 ppm. – **HRMS** (EI, $C_{15}H_{14}INO_2Na$): calc. 389.9961; found: 389.9956.

Axially chiral dibenzo-1,3-diazepines   221

X-ray Crystallographic data

All X-ray crystallographic work in this chapter has been performed by Prof. Dr. John
MCMURTRIE; Chemistry, Physics and Mechanical Engineering, Queensland University of
Technology (Brisbane, Australia). All X-ray images shown in this chapter are depicted with
displacement parameters drawn at 50 % probability level.

N-(3,5-Dimethylphenyl)-N-((N-(2-iodophenyl)acetamido)methyl)acetamide – CCDC 1415528 –

(5.46a)

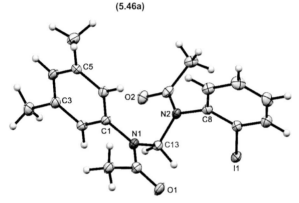

Empirical formula	$C_{19}H_{21}IN_2O_2$	
Formula Weight	436.28	
Temperature	100(2) K	
Wavelength	(Synchrotron) 0.7225 Å	
Crystal system, space group	Triclinic, $P\bar{1}$ (#2)	
Unit cell dimensions	a 8.3540(17) Å	α = 90.10(3)°
	b 9.985(2) Å	β = 90.16(3)°
	c 11.038(2) Å	γ = 99.28(3)°
Volume	908.7(3) Å³	
Z, calculated density	2, 1.595 g cm⁻³	
Absorption coefficient	1.775 mm⁻¹	
F(000)	436	
Crystal size	0.05 × 0.05 × 0.05 mm	
Theta range for data collection	2.10 to 25.17 deg.	
Limiting indices	hkl range -9 to 9, -11 to 11, -12 to 12	
Reflection collected / unique	10895 / 2835 [R(int) = 0.024]	
Completeness to theta = 25.00	91.9%	
Absorption correction	none	
Refinement method	Full matrix least-squares on F²	
Data / restraints / parameters	2812 / 0 / 221	
Goodness-of-fit on F^2	1.053	
Final R indices [I>sigma(I)]	R1 = 0.0256, wR2 = 0.0707	
R indices (all data)	R1 = 0.0263, wR2 = 0.0712	
Largest diff. peak and hole	-0.911, 0.772 e Å⁻³	

1,1'-(1,3-Dimethyl-5H-dibenzo[d,f][1,3]diazepine-5,7(6H)-diyl)bis(ethan-1-one) – CCDC 1415529

– (5.57a)

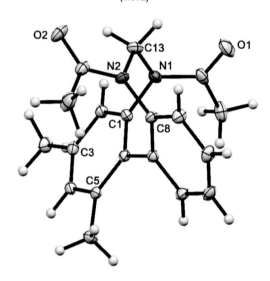

Empirical formula	$C_{19}H_{20}N_2O_2$	
Formula Weight	308.37	
Temperature	100(2) K	
Wavelength	(Synchrotron) 0.7225 Å	
Crystal system, space group	Triclinic, $P\bar{1}$ (#2)	
Unit cell dimensions	a 9.280(2) Å	α = 79.39(3)°
	b 9.500(2) Å	β = 75.05(3)°
	c 9.630(2) Å	γ = 80.83(3)°
Volume	800.6(3) Å³	
Z, calculated density	2, 1.279 g cm⁻³	
Absorption coefficient	0.084 mm⁻¹	
F(000)	328	
Crystal size	0.05 × 0.05 × 0.05 mm	
Theta range for data collection	2.23 to 28.42 deg.	
Limiting indices	hkl range -12 to 12, -12 to 12, -12 to 12	
Reflection collected / unique	26996 / 3812 [R(int) = 0.0213]	
Completeness to theta = 25.00	99.4%	
Absorption correction	none	
Refinement method	Full matrix least-squares on F²	
Data / restraints / parameters	3812 / 0 / 212	
Goodness-of-fit on F^2	1.066	
Final R indices [I>sigma(I)]	R1 = 0.0404, wR2 = 0.1032	
R indices (all data)	R1 = 0.0424, wR2 = 0.1047	
Largest diff. peak and hole	-0.237, 0.320 e Å⁻³.	

5.7 References

[1]   D. Hanson, *C&EN*, **2005**, *83* (25), *The Top Pharmaceuticals That Changed The World: Librium.*

[2]   W. Ried, A. Sinharay, *Chem. Ber.*, **1964**, *97* (4), 1214-1215, *Über heterocyclische Siebenringsysteme, XIII. Notiz zur Ringschlußreaktion mit o.o'-Diamino-biphenyl.*

[3]   A. M. Washton, J. E. Zweben, *Treating Alcohol and Drug Problems in Psychotherapy Practice*; The Guilford Press, 2006.

[4]   H.-w. K. Yang, L. Simoni-Wastila, I. H. Zuckerman, B. Stuart, *Psychiatr. Serv.*, **2008**, *59* (4), 384-391, *Benzodiazepine Use and Expenditures for Medicare Beneficiaries and the Implications of Medicare Part D Exclusions.*

[5]   H. W. Wanzlick, *Angew. Chem. Int. Ed.*, **1962**, *1* (2), 75-80, *Aspects of Nucleophilic Carbene Chemistry.*

[6]   H. W. Wanzlick, H. J. Schönherr, *Angew. Chem. Int. Ed.*, **1968**, *7* (2), 141-142, *Direct Synthesis of a Mercury Salt-Carbene Complex.*

[7]   A. J. Arduengo, R. L. Harlow, M. Kline, *J. Am. Chem. Soc.*, **1991**, *113* (1), 361-363, *A stable crystalline carbene.*

[8]   M. N. Hopkinson, C. Richter, M. Schedler, F. Glorius, *Nature*, **2014**, *510* (7506), 485-496, *An overview of N-heterocyclic carbenes.*

[9]   B. K. Keitz, K. Endo, P. R. Patel, M. B. Herbert, R. H. Grubbs, *J. Am. Chem. Soc.*, **2011**, *134* (1), 693-699, *Improved Ruthenium Catalysts for Z-Selective Olefin Metathesis.*

[10]   M. Iglesias, D. J. Beetstra, J. C. Knight, L.-L. Ooi, A. Stasch, S. Coles, L. Male, M. B. Hursthouse, K. J. Cavell, A. Dervisi, I. A. Fallis, *Organometallics*, **2008**, *27* (13), 3279-3289, *Novel Expanded Ring N-Heterocyclic Carbenes: Free Carbenes, Silver Complexes, And Structures.*

[11]   W. A. Herrmann, M. Elison, J. Fischer, C. Köcher, G. R. J. Artus, *Angew. Chem. Int. Ed.*, **1995**, *34* (21), 2371-2374, *Metal Complexes of N-Heterocyclic Carbenes—A New Structural Principle for Catalysts in Homogeneous Catalysis.*

[12]   G.-F. Du, H. Guo, Y. Wang, W.-J. Li, W.-J. Shi, B. Dai, *J. Saudi Chem. Soc.*, **2015**, *19* (1), 112-115, *N-heterocyclic carbene catalyzed synthesis of dimethyl carbonate via transesterification of ethylene carbonate with methanol.*

[13]   L. He, H. Guo, Y. Wang, G.-F. Du, B. Dai, *Tetrahedron Lett.*, **2015**, *56* (8), 972-980, *N-heterocyclic carbene-mediated transformations of silicon reagents.*

[14]   F. Glorius, S. Bellemin-Laponnaz, *N-Heterocyclic Carbenes in Transition Metal Catalysis*; Springer, 2007.

[15]   W. A. Herrmann, C. Köcher, *Angew. Chem. Int. Ed.*, **1997**, *36* (20), 2162-2187, *N-Heterocyclic Carbenes.*

[16]   D. M. Flanigan, F. Romanov-Michailidis, N. A. White, T. Rovis, *Chem. Rev.*, **2015**, *115* (17), 9307-9387, *Organocatalytic Reactions Enabled by N-Heterocyclic Carbenes.*

[17]   C. C. Scarborough, A. Bergant, G. T. Sazama, I. A. Guzei, L. C. Spencer, S. S. Stahl, *Tetrahedron*, **2009**, *65* (26), 5084-5092, *Synthesis of Pd(II) complexes bearing an enantiomerically resolved seven-membered N-heterocyclic carbene ligand and initial studies of their use in asymmetric Wacker-type oxidative cyclization reactions.*

[18]   C. C. Scarborough, M. J. W. Grady, I. A. Guzei, B. A. Gandhi, E. E. Bunel, S. S. Stahl, *Angew. Chem. Int. Ed.*, **2005**, *44* (33), 5269-5272, *Pd(II) complexes possessing a seven-membered N-heterocyclic carbene ligand.*

[19]   C. C. Scarborough, I. A. Guzei, S. S. Stahl, *Dalton Trans.*, **2009**, (13), 2284-2286, *Synthesis and isolation of a stable, axially-chiral seven-membered N-heterocyclic carbene.*

[20]   C. C. Scarborough, B. V. Popp, I. A. Guzei, S. S. Stahl, *J. Organomet. Chem.*, **2005**, *690* (24-25), 6143-6155, *Development of 7-membered N-heterocyclic carbene ligands for transition metals.*

[21]   S. S. Stahl, C. C. Scarborough; Patent number US20060229448A1, USA, 2006.

[22]   W. Ried, J. Bräutigam, *Chem. Ber.*, **1966**, *99* (10), 3304-3308, *Heterocyclische Siebenringsysteme, XVII. Diazepine aus 2,2'-Diamino-binaphthyl-(1.1') und 2.6.2'.6'-Tetraamino-4.4'-bis-methoxycarbonyl-biphenyl mit Imidsäureestern.*

[23]   R. J. W. Le Fevre, *J. Chem. Soc. (Resumed)*, **1929**, (0), 733-738, *CIII.-Corrections in the chemistry of diphenyl derivatives of the "Kaufler" type, and the formation of dibenzoctdiazines.*

[24]   K. Matsuda, I. Yanagisawa, Y. Isomura, T. Mase, T. Shibanuma, *Synth. Commun.*, **1997**, *27* (14), 2393-
         2402, *Alternative Synthesis of Dibenzo-and Dipyrido-[1,3]Diazepines from Thioamides and
         o,o'-Diaminobiaryls.*

[25]   W. E. Kreighbaum, H. C. Scarborough, *J. Med. Chem.*, **1964**, *7* (3), 310-312, *6-Amino Derivatives of
         5H-Dibenzo[d,f][1,3]diazepine.*

[26]   W. Ried, W. Storbeck, *Chem. Ber.*, **1962**, *95* (2), 459-472, *Über heterocyclische Siebenringsysteme, XI.
         Ringschlußreaktionen mit o.o'-Diamino-biphenyl.*

[27]   W. Ried, A. Sinharay, *Chem. Ber.*, **1965**, *98* (11), 3523-3531, *Über heterocyclische Siebenringsysteme,
         XIV. Weitere Ringschlußreaktionen mit o.o'-Diamino-biphenyl.*

[28]   M. Alajarín, P. Molina, P. Sánchez-Andrada, M. Concepción Foces-Foces, *J. Org. Chem.*, **1999**, *64* (4),
         1121-1130, *Preparation and Intramolecular Cyclization of Bis(carbodiimides). Synthesis and X-ray
         Structure of 1,3-Diazetidine-2,4-diimine Derivatives.*

[29]   A. I. Roshchin, R. G. Kostyanovsky, *Mendeleev Commun.*, **2003**, *13* (6), 275-276, *Configurationally
         stable axially chiral N,N'-dialkyl-2,2'-biphenylene-N,N'-ureas.*

[30]   M. R. Ibrahim, Z. A. Fataftah, O. A. Hamed, *J. Chem. Eng. Data*, **1988**, *33* (1), 69-70, *Synthesis of
         N,N'-diacylated 2,2'-biphenyleneurea.*

[31]   S. Kumar, R. Pratap, A. Kumar, B. Kumar, V. K. Tandon, V. J. Ram, *Beilstein J. Org. Chem.*, **2013**, *9*,
         809-817, *Direct alkenylation of indolin-2-ones by 6-aryl-4-methylthio-2H-pyran-2-one-3-carbonitriles:
         A novel approach.*

[32]   R. Mirabdolbaghi, M. Hassan, T. Dudding, *Tetrahedron: Asymmetry*, **2015**, *26* (10-11), 560-566, *Design
         and synthesis of a chiral seven-membered ring guanidine organocatalyst applied to asymmetric
         vinylogous aldol reactions.*

[33]   Z.-j. Jiang, W. Wang, R. Zhou, L. Zhang, H.-y. Fu, X.-l. Zheng, H. Chen, R.-x. Li, *Catal. Commun.*,
         2014, *57*, 14-18, *6H-Dibenzo[d,f-[1,3]diazepin-6-ylidene,5,7-dihydro-5,7-diphenylphosphinyl]: A new
         ligand for palladium-catalyzed Mizoroki-Heck coupling.*

[34]   M. M. Rogers, J. E. Wendlandt, I. A. Guzei, S. S. Stahl, *Org. Lett.*, **2006**, *8* (11), 2257-2260, *Aerobic
         Intramolecular Oxidative Amination of Alkenes Catalyzed by NHC-Coordinated Palladium Complexes.*

[35]   J. J. Dunsford, K. J. Cavell, B. M. Kariuki, *Organometallics*, **2012**, *31* (11), 4118-4121, *Gold(I)
         Complexes Bearing Sterically Imposing, Saturated Six- and Seven-Membered Expanded Ring
         N-Heterocyclic Carbene Ligands.*

[36]   S. Kumar, R. Pratap, A. Kumar, B. Kumar, V. K. Tandon, V. J. Ram, *Tetrahedron*, **2013**, *69* (24), 4857-
         4865, *Synthesis of dibenzo[d,f]diazepinones and alkenylindolinones through ring transformation of
         2H-pyran-2-one-3-carbonitriles by indolin-2-ones.*

[37]   K.-S. Masters , A. Bihlmeier, W. Klopper, S. Bräse *Chem. Eur. J.*, **2013**, *19* (52), 17827-17835, *Tethering
         for Selective Synthesis of 2,2´-Biphenols: The Acetal Method.*

[38]   J. Barluenga, A. M. Bayón, P. J. Campos, G. Canal, G. Asensio, E. González-Nuñez, Y. Molina, *Chem.
         Ber.*, **1988**, *121* (10), 1813-1816, *A simple and general synthesis of symmetrical and unsymmetrical
         bis(arylamino)methanes. Reactions of N,O-acetals with nitrogen bases.*

[39]   A. G. Giumanini, G. Verardo, E. Zangrando, L. Lassiani, *J. Prakt. Chem.*, **1987**, *329* (6), 1087-1103,
         *1,3,5-Triarylhexahydro-sym-triazines and 1,3,5,7-tetraaryl-1,3,5,7-tetrazocines from aromatic amines
         and paraformaldehyde.*

[40]   K.-S. Masters, S. Bräse, *Angew. Chem. Int. Ed.*, **2013**, *52* (3), 866-869, *The Acetal Concept:
         Regioselective Access to ortho,ortho-Diphenols via Dibenzo-1,3-dioxepines.*

[41]   K.-S. Masters, *RSC Adv.*, **2015**, *5*, 29975-29986, *Multi-bond forming and iodo-selective base-promoted
         homolytic aromatic substitution.*

[42]   F. Ullmann, J. Bielecki, *Chem. Ber.*, **1901**, *34* (2), 2174-2185, *Ueber Synthesen in der Biphenylreihe.*

[43]   F. Ullmann, P. Sponagel, *Chem. Ber.*, **1905**, *38* (2), 2211-2212, *Ueber die Phenylirung von Phenolen.*

[44]   F. Ullmann, *Chem. Ber.*, **1903**, *36* (2), 2382-2384, *Ueber eine neue Bildungsweise von
         Diphenylaminderivaten.*

[45]   S. Zhang, F. Xiong, J. Luo, L. Ran, *Guangdong Huagong*, **2013**, *40* (3), 71-73, *Mechanism of Ullmann
         reaction in synthesis of biaryls and research status in recent years.*

[46]   S. Xiao, J. Zhu, X. Mu, Z. Li, *Youji Huaxue*, **2013**, *33* (8), 1668-1673, *Advances in investigation of
         Ullmann reaction accelerated by microwave irradiation.*

[47]   R. Tanaka, *Yuki Gosei Kagaku Kyokaishi*, **2011**, *69* (8), 937-938, *Recent advancements in Ullmann reaction.*

[48]   E. Sperotto, G. P. M. van Klink, G. van Koten, J. G. de Vries, *Dalton Trans.*, **2010**, *39* (43), 10338-10351, *The mechanism of the modified Ullmann reaction.*

[49]   C. Sambiagio, S. P. Marsden, A. J. Blacker, P. C. McGowan, *Chem. Soc. Rev.*, **2014**, *43* (10), 3525-3550, *Copper catalysed Ullmann type chemistry: from mechanistic aspects to modern development.*

[50]   X. Ribas, I. Gueell, *Pure Appl. Chem.*, **2014**, *86* (3), 345-360, *Cu(I)/Cu(III) catalytic cycle involved in Ullmann-type cross-coupling reactions.*

[51]   Z. Li, Z. Wu, H. Deng, X. Zhou, *Youji Huaxue*, **2013**, *33* (4), 760-770, *Progress in copper-catalyzed Ullmann-type coupling reactions in water.*

[52]   X. Guo, A. Xu, J. Wang, M. Jia, Z. Bao, *Gongye Cuihua*, **2015**, *23* (3), 172-177, *Research advances in the catalysts for Ullmann coupling reactions.*

[53]   D. A. Everson, D. J. Weix, *J. Org. Chem.*, **2014**, *79* (11), 4793-4798, *Cross-Electrophile Coupling: Principles of Reactivity and Selectivity.*

[54]   B. Panunzi, A. Tuzi, M. Tingoli, *Inorg. Chem. Commun.*, **2010**, *13* (1), 153-156, *Synthesis and resolution of 1,11-diamino-dibenzo[d,f][1,3]dioxepine: a route to new asymmetric ligands and their complexes.*

[55]   C. S. Yeung, V. M. Dong, *Chem. Rev.*, **2011**, *111* (3), 1215-1292, *Catalytic Dehydrogenative Cross-Coupling: Forming Carbon-Carbon Bonds by Oxidizing Two Carbon-Hydrogen Bonds.*

[56]   B. M. Trost, *Angew. Chem. Int. Ed.*, **1995**, *34* (3), 259-281, *Atom economy - a challenge for organic synthesis: homogeneous catalysis leads the way.*

[57]   K. C. Nicolaou, P. G. Bulger, D. Sarlah, *Angew. Chem. Int. Ed.*, **2005**, *44* (29), 4442-4489, *Palladium-catalyzed cross-coupling reactions in total synthesis.*

[58]   J. Hassan, M. Sévignon, C. Gozzi, E. Schulz, M. Lemaire, *Chem. Rev.*, **2002**, *102* (5), 1359-1470, *Aryl–Aryl Bond Formation One Century after the Discovery of the Ullmann Reaction.*

[59]   A. Dewanji, S. Murarka, D. P. Curran, A. Studer, *Org. Lett.*, **2013**, *15* (23), 6102-6105, *Phenyl Hydrazine as Initiator for Direct Arene C-H Arylation via Base Promoted Homolytic Aromatic Substitution.*

[60]   M. Hartmann, C. G. Daniliuc, A. Studer, *Chem. Commun.*, **2015**, *51* (15), 3121-3123, *Preparation of phenanthrenes from ortho-amino-biphenyls and alkynes via base-promoted homolytic aromatic substitution.*

[61]   D. Leifert, C. G. Daniliuc, A. Studer, *Org. Lett.*, **2013**, *15* (24), 6286-6289, *6-Aroylated Phenanthridines via Base Promoted Homolytic Aromatic Substitution (BHAS).*

[62]   D. Leifert, A. Studer, *Org. Lett.*, **2015**, *17* (2), 386-389, *9-Silafluorenes via Base-Promoted Homolytic Aromatic Substitution (BHAS) - The Electron as a Catalyst.*

[63]   S. Wertz, D. Leifert, A. Studer, *Org. Lett.*, **2013**, *15* (4), 928-931, *Cross Dehydrogenative Coupling via Base-Promoted Homolytic Aromatic Substitution (BHAS): Synthesis of Fluorenones and Xanthones.*

[64]   S. Lindner, S. Bräse, K.-S. Masters, *J. Fluorine Chem.*, **2015**, *179*, 102-105, *Simple and expedient metal-free C-H-functionalization of fluoro-arenes by the BHAS method - Scope and limitations.*

[65]   W. Liu, L. Xu, *Tetrahedron*, **2015**, *71* (30), 4974-4981, *Ligand-controlled chemoselectivity in potassium tert-amylate-promoted direct C-H arylation of unactivated benzene with aryl halides.*

[66]   Q. Song, D. Zhang, Q. Zhu, Y. Xu, *Org. Lett.*, **2014**, *16* (20), 5272-5274, *p-Toluenesulfonohydrazide as Highly Efficient Initiator for Direct C-H Arylation of Unactivated Arenes.*

[67]   A. Studer, D. P. Curran, *Nature Chem.*, **2014**, *6* (9), 765-773, *The electron is a catalyst.*

[68]   G. Bringmann, M. Breuning, S. Tasler, *Synthesis*, **1999**, *1999* (4), 525-558, *The Lactone Concept: An Efficient Pathway to Axially Chiral Natural Products and Useful Reagents.*

[69]   G. Bringmann, M. Breuning, R.-M. Pfeifer, W. A. Schenk, K. Kamikawa, M. Uemura, *J. Organomet. Chem.*, **2002**, *661* (1), 31-47, *The lactone concept—a novel approach to the metal-assisted atroposelective construction of axially chiral biaryl systems.*

[70]   G. Bringmann, D. Menche, *Acc. Chem. Res.*, **2001**, *34* (8), 615-624, *Stereoselective Total Synthesis of Axially Chiral Natural Products via Biaryl Lactones.*

[71]   B. S. Bhakuni, A. Kumar, S. J. Balkrishna, J. A. Sheikh, S. Konar, S. Kumar, *Org. Lett.*, **2012**, *14* (11), 2838-2841, *KOtBu Mediated Synthesis of Phenanthridinones and Dibenzoazepinones.*

[72]   R. H. Crabtree, *Chem. Rev.*, **1985**, *85* (4), 245-269, *The organometallic chemistry of alkanes.*

[73]   T. W. Lyons, M. S. Sanford, *Chem. Rev.*, **2010**, *110* (2), 1147-1169, *Palladium-Catalyzed Ligand-Directed C–H Functionalization Reactions.*

[74]   L.-C. Campeau, M. Parisien, M. Leblanc, K. Fagnou, *J. Am. Chem. Soc.*, **2004**, *126* (30), 9186-9187, *Biaryl Synthesis via Direct Arylation: Establishment of an Efficient Catalyst for Intramolecular Processes.*

[75]   M. Lafrance, N. Blaquière, K. Fagnou, *Chem. Commun.*, **2004**, (24), 2874-2875, *Direct intramolecular arylation of unactivated arenes: Application to the synthesis of aporphine alkaloids.*

[76]   L.-C. Campeau, S. Rousseaux, K. Fagnou, *J. Am. Chem. Soc.*, **2005**, *127* (51), 18020-18021, *A Solution to the 2-Pyridyl Organometallic Cross-Coupling Problem: Regioselective Catalytic Direct Arylation of Pyridine N-Oxides.*

[77]   M. Lafrance, K. Fagnou, *J. Am. Chem. Soc.*, **2006**, *128* (51), 16496-16497, *Palladium-Catalyzed Benzene Arylation: Incorporation of Catalytic Pivalic Acid as a Proton Shuttle and a Key Element in Catalyst Design.*

[78]   S.-D. Yang, C.-L. Sun, Z. Fang, B.-J. Li, Y.-Z. Li, Z.-J. Shi, *Angew. Chem. Int. Ed.*, **2008**, *47* (8), 1473-1476, *Palladium-catalyzed direct arylation of (hetero)arenes with aryl boronic acids.*

[79]   G. P. McGlacken, L. M. Bateman, *Chem. Soc. Rev.*, **2009**, *38* (8), 2447-2464, *Recent advances in aryl-aryl bond formation by direct arylation.*

[80]   M. Lafrance, N. Blaquiere, K. Fagnou, *Eur. J. Org. Chem.*, **2007**, *2007* (5), 811-825, *Aporphine Alkaloid Synthesis and Diversification via Direct Arylation.*

[81]   R. Cano, A. F. Schmidt, G. P. McGlacken, *Chem. Sci.*, **2015**, *6* (10), 5338-5346, *Direct arylation and heterogeneous catalysis; ever the twain shall meet.*

[82]   L.-C. Campeau, D. R. Stuart, K. Fagnou, *Aldrichimica Acta*, **2007**, *40* (2), 35-41, *Recent advances in intermolecular direct arylation reactions.*

[83]   L. Ackermann, R. Vicente, A. R. Kapdi, *Angew. Chem. Int. Ed.*, **2009**, *48* (52), 9792-9826, *Transition metal-catalyzed direct arylation of (hetero)arenes by C-H bond cleavage.*

[84]   D. Alberico, M. E. Scott, M. Lautens, *Chem. Rev.*, **2007**, *107* (1), 174-238, *Aryl–Aryl Bond Formation by Transition-Metal-Catalyzed Direct Arylation.*

[85]   F. Shibahara, T. Murai, *Asian J. Org. Chem.*, **2013**, *2* (8), 624-636, *Direct C–H Arylation of Heteroarenes Catalyzed by Palladium/Nitrogen-Based Ligand Complexes.*

[86]   L.-C. Campeau, M. Parisien, A. Jean, K. Fagnou, *J. Am. Chem. Soc.*, **2006**, *128* (2), 581-590, *Catalytic Direct Arylation with Aryl Chlorides, Bromides, and Iodides: Intramolecular Studies Leading to New Intermolecular Reactions.*

[87]   A. Kimishima, H. Umihara, A. Mizoguchi, S. Yokoshima, T. Fukuyama, *Org. Lett.*, **2014**, *16* (23), 6244-6247, *Synthesis of (–)-Oxycodone.*

[88]   G. Bringmann, A. J. Price Mortimer, P. A. Keller, M. J. Gresser, J. Garner, M. Breuning, *Angew. Chem. Int. Ed.*, **2005**, *44* (34), 5384-5427, *Atroposelective synthesis of axially chiral biaryl compounds.*

[89]   M. C. Kozlowski, B. J. Morgan, E. C. Linton, *Chem. Soc. Rev.*, **2009**, *38* (11), 3193-3207, *Total synthesis of chiral biaryl natural products by asymmetric biaryl coupling.*

[90]   M. Leblanc, K. Fagnou, *Org. Lett.*, **2005**, *7* (14), 2849-2852, *Allocolchicinoid synthesis via direct arylation.*

[91]   H.-Y. Sun, S. I. Gorelsky, D. R. Stuart, L.-C. Campeau, K. Fagnou, *J. Org. Chem.*, **2010**, *75* (23), 8180-8189, *Mechanistic Analysis of Azine N-Oxide Direct Arylation: Evidence for a Critical Role of Acetate in the Pd(OAc)₂ Precatalyst.*

[92]   D. Lapointe, T. Markiewicz, C. J. Whipp, A. Toderian, K. Fagnou, *J. Org. Chem.*, **2011**, *76* (3), 749-759, *Predictable and Site-Selective Functionalization of Poly(hetero)arene Compounds by Palladium Catalysis.*

[93]   M. Livendahl, A. M. Echavarren *Isr. J. Chem.*, **2010**, *50* (5-6), 630-651, *Palladium-Catalyzed Arylation Reactions: A Mechanistic Perspective.*

[94]   S. I. Gorelsky, D. Lapointe, K. Fagnou, *J. Am. Chem. Soc.*, **2008**, *130* (33), 10848-10849, *Analysis of the Concerted Metalation-Deprotonation Mechanism in Palladium-Catalyzed Direct Arylation Across a Broad Range of Aromatic Substrates.*

[95]   M. Lafrance, C. N. Rowley, T. K. Woo, K. Fagnou, *J. Am. Chem. Soc.*, **2006**, *128* (27), 8754-8756, *Catalytic Intermolecular Direct Arylation of Perfluorobenzenes.*

[96]   D. García-Cuadrado, A. A. C. Braga, F. Maseras, A. M. Echavarren, *J. Am. Chem. Soc.*, **2006**, *128* (4), 1066-1067, *Proton Abstraction Mechanism for the Palladium-Catalyzed Intramolecular Arylation.*

[97]   M. Lafrance, D. Lapointe, K. Fagnou, *Tetrahedron*, **2008**, *64* (26), 6015-6020, *Mild and efficient palladium-catalyzed intramolecular direct arylation reactions.*

[98]   D. W. Old, J. P. Wolfe, S. L. Buchwald, *J. Am. Chem. Soc.*, **1998**, *120* (37), 9722-9723, *A Highly Active Catalyst for Palladium-Catalyzed Cross-Coupling Reactions: Room-Temperature Suzuki Couplings and Amination of Unactivated Aryl Chlorides.*

[99]   C. Han, S. L. Buchwald, *J. Am. Chem. Soc.*, **2009**, *131* (22), 7532-7533, *Negishi Coupling of Secondary Alkylzinc Halides with Aryl Bromides and Chlorides.*

[100]   A. Albini, E. Fasani, V. Frattini, *J. Chem. Soc. Perkin Trans. 2,* **1988**, (2), 235-240, *Medium and substituent effects on the photochemistry of phenanthridine N-oxides. Is an intermediate of diradical character involved in the photorearrangement of heterocyclic N-oxides.*

[101]   S. Shang, D. Zhang-Negrerie, Y. Du, K. Zhao, *Angew. Chem. Int. Ed.*, **2014**, *53* (24), 6216-6219, *Intramolecular Metal-Free Oxidative Aryl–Aryl Coupling: An Unusual Hypervalent-Iodine-Mediated Rearrangement of 2-Substituted N-Phenylbenzamides.*

[102]   M. Hori, T. Kataoka, H. Shimizu, K. Matsuo, *Tetrahedron Lett.*, **1979**, *20* (41), 3969-3972, *Synthesis of novel cyclic sulfilimines, 2-azathiabenzene derivatives.*

[103]   H. Shimizu, K. Hamada, M. Ozawa, T. Kataoka, M. Hori, K. Kobayashi, Y. Tada, *Tetrahedron Lett.*, **1991**, *32* (34), 4359-4362, *Reactions of azathiabenzene anion with electrophiles. Formation and x-ray analysis of novel heterocyclic compounds from the reaction with carboxylic ester.*

[104]   H. Shimizu, T. Hatano, T. Matsuda, T. Iwamura, *Tetrahedron Lett.*, **1999**, *40* (1), 95-96, *Novel polar cycloaddition of 1,2-thiazinylium salt.*

[105]   J. Ryan, C. Hyland, J. Just, A. Meyer, J. Smith, C. Williams, *Prog. Heterocycl. Chem.*, **2013**, *25*, 455-495, *Seven-Membered Rings.*

[106]   I. Karamé, M. Lorraine Tommasino, M. Lemaire, *J. Mol. Catal. A: Chem.*, **2003**, *196* (1–2), 137-143, *N,N- and N,S-ligands for the enantioselective hydrosilylation of acetophenone with iridium catalysts.*

[107]   A. Aranyos, D. W. Old, A. Kiyomori, J. P. Wolfe, J. P. Sadighi, S. L. Buchwald, *J. Am. Chem. Soc.*, **1999**, *121* (18), 4369-4378, *Novel Electron-Rich Bulky Phosphine Ligands Facilitate the Palladium-Catalyzed Preparation of Diaryl Ethers.*

[108]   Á. Mosquera, M. A. Pena, J. Pérez Sestelo, L. A. Sarandeses, *Eur. J. Org. Chem.*, **2013**, *2013* (13), 2555-2562, *Synthesis of Axially Chiral 1,1'-Binaphthalenes by Palladium-Catalysed Cross-Coupling Reactions of Triorganoindium Reagents.*

[109]   Y. Wang, B.-M. Ji, K.-L. Ding, *Chin. J. Chem.*, **2002**, *20* (11), 1300-1312, *Synthesis of Aminophosphine Ligands with Binaphthyl Backbones for Silver(I)-catalyzed Enantioselective Allylation of Benzaldehyde†.*

[110]   F. Seng, *Synthesis*, **1977**, (11), 753, *A facile method for preparing cyclic amidinium salts.*

[111]   H. E. Gottlieb, V. Kotlyar, A. Nudelman, *J. Org. Chem.*, **1997**, *62* (21), 7512-7515, *NMR Chemical Shifts of Common Laboratory Solvents as Trace Impurities.*

Chapter 6. Summary

Sulfonyl ynamides are highly versatile and synthetically useful reagents. This thesis describes recent advances in their synthesis, modification and use in heterocyclic organic synthesis.

A wide array of sulfonyl ynamides was prepared via copper-catalyzed amidative cross-couplings of sulfonyl amides with bromo acetylenes or from dichloroenamide precursors via an elimination, lithiation and quenching procedure. Additionally, terminal (unsubstituted) ynamides could be further diversified via Sonogashira couplings. Investigations into the preparation of solid-supported ynamides found that carboxylic acid-bearing ynamides could be coupled to a Rink amide resin and that dichloroenamides could be attached to a Merrifield triazene resin using CuAAC reactions. The loading of the amide-coupled Rink-ynamide resin was found to be unpractically low, but the coupling of the dichloroenamide to the resin worked very well with an overall of 80% yield. Investigations concerning the final elimination reaction to the ynamide are currently ongoing.

Electrophilically activated amides were reacted with sulfonyl ynamides in order to access highly functionalized 4-aminoquinolines. The amide activation procedure consists of treating the amides with triflic anhydride and 2-chloropyridine at −78 °C to form a highly reactive nitrilium species that was then reacted with the sulfonyl ynamide. After annulation the final 4-aminoquinoline was formed. The advantage of this approach over pre-existing synthetic strategies towards 4-aminoquinolines is that this method allows for full synthetic flexibility regarding substituents on all the positions on the quinoline ring. The tosylated and benzylated 4-amino group was deprotected using potassium diphenylphosphide in THF with subsequent hydrogenation. All obtained quinolines are currently being screened for their biological activity.

Sydnones are 5-membered meso-ionic heterocycles that are known for their use in pyrazole syntheses via cycloaddition reactions with electron-deficient alkynes. Interestingly, it was found that the electron-rich sulfonyl ynamides also reacted with sydnones under copper catalysis to form 4-aminopyrazoles. However, choice of the copper catalyst was of crucial importance as hydrolysis of the ynamide was a persistent side-reaction. This previously unreported pyrazole synthesis employed copper sulfate and a reducing agent and was found to be only compatible with C-4 unsubstituted sydnones. In order to avoid the copper-catalyzed hydrolysis of the terminal ynamides, strain-promoted cycloadditions were investigated. An *in situ* prepared 3-azacyclohexyne was found to tolerate a wide array of C-4 substituted sydnones, producing a mixture of both the 3,4- and 4,3-regio-isomers in good yields. Alternatively, a 3-azacyclooctyne was prepared *in situ* and directly reacted with a sydnone, leading to both pyrazole regio-isomers in modest yield. Isolation and characterization of the cyclic 8-membered ynamide is currently ongoing.

Additional investigations into heterocyclic methodology led to the development of highly sophisticated, non-symmetrical and axially-chiral dibenzo-1,3-diazepines, -oxazepines and -thiazepines from simple, commercially available anilines. The anilines were coupled to the corresponding reaction partner via a chloromethyl intermediate and the 7-membered ring was subsequently formed using direct arylation. Additional investigations, including biological tests, development of a chiral synthetic route and conversion of the dibenzo-1,3-diazepines into *N*-heterocyclic carbenes, are currently ongoing.

Annexes

List of abbreviations

)))	sonication	CuAAc	copper-catalyzed alkyne-azide cycloaddition
1,10-phen.	1,10-phenanthroline	CuSAC	copper-catalyzed sydnone-alkyne cycloaddition
3-NBA	3-nitrobenzylalcohol		
Å	Ångström	Cy	cyclohexyl
Ac	acetyl	δ	chemical shift
ACN	acetonitrile	d	doublet (NMR), day(s)
AIBN	2,2'-azobisisobutyronitrile	*D*	*dexter* (right)
app.	apparent	DAD	diode array detector
aq.	aqueous	DBAD	di-*tert*-butyl azodicarboxylate
Ar	arene	dd	doublet of doublets,
asc.	ascorbate	ddd	doublet of doublet of doublets
ATR	attenuated total reflection	ddt	doublet of doublet of triplets
BHAS	base-promoted homolytic aromatic substitution	DEPT	distortionless enhancement by polarization transfer
BINAP	(2,2'-bis(diphenylphosphino)-1,1'-binaphthyl)	DFT	density functional theory
		DIPEA	*N,N*-diisopropylethylamine
BINOL	1,1'-bi-2-naphthol	DMA	dimethylacetamide
Bn	benzyl	DMF	dimethylformamide
Boc	*tert*-butyloxycarbonyl	DMSO	dimethylsulfoxide
bs	broad singlet	DNA	deoxyribonucleic acid
Bu	butyl	dq	doublet of quintets
°C	degrees Celsius	dt	doublet of triplets
calc.	calculated	*E*	*entgegen* (opposite)
*c*Hex	cyclochexane	E	electrophile
cm	centimeter	EDG	electron-donating group
CMD	concerted metalation-deprotonation	EDTA	ethylenediaminetetraacetic acid
conc.	concentrated	e.g.	*exempli gratia* (for example)

EI	electron impact ionization	$[M]^+$	molecular ion
Equiv.	equivalent	mL	milliliter
ESI	electrospray ionization	Me	methyl
Et	ethyl	Mes	mesityl
et al.	*et alia* (and others)	MHz	megahertz
etc.	*et cetera* (and so forth)	mg	milligram
eV	electronvolt	min	minute
EWG	electron-withdrawing group	m.p.	melting point
FAB	fast atom bombardement	MS	mass spectrometry
g	gram	MS4Å	molecular sieves with 4Å pores
GC	gas chromatography		
gen	generation	μW	microwave irradiation
GP	general procedure	*m/z*	mass/charge ratio
h	hour	ṽ	wave number
HOBt	1-hydroxybenzotriazole	*n*	*normal*
HPLC	high performance liquid chromatography	Na-asc.	sodium ascorbate
		NBS	*N*-bromosuccinimide
HRMS	high resolution mass spectrometry	NHC	*N*-heterocyclic carbene
		NMR	nuclear magnetic resonance
Hz	Hertz	NMP	*N*-methyl-2-pyrrolidone
i	iso		
IR	infrared	Nu	nucleophile
J	coupling constant	n.d.	not determined
KHMDS	potassium bis(trimethylsilyl)amide	*o*	*ortho*
		*o*DCB	orthodichlorobenzene
L	*laevus* (left)	org.	organic
L	liter(s)	oxid.	oxidative
λ	wavelength	*p*	*para*
LDA	lithiumdiisopropylamide	*p.a.*	*pro analysi*
M	molar	PG	protecting group
μ	micro	Ph	phenyl
m	*meta*	phen.	1,10-phenanthroline
m	multiplet (NMR), medium (IR), milli-	PMP	*para*-methoxyphenyl
		ppm	parts per million

Pr	propyl	*t, tert*	*tertiary*
PS	polystyrene	t	triplet
*p*TsOH	*para*-toluenesulfonic acid	T, Temp.	temperature
pyr	pyridine	TBAF	tetra-*n*-butylammonium fluoride
q	quartet		
quant.	quantitative	td	triplet of doublets
quin	quintuplet	Tf	triflyl
R	*rectus* (right)	TFA	trifluoroacetic acid
R	radical	TFAA	trifluoroacetic acid anhydride
rac	racemic	THF	tetrahydrofuran
Ref.	reference	TIPS	triisopropylsilane
R_f	retention factor	TMS	trimethylsilyl
RT	room temperature	TLC	thin-layer chromatography
S	*sinister* (left)	tol	toluene
s	singlet (NMR), strong (IR)	Ts	tosyl
sat.	saturated	tt	triplet of triplets
SET	single electron transfer	UV	ultraviolet
sol.	solution	vs	very strong (IR)
SPAAC	strain promoted alkyne-azide cycloaddition	vw	very weak (IR)
		W	Watts
SPOS	solid-phase organic synthesis	w	weak (IR)
symm.	symmetrical	Z	*zusammen* (together)

Curriculum Vitae

Tim Wezeman was born on the 13th of April, 1989 in Essen, Germany. After finishing high school in Almere, the Netherlands, Tim went to Wageningen University in September 2007 to study Molecular Life Sciences. Already early on in his bachelor degree he found himself being drawn to organic chemistry and visited the Eindhoven University of Technology for a 6 month in-depth minor in Polymer Chemistry. In November 2010 he concluded his Bachelor of Science with a thesis in a joint-venture between the Physical Chemistry & Colloid Science and Organic Chemistry Departments working on novel surfactants for CO_2-based dry-cleaning.

Tim then continued with his Masters degree in Molecular Life Science, focused fully on organic synthesis. In the laboratory of Prof. Dr. Han Zuilhof he worked with Assistant Prof. Dr. Tom Wennekes on the synthesis of pseudaminic acid and related azidosugars from amino acids. From October 2011 until April 2012 he joined the laboratory of Prof. Dr. Marco A. Ciufolini at the University of British Columbia in Vancouver, Canada and investigated the synthesis of (R)-Telomestatin, a heterocyclic, polycyclic broad-spectrum anti-cancer candidate. In November 2012 he obtained his Master of Science degree.

For his doctoral studies, Tim joined the laboratory of Prof. Dr. Stefan Bräse in March 2013 and joined the ECHONET training network as a Marie Curie Early Stage Researcher. During his PhD he joined the laboratories of Dr. Kye-Simeon Masters at the Queensland University of Technology in Brisbane, Australia and Dr. Stephen D. Lindell at Bayer CropScience GmBH in Frankfurt-Höchst for short interdoctoral research stays. As part of his Marie Curie Program he attended several workshops and summer schools covering various subjects. Parts of his PhD work were presented at several meetings and conferences, most notably the 2015 International Society of Heterocyclic Chemistry Congress in Santa Barbara where he presented a short talk on his quinoline syntheses.

Tim can most conveniently be reached by email: **timwezeman@gmail.com**

List of publications

Scientific articles & Book Chapters

- Tim **Wezeman**, Júlia Comas-Barceló, Martin Nieger, Joseph P.A. Harrity and Stefan Bräse, *submitted.*

- Tim **Wezeman**, Sabilla Zhong, Martin Nieger and Stefan Bräse, Angew. Chem. Int. Ed. *2016*, **55**, 3823 – 3827, http://dx.doi.org/10.1002/anie.201511385 and Angew. Chem. *2016*, **128**, 3888 – 3892, http://dx.doi.org/10.1002/ange.201511385

- Tim **Wezeman** and Kye-Simeon Masters, Book Chapter "Xanthones are Privileged Scaffolds in Medicinal Chemistry – but are they Over-privileged? Ed. S. Bräse, *2016* http://dx.doi.org/10.1039/9781782622246-00312

- Tim **Wezeman**, Yuling Hu, Stefan Bräse, John McMurtrie and Kye-Simeon Masters, Aust. J. Chem. *2015*, **68**, 1859 – 1865, http://dx.doi.org/10.1071/CH15465.

- Tim **Wezeman**, Stefan Bräse and Kye-Simeon Masters, Nat. Prod. Rep., *2015*, **32**, 6 – 28, http://dx.doi.org/10.1039/C4NP00050A, Front Cover Article.

- Tim **Wezeman**, Kye-Simeon Masters and Stefan Bräse, Angew. Chem. Int. Ed. *2014*, **53**, 4524 – 4526, http://dx.doi.org/10.1002/anie.201402384 and Angew. Chem. *2014*, **126**, 4612 – 4614, http://dx.doi.org/10.1002/ange.201402384.

Scientific presentations

- (Poster) **Tim Wezeman**, Stefan Bräse (August 2016) *"Sulfonyl ynamides: useful tools for N-heterocyclic chemistry"* presented at the Swiss Summer School on Chemical Biology in Villars, Switzerland.

- (Talk) **Tim Wezeman**, Stefan Bräse (July 2016) *"Pyrazole Synthesis via Strained Cyclic Sulfonyl Ynamides"* presented at "Marcial Moreno Mañas" XI International Summer School on Organometallic Chemistry in Donostia/San Sebastián, Spain

- (Poster) **Tim Wezeman**, Stefan Bräse (July 2016) *"Sulfonyl ynamides: useful tools for N-heterocyclic chemistry"* presented at "Marcial Moreno Mañas" XI International Summer School on Organometallic Chemistry in Donostia/San Sebastián, Spain

- (Talk) Tim **Wezeman**, Sabilla Zhong, Stefan Bräse (August 2015) *"Synthesis of highly functionalized 4-aminoquinolines"* presented at the 25th International Society for Heterocyclic Chemistry Congress in Santa Barbara, USA.

- (Talk) Tim **Wezeman**, Stefan Bräse, Kye-Simeon Masters (February 2015) *"Synthesis of dibenzo-1,3-diazepines"* presented at the Nanotechology and Molecular Science HDR Symposium in Brisbane, Australia.

- (Poster) Tim **Wezeman**, Stefan Bräse (September 2014) *"Synthesis of heterocycles using solid-phase sulphonynamide based chemistry"* presented at the 4th KIT PhD Symposium.

- (Poster) Tim **Wezeman**, Stefan Bräse (June 2014) *"Synthesis of heterocycles using solid-phase sulphonynamide based chemistry"* presented at *"A Corbella"* International Summer School on Organic Synthesis.

Acknowledgements

First of all I would like to thank my supervisor, Prof. Dr. Stefan Bräse, for welcoming me in his group, allowing me to work on this interesting topic and always giving me the opportunity to explore and learn more about chemistry. I would like to thank him for his guidance as well as the support and freedom to work on my own ideas during my time in his group.

Dr. Kye-Simeon Masters is thanked for his continued helpful advice and support, enjoyable collaborations and supervision during my visit to his laboratory at the QUT.

Dr. Stephen D. Lindell is thanked for the valuable discussions, helpful advice, and supervision during my visit to his laboratory at Bayer CropScience.

Prof. Dr. Joachim Podlech is kindly acknowledged for the acceptance of the co-reference of this thesis.

I would like to express my gratitude to everybody at the Karlsruhe Institute of Technology that has helped me during my research, especially Pia Lang, Tanja Ohmert, Angelika Kernert, Ingrid Roßnagel, Andreas Rapp and Patrick Weis.

Thanks to everybody that contributed to my research, especially to all the students that I supervised: Arnaud Barbier, Patrick Hodapp, Robert Vogt, Thomas Sattelberger, Valentin Beyer, Sarah Forcier and Nicolai Wippert, and to my collaborators: Sabilla Zhong, Yuling Hu, Martin Nieger, John McMurtrie and Júlia Comas-Barceló.

Thomas Hurrle and Patrick Hodapp are deeply thanked for the detailed proofreading of this manuscript.

I would like to thank the entire M6 laboratory, and especially Pabhon Poonpatana, Anthony Verderosa, Colin Schiemer, Jesse Allen, Liam Walsh and of course Kye, for making my time at the QUT unforgettable.

Thanks to all my colleagues at Bayer CropScience for always being helpful and providing such a kind and productive environment during my visit.

The European Union is thanked for their financial support of my thesis through the funding of the MSCA-FP7-ECHONET-ITN. All my ECHONET colleagues are thanked for their support and pleasant collaboration, be it of scientific nature or at social occasions.

The entire Bräse research group, past and present, is thanked for the friendly and open working atmosphere as well as for making my time in Karlsruhe so much fun: you are all great! Special thanks go to Thomas Hurrle, Patrick Hodapp, Alexander Braun, Stephan Münch and Isabelle Wessely.

Verder wil ik graag mijn ouders, broers en vrienden in Nederland bedanken voor hun standvastige support en aanmoedigingen om te doen wat ik graag wil doen. In het bijzonder Emil den Bakker, Jaap Steenkamer, Jeroen Berg, Casper Gerritsen, Aline Welzen, Daniël Gerritsen en alle (oud)-leden van de Straat wil ik bedanken voor hun aanhoudende vriendschap en voor de welkome afleidingen.

Mijn grootse dank gaat uit naar Hetty. Zonder haar was dit avontuur duizend minder leuk geweest.